# THE PRECARIOUS HUMAN ROLE IN A MECHANISTIC UNIVERSE

# THE PRECARIOUS HUMAN ROLE IN A MECHANISTIC UNIVERSE

---

## The Enigma and Stigma of Imaginative Thought in an Era of Understanding

John F Brinster

Library of Congress Control Number: 2010917913
ISBN: Hardcover 978-1-4568-2683-3
Softcover 978-1-4568-2682-6
Ebook 978-1-4568-2684-0



This book was printed in the United States of America.

**Note: The original study on which this book is based was published in draft form under the title "The Enigma and the Stigma" (ISBN: 978-1-4415-8298-0) and distributed for peer review.**

**To order additional copies of this book, contact:**
Xlibris Corporation
1-888-795-4274
www.Xlibris.com
Orders@Xlibris.com
87571

# Contents

# Acknowledgement

I am indebted to my family for their understanding of my forward-looking views and to Dr James Tietjen, Professor Marvin Bressler, Professor Renee Weber, among others who read portions of the first draft and to Lor Gehret and Karen Verde for editorial suggestions.

# Introduction

## Apologia Pro Vita Sua

Not long after Friedrich Nietzsche declared that "God is Dead", William James, the founder of psychology of religion, ushered in the twentieth century with his outstanding treatise on *The Varieties of Religious Experience*. It was a classic examination of belief in the existence of an all powerful god that created man and his world with a promise of life in the hereafter. His book was based on the Gifford Lecture Series delivered at the University of Edinburgh, the first of which was titled "Religion and Neurology", clearly an effort to relate spirituality to human mind function at the very outset. But after five hundred or more pages of his book describing details of his study, he seemed compelled to add a postscript, for he had yet registered no defining conclusion about his own views. He wrote, "Notwithstanding my own inability to accept either popular Christianity or scholastic theism . . . subjects me to being classed among the supernaturalists of the piecemeal or crasser type." In a letter to a friend he further wrote, " . . . although all the special manifestations of religion may have been absurd (I mean creeds and theories), yet the life of it as a whole is mankind's most important function."

More than a full century of unprecedented human advancement has since taken place and, although nothing new has been learned about God in the interim, man has learned infinitely more about his universe and about the function of his own spiritual mind. The spectacular advances of the past century opened a new era of intellectual capacity, the effects of which are only now spreading throughout society with sufficient understanding to be meaningful. God is apparently not yet dead for most people, but for many others, it seems that his humanlike image is beginning to suffer gradual dilution and heading for eventual abandonment. This book not only explains why the concept of an anthropomorphic god remains pervasive but also why the slow trend toward abandonment will grow stronger over future generations, with extraordinary social consequences.

After a millennium or more of widespread theological study, overwhelming documentation of religious history and, after many scholarly volumes about beliefs and practices, there is little need here for another analysis of specific traditional religions. The emotions of spiritual man have now ventured about as far as they can go with religious beliefs, and religions have reached a point where belief is simply a "given" offered and taken for granted by the masses without serious question. There are many different interpretations of human existence and purpose throughout the world, and promises made by man to fellow man on behalf of his assumed gods have not only approached a limit but are expected to be increasingly ineffective. The religion of future, further developed man will be cloaked in greater realism, in unprecedented appreciation of the beauty and harmony of a measurable, more fully understood mechanistic universe. It will be based on unprecedented appreciation of the function of the human mind and its emotions, on greater interest in man's position on his planet and on increased desire for earthly unity and cooperation.

The historical path of spiritual development suggests that, in the course of time, religion as we have known it may even become an anachronism. Decreasing acceptance on "faith" alone and increasing exercise of rationality, generation by generation, are the consequences of accelerated cultural evolution. This book deals with the recognition and explanation of an entirely new direction, not just a rehash of visions of the past. It sees an expansion of greater reasoning that is slowly beginning to penetrate the traditional theological world, a trend that is destined to change society as never before. Together with other irreversible characteristics of man, it is a trend that will finally carry man beyond existing spiritual bounds into a world of greater reality, never to return.

As he ended his famous book, William James suggested that he would like to return to the unexplained experiences in another book, but one was never written to address those particular concerns. I propose to make that journey with the reader in this book and, in so doing, humbly request that it be considered a valid James sequel, the causal foundation of his religious expositions. Despite the spiritual stranglehold on current man, the natural forces of change will eventually replace traditional religion with "reality" as "mankind's most important function." However, understanding of spiritual change first requires entry into the mind where religion is exclusively created, maintained, and modified.

Since man's very emergence from ancestral animal roots, his developing mind has created all existing human knowledge and understanding as well as means for intelligent communication among peoples of the Earth. Understandably, religious beliefs and religions remain part of his past creations and interpretations of life. The natural responses of the evolving mind have since brought man very much further and they are destined to continue to build and to produce newness and change with increasingly greater understanding as long as man survives. But we now know that human minds have a somewhat feral base and have evolved from animal ancestors with an inclination to desire and imagine the existence of a supreme humanlike being, a protective figure that created the world and controls the fate of man with an offer of eternal life. However, mind development is destined to move far beyond such concepts to concepts based on greater reality. This "greater reality" will most certainly follow the dictates of advancing factual knowledge and increasing capability of human reasoning. Indeed, man is destined to be a different man. It was the dictum of Nietzsche that declared man to be a "bridge" and not an "endpoint."

The term "imaginative" is used quite emphatically in this book because traditional spiritual vision is based, not on any credible evidence, neither on observation nor experience, but solely on emotional feeling and hearsay about the way man would like it to be. It is what man has imagined to be the case based purely on what he calls "faith." If evolving man were naturally absent neural mechanisms capable of strong emotions of fear and desire, the concept of god might never have arisen. The religious descriptions of the William James study represented the strongest spiritual emotions that could be experienced by man, but yet, they left James unconvinced of his need for theism.

Because man has long wrapped his primitive spiritual visions in a solid framework of idealism and promise, they not only persist today, spread throughout much of the world, but they have become "virtual reality" for the masses. A large percentage of humanity has been taught accordingly and believes in the existence of such a humanlike god, one that it can access at any time through supplication called prayer without any physical means, and one that benevolently guides its lives. Some also believe that humans are possessed by a ghostlike soul that continues to live after them. Spiritual desire is often so strong in the minds of man that logical explanations appear to be simply avoided or overlooked. Characteristic of emotional conviction, such belief is often rigid, extreme, and virtually intractable. However, a lesser, but increasing, percentage of world population firmly believes that no such personal god exists and that the universe and everything in it is guided by an observed mechanistic force that has no recognition of man. This book is about the reconciliation and eventual resolution of those conflicting ideologies and how society is affected.

Unresolved differences among interpretations of existence have arisen over the centuries to create a fundamental human enigma of gigantic proportions and they have led to widespread social stigma, the basis of untold human friction, name-calling, conflict, and separatism. Because of ongoing uncertainty and confusion about belief, and because of its imaginative association with divine purpose, security, and immortality, conflicts regarding spiritual understanding have indeed invaded and strongly influence every aspect of society and individual life. Although increased new concern has recently been publicly expressed, no effective move toward resolution of this most pervasive, outstanding enigma has, to now, been offered to society in modern terms with sufficient explanation and credibility. The time has now arrived when beliefs in advancing and globalizing modern society can, and must, be clarified more specifically and the course of resolution must be put on record in the best interest of mankind. Changes do not entirely rest in the hands of man but are largely the natural consequence of powerful universal forces that subtly act on all material and evolving life and provide the main theme of this book.

When the literary critic, Belinsky told the Russian writer, Ivan Turgenev that it was not appropriate for them to proceed to lunch because they hadn't yet solved the "god problem", America was in the initial throes of freeing slaves. But today, they could dine more promptly, for the problem is no longer the mystery they then thought it to be. Informed men of vision now sufficiently understand the human mind and the universe around it to be able to explain early spiritual visions in terms of modern knowledge and, despite predictions of expanding religion by aggressive religious factions, the new segment of society now detects unmistaken signs of relative freedom from religious enigma and stigma on the distant horizon.

Although modern minds still differ widely on details of beliefs, an increasing number now regard traditional concepts to be somewhat primitive, inconsistent with observed reality, and contrary to established factual knowledge. The first page of the April 13, 2009 issue of *Newsweek* read, "The Christian God is not dead in American life, but he is less of a force in our politics and culture than at any other time in recent memory" (1). A study showed that "no religion" was the only demographic group that recently grew in all fifty states. The president of the Southern Baptist Theological Seminary suggested that America's religious culture is "cracking" (2). The modern world has indeed become less reliant on religious history and teaching. Mysticism is bowing to human understanding and explanation. Europe is tending toward increased secularism (3). Sweden and Denmark claim to be largely without a god or prayer (4). England and Syria are two of the most secular of nations. Islamic regions around the world remain captive by tradition with sharply increasing unease. Russia is struggling with recently reworked Orthodoxy (5). The Anglicans are sorely divided by encroaching social issues (6). Modern China continues to resist Christian growth. Over half of the Jewish populations of America and Israel (7) continue to be secular. Buddhism is dying out in Japan (8). The moving picture industry (9) and universities (10) are being newly nudged by the secular spirit. Much of the advancing undeveloped world is engaged in uncertain spiritual struggle with fewer primitive belief systems. The relentless forces of modernity, on the heels of retreating tradition, are now quietly penetrating all the nooks and crannies of human spirituality with noticeable effect. Many believe that it cannot come too soon for world behavior cannot long tolerate expanding ideological conflict or dependence on imagined influences for assistance, guidance, or restraint.

Highly organized religions now appear to be facing a strong backlash by determined elements of modern society. But the religious not only continue to pursue theological discipline with great fervor, they also maintain aggressive efforts in the role of self-appointed missionaries citing essential human needs and benefits, frequently instilling fear of divine penalties in those who do not favorably respond. Education, even at the university level (11), still encourages the application of "faith" to life's pursuits while the advancement of university science simultaneously teaches conflicting factual information. Such opposing views always reflect certain pending change. As potential new believers dwindle in the developed world, missionaries must ferret out new emotional and vulnerable minds. Further globalization, relentlessly driven by natural forces, will eventually neutralize the entire world of man. Traditional religion and secularism are indeed immiscible ingredients of modern planetary intelligence, and this book envisions a point in time when only one will survive.

Although religion, at times, still tends to spread like a contagious disease, underlying traditional tenets are gradually being diluted by human indifference and are now perceived as less profound. For the most part, the overall behavior of the expanding modern world is anything but "religious." Despite religious declaration, behavior generally remains a mixture of ancestral animal instincts with only moderate refinement and scattered recognition of ideal religious prohibitions. Religiosity seems to be retreating from heaven and hell and thought and behavior everywhere are losing their spiritual meaning in favor of greater earthly meaning. There is increased pursuit of social and personal interests in lieu of purely religious celebration. Memberships in megachurches and other houses of worship represent greater social involvement and less serious personal commitment. The fear of breaking with the long promoted "god notion" is waning among much of society as more of its self-appointed intellectual spokesmen cry out for broad new rational understanding. At the well educated level many more people are now saying, "I am spiritual but not religious." Such new distinction represents a significant milestone in changing religious thought discussed in these chapters.

Less than a hundred years after William James, author Robert Coles, in *The Secular Mind* (12) suggested that daily living might now separate man from life's "sacred" mysteries. He questioned current spirituality, suggesting that it may be a "hobby rather than an all-encompassing approach to life." He questioned serious future continuity of the concept of the soul and considered how "secular culture" has settled in the hearts and minds of Americans. He pondered a shift from religious control over western society to "scientific dominance of the mind." In over fifty books, Coles, a Harvard psychiatrist and academic, contributed a lifetime of observations about human growth and development, indeed, a unique vision of intellectual, emotional, and moral intelligence.

This book takes the mind a giant step further. It not only discusses the significance of "imaginative spiritual thought" as it exists today, but also may be the first to envision prospects of a steady, long-term movement away from traditional beliefs toward new realism and to explain the reasons behind it. Based on modern knowledge, this writing examines the issue at its underlying roots and presents a credible basis for ultimate spiritual unity. It may also have the effect of bringing comfort to minds that have already sensed greater reality amid the overwhelming pressure of spiritual promises and of offering support for those minds that may still be in question.

This frank presentation is likely to instill a stigma of "godlessness", "atheist", or "disbeliever" in the minds of serious believers. Many who believe in the findings of this study might even feel uncomfortable in agreeing with them publicly. Publishers and reviewers might be negatively influenced because of personal beliefs and some may even feel its theme to be unrealistic and its arguments fruitless in view of strong current tradition. For those reactions, I offer my apologies or forgiveness, whichever is applicable. However, based on the underlying mechanism of belief and the overwhelming power of change, this book envisions eventual reversal of such stigma, perhaps beginning as soon as the next century or so, when the stigma of "imaginative belief" may be expected to rise to replace the stigma now rooted in accusations of "secularist", "agnostic", "pagan", "heathen", "humanist", or "atheist", as it paves the way to eventual elimination of all religious stigma.

This book is largely based on a study of how the modern mind is stimulated, responds, and functions judgmentally in spiritual matters and of consequences concerning those specific issues. It suggests that the current mind of man, still very much confused about the meaning and satisfaction of his emotional needs, will increasingly recognize and acknowledge the reality it has long measured and observed without appropriate response. It is not a book on religion-bashing nor does it level criticism or scorn against those who believe differently, nor is it an indirect attempt to effect conversion to godlessness. It is simply an interpretation of an observed long-term trend and its implications relative to advancing society with many supporting references. It sees a natural, more rational response to modern reality in maturing man, just as deep spirituality became the natural response to imaginary visions in earlier formative man.

Four of my earlier publications of limited distribution recognized the beginning signs of changing spirituality but, with this fifth book, it has been my intention to better organize prior observations and bring them up to date in a more comprehensive and scholarly fashion. Well before the turn of the century, I felt it desirable and even essential to call attention to the transformation of imaginative religious ideas to a greater sense of reality, the long overlooked "theology of nature". My first book in this series, *The Way Things Are* (13) was a simplified introduction to the reality of man. The second, *The Natural Bible for Modern and Future Man* (14), represented how a pro-modern Bible might, and should read. The third, *The Man Who Created God* (15) was a fictional story about one man's attempt to establish the Einsteinian form of cosmic religion, and *The Abduction* (16), retaining some of the same characters, was a fictional story with strong satirical flavor about the abduction of the Christ Child from the manger by the first Chinese ambassador, suggesting alternative, more realistic religious events in first century Judaea at the origin of Christianity. The concept of resurrection, currently in dispute (17), was particularly readdressed. Among other timely related works was an op/ed on the merits of Albert Einstein's "cosmic reverence" written in celebration of the Einstein Centennial and published in the *Philadelphia Inquirer* (18). Albert Einstein is discussed in chapter six.

To properly set the stage, the first two chapters discuss how the modern mind uniquely handles religion. Faith and belief, by their very nature, are considered to be entirely within the mind, a particular state of mind, and as such, their current status and ongoing changes cannot be properly explained or their trends predicted by either philosophy or theology, but only in terms of neural function. Mind research, an explosive new field, is still an immature and extremely complex subject. However, in simplifying its presentation using only interpretive neuroscientific descriptions, this text has avoided complex words and ideas that might be familiar only to students of applicable sciences.

An appendix of appropriate references is provided including references that might be usefully scanned separately for they represent much of the current molding of developing society surrounding this subject. Those in parentheses are appropriately numbered to correspond with numbers included in the text. Although not all references were cited in the final text after editing, they represent significant reader information applicable to the theme at hand. Its bibliography largely covers recent events rather than thought and behavior of the more distant past. It includes many new concerns about imaginative belief shared, not by demonic agents of contrariety, but by well informed modern minds of reason. It not only represents recognition of an impending sociological change, but awareness of this trend may be useful in dealing more effectively with other human issues in the process of further globalization in the immediate decades ahead.

Listings particularly include hundreds of references to current newspaper articles representing a wide variety of relevant information. They represent much of what the public is now reading about religious thought and behavior and about ongoing change in popular understanding. Such news articles are more effective in reaching the average public than are research papers and textbooks, and their inclusion on the Internet in most instances assures widespread distribution of the nature of information on which this book is largely based.

Intended for a wide range of readership, this work is suitable for old and young alike, for minds of widely different beliefs, and especially for minds that seem intractably cemented to traditional religions. Although human direction will not be deflected or assured by this or by any other single writing, it represents a lifetime of observation and forward-looking thought about religiosity, indeed, a scholarly effort to interpret its latest understanding and direction. It attempts to set the religious record straight for modern man based on his own factual knowledge, knowledge skillfully determined, confirmed, and universally accepted. It describes how hard reasoning is "knowingly" clouded and often blocked by strong emotions. The currently disturbed social condition of the world, clearly without effective central guidance of any divine power, suggests that remedial measures applied to human spiritual thought are in urgent need of attention.

The results of this study should be viewed as only the initial entry to man's ordained future of spiritual reality. It speaks of freedom from his most troublesome and fearful emotions and the path to greater world unity of ideology, a goal that may perhaps constitute the real "savior" and "salvation" of mankind. Surely, if Turgenev and Belinsky were to dine today, and one suggested the existence of a controlling humanlike god, one that is said to exist everywhere but can never be seen, one that is said to communicate with man but is never heard, one that is said to monitor man's behavior but never responds, one that is said to aid and protect man but never does, and one that is said to make claims to immortality contrary to the established principle of cellular death, the other might simply blame it on the vodka and enjoy a good lunch.

For readers whose cherished beliefs have been firmly held and defended for many long years and for committed students of theology, its theme will tend to fall on infertile ground for the stigmatic reasons enumerated. Its frankness and openness might be better understood and appreciated by the younger, more supple, untarnished minds of new generations that may be confronted with natural emotions of developing spirituality; more alert informed minds especially capable of assimilating a wider and deeper range of modern understanding. It is important reading for all who share belief in the gradual onset of secularism as explained in these chapters, but it is even more essential for those who think otherwise.

A neutral approach was initially assumed in this study in order to produce an unbiased representation of spiritual trends. However, it became clear that traditional beliefs are still based on imaginative thought and would eventually have to be modified to reflect compatibility with factual information. Any other route would be a path of hypocrisy and irrationality. For maximum appreciation, the reader should therefore maintain an open mind and a critical view of narrow education. He should, moreover, require a balancing disclaimer with respect to all expressions of religious dogma. Efforts such as those of the "congregatio de propaganda fide" (referring to the congregation for spreading the faith instituted by Pope Leo XV) are simply no longer suited to the modern mind.

This vision of eventual world secularism is based on a study of the fundamental mechanism of the human mind of belief, on the temporal pattern and direction of changing human thought processes and on the observed material nature of the world in which man lives, all explained in simple language of common understanding. No other forward-looking trend with respect to spirituality seems equally possible. The reader should not be deluded by the popular notion that man will "always" require and search for spirituality, but should recognize the more applicable imperative that developing man will, above all else, seek "rational" understanding, rejecting that which its senses cannot establish as factual reality. More and more people will therefore, admit a sense of spirituality but not religiosity.

The trend is based on the unstoppable and ungovernable natural forces of change slowly creeping over man's aging planet. It is perhaps the "unintended consequence" of the ongoing forces of the transcendent "god of nature" discussed at length in this study. These are forces that rule chemistry and motion but recognize no relationship with man or his emotions. Imaginative belief with all its variation and conflicts has at times been a great healer of fearful visions found in man's life of reality and an instrument of human happiness and satisfaction. However, this study claims that only factual recognition of the basis of existence can lead to true satisfaction, to eventual social unity and to eventual peaceful coexistence.

Again, this book should not be interpreted as religion-bashing in any sense, for it maintains the highest continuing respect for many well educated persons who are firm believers, persons of substantial contribution and righteousness. It fully respects "believer rationale", a different natural interpretation of being that forms an alternative structure of "neural wiring", one that conflicts with neural wiring formed in recognition of established factual knowledge. However it may be formed in the learning process, it is the neural wiring of man that determines all his thoughts and behavior. The following chapters explain how such wiring is the result of neural processes of learning, memory, and emotion, not the result of divine inspiration.

Finally, this discourse also reflects the "cosmic" philosophy of my former neighbor, the late Albert Einstein, and perhaps, the fulfillment of my promise to the late Sir John Templeton, financier and promoter of traditional belief, a promise made many years ago at the Princeton Theological Seminary that I would publish views that more fully explain the divergent relationship between science and religion. At that time, I was not aware that William James had promised his own father at the Princeton Theological Seminary that someday he would also deal in a "sustained way" with the issue of religion. This writing should satisfy both promises.

I took the liberty of adding an appropriate title to this introduction, *Apologia Pro Vita Sua*, which is simply translated "A written justification for one's beliefs." In contrast to many other studies of social significance with similar depth of inquiry, I do not anticipate immediate recognition or popular acceptance, for it represents a view of substantial contrariety beyond widespread current spiritual teaching and established popular tradition. Although this controversial route of study has proven a most difficult task, I did not suffer any of the fears experienced by the late John Updike, whose writing he called his sole addiction, "an illusory release and a presumptuous taming of reality." It was said that Updike could write "breezily in the morning of what he could not contemplate in the dark without turning in panic to God."

John F Brinster
Princeton, NJ

# Prologue

CASE 1. He kneels at his place of worship, bows his head, and is soon led into a world of imagination, the traditional spiritual realm of his Lord. He has visions of an ideal heaven and perhaps clouds of angels. He is told that he has a soul, an undetectable shadow that may live beyond his body and that he can communicate with his Lord by mere thought with expectation of response. He is told that his behavior is monitored and judged divinely. He reacts with a mixture of pleasure and fear. He loosely remembers this ideal world of hope and mystery, a dream without practical definition, as his vision returns to his world of reality, the world in which he was born and knows through his five senses, to his planet floating among an infinite number of barren worlds in space, to the world which man himself has made to function by his culture, his communication, his technology, and his economics. It is his "now" world, the only world in which he can sense happiness and satisfaction in coexistence with fellow man, but an imperfect world in which death will end all his visions forever. He reflects deeply on his imaginative experience and ponders the doubts of Mother Teresa. He wonders if his mind of hope is of primitive structure, if this paradigm might be explained with twenty-first century understanding.

**CASE 2.** In his third decade of research and teaching, a science professor of international renown stands before his students who represent traditional beliefs, Catholics, Protestants, Muslims, Buddhists, and Jews. Together with his departmental colleagues, all labeled "atheists" by the student body, his work had led to the conclusion that there can be no anthropomorphic god and that there can be no life beyond death. He lectures about the observed mechanistic universe, the origin of the solar system and its planets, the chemical origin of life on Earth, and about genetic structure and its evolution. His cosmic and natural religion, based on deep appreciation of the harmony and beauty of transcendent nature and the uniform pattern of its functional laws, can only envision the existence of a supreme force that functions independently of man. Meanwhile, his university promotes faith-based teaching of a purely imaginative nature to satisfy the emotions of students and trustees. He contemplates his dedication to science to explain the world in conflict with the spirituality of his students. He ponders his commitment to truth and the university responsibility to future society for the advancement of factual knowledge.

This book was written to explain the world's most pervasive current enigma, the "god problem" as represented by these two classic examples. In doing so, it provides the basis for reconciliation and resolution of perceived conflicts in the disciplines of science and religion.

# Chapter One

## Forces of Change

I have struggled to comprehend
the bell,
to grasp why its resonance
elevates the Host and ignites
the incense of the Mass.

I have wanted to find a fair
explanation
of "Very God of Very God",
flesh and non-flesh striving
toward substance

I have so little to show for all
my efforts
a recognition of Catholicism
as profoundly false yet gesturing
towards truth.

—Excerpts from *ORATORIO*,
by Theodore Meth, Princeton, NJ

**Secularism and peaceful coexistence are among imperatives subtly contained in the natural forces that constantly impact developing human life. Ascribed to an imagined humanlike god by early man, these forces have increasingly given mankind fundamental understanding and a new sense of meaning and purpose:** This book is largely concerned with powerful natural forces that, among other effects, continually influence change in human thought and behavior. They are the forces of nature that man rarely detects in the blinding clouds of advancement but are accepted and understood only when they become history. History is indeed woven from the changing fabric of human emotion and from the continuous advancement of human capability and knowledge. Beyond human emotion, the "god problem" seems to be largely a problem of lack of human understanding keyed to limited education, teaching, and distribution of factual information.

It is complicated by the widespread encouragement of irrational spiritual response that only a limited number of minds are yet inclined to recognize, interpret, and understand. However, the forces that move man forward are expected to eventually inundate ignorance of reality as well as the fear of mortality as human understanding continues its inevitable rise. These powerful natural forces will indeed move man far beyond the most improbable vision of most current readers and well beyond the vision of this book. Individual lives, however, represent small elements of a virtually infinite process and their integrated future linkages can only be envisioned in current terms. Martin Rees recently reminded us in Science: *The Coming Century*, published in the New York Review of Books, November 20, 2008, that those who witness the sun's demise will not any longer be human, but will "be as different from us as we are from bacteria."

Promotion by spiritual leaders of an all-powerful, benevolent, father-like God might be expected to have influenced early emotional man, but it is difficult to understand how the notion of a humanlike form of God grew so central and remains so fixed in the normally skeptical and suspicious mind of modern man. But then, man is relatively new in the universe with an imperfect functional mind, a mind fueled by desire and emotion, one still developing in rational understanding. It does not seem so much a question of limited intellectualism, but one of latent spiritual fear and lack of understanding of the mechanisms of the emotional mind itself.

Man's willingness to accept a humanlike "god notion" without question in the light of current knowledge is indeed something of a neural phenomenon and most certainly will occupy only a limited period in further civilization. Although earlier intellectual minds began to question the god notion, the reality of human origin, makeup, and function is only now being recognized by a significant learned portion of society. Appreciation by the masses may well require an extended period of more widespread further human development. Whatever the timeframe, change is already written clearly in the latent structure and function of the present neural mechanisms of constantly evolving man.

**Godless without human life?** The Earth was occupied by only plants and animals for billions of years after its formation when there was no human thought, no ideology or spirituality, and certainly no vision of a humanlike god. If basic life had not been naturally formed biochemically, humans would not even exist today (19). That means there would then have been no concept of a humanlike god or no need for such a god. Clearly, man came before his concept of God. But then, no continuing concept of a traditional god would be expected after the supporting sun ends its life of nuclear fusion as predicted, at which time the people of Earth will become extinct forever more (20). There would then hardly be a need for God. Theology may provide easy answers to these questions but the intellectual and scientific communities remain concerned with factual explanation.

When human mammalian form emerged from lesser life over a million years ago, it retained the primitive awareness and emotions of animal ancestors (21). Because neural systems of advancing life gradually acquired "awareness" and finally "awareness of awareness", man began to search for meaning and purpose, for a "creator", and for an explanation of both life and death. As language developed among peoples of the Earth, the primitive mind created many different belief systems relating human values and behavior to a god image. Separate civilizations produced different interpretations of their fears and inner feelings, some even imagining direct messages from their gods of varying descriptions. It is important to be aware that human uncertainty and search for spiritual meaning does not in itself constitute the existence of a god.

Spiritual leaders periodically arose in past millennia to stimulate vulnerable minds, some even suggesting personal knowledge of the character and intentions of their god. But man could not decide then which belief, if any, was the correct one and cannot decide which one is, even today. The spontaneous formation of life, the mechanisms of the modern mind, and the nature of the universe around it, have now become increasingly understood in functional terms. For the good of man, it is therefore necessary to take stock of his modern state of understanding to determine whether his retained earlier spiritual visions fashioned from "feeling" and relatively limited knowledge, were right or wrong.

Enhanced recognition of reality: Entering the twenty-first century, we see that man, the most advanced "evolved" mammal to date, indeed, has continued to undergo change. He has not only assumed a more realistic vision of life, but many more informed members of advancing society have now begun in earnest to criticize and question his earlier, solidly entrenched spiritual visions as being primitive and unrealistic. Modern enhancement of reality perception is now reaching a sufficient level to begin to detect the true nature of imaginative perception. The induced fear of a humanlike god and the natural fear of human mortality, the principal concerns of imaginative beliefs, seem to have already peaked and have begun to fade in many minds. Based on careful observation, select man is beginning to adopt a more factual philosophy of a natural mechanistic world origin and function in which a human mind form played no role at all. Of course, it could not have, for it came only billions of years after the universe was formed.

**Change is inevitable:** Religion is not a matter of the consensus of a number of believers or non-believers but a matter of depth of absolute understanding and rationality. The new movement of realism has begun to take root however, and is now destined to become a permanent avenue of man's spirit of change in his continuing natural advancement. Meanwhile, the weaknesses and slowly fading intensity of traditional beliefs suggest major changes in the social mind. Although broad surveys on religion reflect the immediate position of particular social groups (22), they do not adequately reflect the pace and influences of ongoing dynamic change. Historically, we know that significant changes first take place among concerned visionaries, a relatively small informed element of a population, and then eventually reach out to all. The reader must remember that, despite overwhelming doubts, the once flat world became round. Similarly, the sun eventually replaced the Earth as the center of our early cosmos and will long remain there. Symphonic music, an important human emotion, did not even exist prior to a few centuries ago, but will now continue forward as a major part of man's life. The irreversible processes of positive change become more pronounced in each successive generation because of constantly expanding human experience and mind function, but it is often little noticed during the time that change is taking place.

Even in this new millennium, amid all the distractions by human conflict and complex interaction, man still seems to be quietly undergoing transformation, not only in his relationship with planetary fellow occupants, but also in his spiritual visions that are largely based on an assumed relationship with an invisible controlling humanlike god. However, for modern man, relationships have not become as deeply metaphysical as in the past. The underlying tenets of organized religions are not only fading but imaginative spiritual thought is increasingly being regarded with indifference and now, more frequently challenged, particularly by those minds that are supported by the very products of accelerating human progress. There is increasing consensus among informed intellectuals that man may well have had it wrong with respect to his most common spiritual tradition. Clearly, an extended race is developing between religious promotion and factual education.

As the world population expanded and became more informed, monastic spiritual emotions became somewhat distorted and have been increasingly transformed into more social, political, and even militaristic entities. Accordingly, some religions have assumed a decided cultural base. But even in their modern complexity, it appears that the spiritual enigma and stigma are approaching a zenith among men and society can anticipate a descending turn toward an eventual nadir. Notable concern emerging from some of the most respected informed minds promises a movement of considerable social significance, one that is destined to form a basis for spiritual resolution, perhaps in the next century. Such prediction is no longer a speculation but, based on broad assimilation of modern knowledge, represents certain eventuality. Typically, such currently unpopular proposals are at first rejected and written off, for they swim against the public current, but if there is enough substance to swim with advancing human knowledge, and in the same direction as human development itself, they can be expected to endure and eventually succeed.

As much as we all might like it to be different, factual science knows with certainty that life is an insignificant, spontaneous product of the vast chemical universe and that man is its most advanced form on planet Earth (23). We know with certainty that the emerged character of developed man is largely the product of his learning and desires, and that his emotions are responses to his thoughts and feelings and to stimuli through his five senses (34). Among his most primitive desires is a protective relationship of affection with an imagined superior god. But with his newly gained modern knowledge many men no longer require an imaginative "crutch" but look forward to a full life in "reality." His preference is expected to slowly turn to greater pursuit of that interpretation. Neurally, only fresh generations can readily form minds of expanded spiritual freedom and absorb the meaning of such new knowledge. However, the forward vision of man is easily obscured by current emotion and by new technology with which he is increasingly surrounded and continually buoyed.

**Beyond the religious era:** This book does not see belief so much as a Dennett "spell" (24), a Dawkins "delusion" ( 25) or a Hitchens "poison" (26), nor does it see traditional belief immediately as a Grayling "last gasp" (27), nor abruptly subject to a Harris "end of faith" (28). Those references, collectively referred to as the new atheism, are thought to be justified knee-jerk reactions to the unsettled spiritual condition and religion related concerns of current society. This book looks to the changing nature of man, not God. It does, however, share implications of a protracted natural change in spiritual thought based on two prime factors; first, on the ongoing evolutionary-like transformation of the reasoning structure governing man's neural ability to recognize imagination and to interpret reality, and second, on his increasing sense and acceptance of observed reality as influenced by accelerating processes of communication, advancing factual knowledge, and by greater individual interaction with his surrounding world.

These are indeed powerful new ingredients that ensure inevitable change in spiritual thought but yet without accurately predictable intensity or time scale. Because of the strong past desire for a commanding god and consequential deep religious entrenchment, the entire process of predicted spiritual inversion might consume many generations but the end result now seems assured. Although new knowledge of human brain function suggests beneficial neuronal regeneration and substantial plasticity in adaptation over one's lifetime, existing deep believers will not likely be sufficiently able to extricate themselves from a semi-locked state of positive neural feedback (57) to appreciate these comments, as explained more fully in the following chapter on spiritual mind function.

**In defense of imagination:** In his new book, *God and the New Atheism*, John Haught commented on "the outbreak of atheistic treatises." He said, "I must confess that it has been disappointing for me to have witnessed the recent surge of interests in atheism. It's not that my own livelihood, that of a theologian, is at stake, although the authors in question would fervently wish that it were so. Nor is it that the treatment of religion in these tracts consists mostly of breezy over-generalizations that leave out almost everything that theologians would want to highlight in their own contemporary discussion of God."

The "highlights" offered by most theologians in discussions of God do indeed constitute the very imaginative visions to which this book largely refers. Except for aggressive ministries, the "livelihood" of persons in the imaginative religious loop is already being negatively affected (29) and is expected to be further affected in the long course of time. I trust that the open treatment of spiritual belief in this book is absent "breezy over-generalizations" and will not be regarded as "harsh or unfair criticism" but viewed only as a careful examination of the subject in real terms with full appreciation of its sensitivity to the long standing principles and depth of sincere belief.

According to an Associated Press release authored by Rachel Zoll in May 2007, Richard Mouw, president of Fuller Seminary in California, also commenting on the sudden appearance of atheistic publications, was quoted as saying, "It's almost like they all had a meeting and said, "Let's counterattack." He appropriately added, "Whatever may be wrong with Christopher Hitchens' attacks on religious leaders, we have certainly already matched it in our attacks." It is indeed unfortunate that the new publications that simply recognize imaginative generation of religious notions are regarded as "attacks." Although this book may also be considered a reaction to the current state of imaginative religion, it is intended as confirmation and a more in-depth development of my earlier published observations of ongoing natural changes in religious thought.

The appearance of many more atheistic-leaning books also supports the principal theme of this book suggesting humanism and secularism (30). One notable recent book in this category, written by Victor Stenger, entitled, *God, the Failed Hypothesis* (31), described how science shows that God does not exist. This is the work of a recognized physicist who combined arguments from physics, astronomy, biology and philosophy in defense of reason. Professor Emeritus, Mark Perakh of California State, commented on it as follows: "Stenger convincingly shows in his book that a combination of factual evidence with simple logic makes belief in supernatural entities untenable." Dawkins also commented on the Stenger book, pointing out that, despite its lucidity, "the faithful won't change their minds, of course, that is what faith means . . . . god is running out of refuges in which to hide."

**The frontiers of life:** Dr. Lee Silver, a recognized Princeton biophysicist, recently wrote about "the clash of science and spirituality at the new frontiers of life" (32). His courageous examination of life as an element of nature is also consistent with the theme of this book and contributes much to explain important, if not essential facts to which many men are yet oblivious. Michael Gazzaniga, director of the SAGE Center for the Study of the Mind, commented that Silver's book "puts those, who reason by assertion of prior traditions, on the run", that it "makes you rethink the most basic questions about the nature of human existence." Robert May, President of the Royal Society and advisor to the British Government, suggested that it provides a sensitive account of facts derived from advancing science about "unintended consequences and religious beliefs."

**The forces of reality:** Relative to the theme of this book, Mark Taylor's *After God* (33) also suggests that contemporary notions of atheism and the secular are "already implicit in classical Christology and Trinitarian theology" and if we are to negotiate the perils of the new millennia we must refigure the networks that inform our policies and guide our actions. With strong fundamental expressions of faith and the new equally strong arguments of the scientific and intellectual community, claiming the biological and neurological basis of belief, Taylor suggests that changing beliefs are not only more complicated than society thinks, but that the influence on society will be unprecedented. He apparently believes that religion without God will lead to ethics without absolutes and the promotion of new life in an increasingly fragile world. While some of these issues are discussed in later chapters, this book sees man continuing to be driven by a strong natural force directed toward the acquisition of knowledge, the recognition of reality and finally to uniform secularism. As always in the past, society with its ongoing natural creativity over an endless future will have ample time and opportunity to adjust to ethics with positive absolutes as well as to the new spiritual unity suggested in this book. Conflict will continue largely because of the natural strength of the emotional mind of desire and survival, experienced and recognized but yet, little understood by man. Adjustment to the imaginative is often easier than adjustment to reality.

Arguments based on factual science, however applicable and well meant, are still sparingly read, much less understood by the average citizen. For that reason this book does not intend to rehash arguments that might simply fall into the popular age-old public debates between "science and religion." It is largely based on the expanding application of the most recent state of factual knowledge. Although it comments on appropriate scientific disciplines in passing, it concentrates primarily on the all-important natural function of serial advancing minds, emphasizing processes of learning, memory, and judgment, as they concern belief and behavior. It is there that belief begins and is habilitated, and it is there that belief must face the indefatigable opposing strength of increasing human reasoning.

Imaginative religious belief has long been claimed to be the "opiate of man." Some now call it a "placebo" providing relief without genuine substance. But the underlying issue more properly concerns the seminal breakdown of perceived reality through the neural strength of over-powering selective emotions, breakdown that man has similarly experienced in other eras of its long history in the process of ongoing mind development, each era involving appropriate change and recovery to a sense of greater realism. For example, even the venerable pages of the Talmud are said to have abounded in descriptions of the realm of magic, demons, and other manifestations. A true Jewish believer accepted these things as existing between heaven and Earth (51), a notion long since abandoned. The traditional concept of a humanlike god, however persistent in much of current society, is similarly expected to undergo change in successive future minds.

**Desire vs. reason:** What mechanism of the mind allows perception, formed through the five reliable senses of man, to be interpreted as other than reality? Such perception seems to want to exceed reality. One can claim a sixth sense, but in so doing, is knowingly engaging in imagination. A sixth sense is only the uncontrolled interaction of the five senses of reality, incapable of introducing new reality. What induces man to assume an imaginative vision crafted to one's own desires? The strong emotional forces simply overwhelm human capability of exercising normal logic and reason without serious further analysis, reflecting just what it emotionally desires. The meaning of information from the five physical senses is effectively distorted deep within the mind by visions of personal need and desire, perhaps in an "ancient" area of the mind whose chemistry was innately prepared for attachment and loving care upon newborn arrival into the world (34). Spirituality must certainly represent a temporary perception in the evolving mechanism of the emotional mind, a sensitive semi-stable mechanism subject to gradual corrective development and change.

The predicted spiritual change is expected to involve a relatively coarse time scale, generation by generation, in contrast to change over the period of an existing lifetime. Continued traditional belief is therefore seen only as a temporary but extended natural function of the earlier emotional and imaginative mind, subject to gradual change toward increasing recognition of reality. A new book, *Across the Secular Abyss: From Faith to Wisdom* by Bainbridge (35), is said to draw upon the best evidence "to understand profound religious changes occurring in advanced societies today, as previously separate fields of science converge, leaving progressively less room for faith." The natural trend, as described in this book, is bound to transcend causal views of faith-based religions despite their firm place of human importance in current society and their occasional spurts of temporary expansion often likened to the outbreak and spread of human disease.

**Beyond human control:** The increased number of frank and pointed publications by intellectuals, philosophers, and scientists that have recently appeared in print, all tend to support my earlier seminal views that seem to have marked a turning point in mind development and reasoning with respect to spirituality. However, I should note that sensitive and vulnerable minds cannot be entirely blamed for experiencing strong natural emotions, responses to the overwhelming influences of idealistic imaginary information with which they have been constantly bombarded as the very basis for their lives.

This, somewhat softer treatment of increasing secularism, viewed as a natural consequence of reality-based modern reasoning, sees traditional spiritual development as a prolonged, separable phase of early mind formation, hidden within the continuum of its ongoing development. Beliefs are seen as traces of an obscured primitive phase of mind function, one that began soon after emergence with the awe, fear, and desire that first inspired spirituality and in its perceived needs and ignorance had to rely on imaginative interpretations, interpretations that are now destined to be corrected with new vision. Still developing minds will certainly continue to respond to religious promotion but, generation by generation will gradually have to abandon the tenuous and fluid faith-based world to seek sounder, more fact-based understanding.

**What keeps it alive?** Once informed of possible immortality and the existence of a benevolent humanlike god believed to be in contact with people of the planet, the notion becomes a very powerful central point of life for many uninformed people, for no sound explanation, competing spiritual alternative, or forceful counter-argument has been generally offered them. It is something they should and must submit to their own knowledge and reason. Freeman Dyson once quoted his mother as saying, "If you throw religion out of the door, it will come back through the window." I remember my parents saying the very same thing. But, the notion that man is inherently spiritual is only true to a point. Although this book claims that religion is fading in the civilized intellectual world, it is still very prevalent elsewhere, but then it clearly requires constant reminding, preaching, promotion, ritual, marketing, public prayer, and advertising to be reasonably sustained. Religious social membership is perhaps more significant today than is deep subscription to the underlying tenets of belief. If the mechanisms of spiritual influence were explained to the early mind in the manner that I do here, and if, because of its basic uncertainty, appropriate disclaimers were required by law, as they should be for all uncertain idealized claims and promises, and if full education based on factual human knowledge were made mandatory and available to all, traditional imaginative belief would have faded more noticeably.

**Evolving human reasoning:** Based on the expanding knowledge of man, it therefore seems clear that the natural evolution of his complex developing mind will continue to move monotonically on track in the direction of confirmed reality, slowly separating from unsupportable imaginative faith-based interpretations of the past. Although, perhaps not yet at a clear tipping point, literature and expressed thought show sure signs of spiritual uncertainty, for many minds of traditional bent increasingly retain simultaneous ideas of both belief and non-belief. Reliance on faith alone for any matter of vital human importance is becoming less satisfying and indeed, more acutely incompatible with accepted modern knowledge and reason. A recent news article (36) spoke of Italian anxieties about the future of Europe citing "the loosening of national identities, the rise of immigration, and the decline of Christian belief." Other sources suggested that only one out of ten of its Catholics now attend required weekly church services.

Human brain function has been a popular topic of journalism over the past half century. There have been many informative issues of popular magazines, even major cover-issues, on brain structure and function. It is now quite clear to researchers that the brain was not the product of purposeful design (37). Even the famous Francis Crick, a discoverer of the genetic helix, once facetiously said that he considered God to be a "hacker" with respect to brain design. However, since the "decade of the brain" at the end of the twentieth century, man has begun to understand much more about the origin, structure, and function of the advancing human mind and the response of his five senses to reality. Select man is now simply experiencing a more advanced stage in human "awareness of awareness", more reliant sensing, and increased understanding of emotional response. This condition and the eventuality of the underlying spiritual mind are discussed more fully in the next chapter.

For various reasons, including the widespread influences of religious shrines and biblical lore, the uninformed naive mind acquired the notion that religion can even cure the medical problems of man (38). As much as we would like to think otherwise, man has now determined that prayer has no medical benefit with the exception of the strong placebo effects discussed elsewhere. This notion would then place spirituality in the same category as drugs, that is, the formulation of an imaginative god as a medicine designed to make people feel better. The conflict between the findings of science and the claims of religions will not much longer go unresolved in the advancing human mind, but a sense of resolution has thus far largely reached only a relatively small, sophisticated and intellectual portion of society.

As more factual knowledge begins to penetrate all of society, man will most likely first turn away from the notion of a humanlike god and its heavenly framework, and then the notion of immortality of the soul, and will fix his mind on forces underlying the more mechanistic formation and function of the universe as observed and described by accepted methods of modern science. Religious leaders, eventually acknowledging reality, may then turn to universal worldly improvement and churches may become centers of factual communication among people, vocal sources of modern understanding and more acceptable behavioral principles that enhance coexistence.

**The pace of change:** Given explanations and corrections of belief processes and the prediction of greater realism, science itself has not yet fully mastered the origin of the universe including fundamentals of gravitation, magnetism, and space-time. Although science also assumes a kind of "faith", it is likely that parasitic man may never enjoy full privity with his universe on all these fundamentals. Meanwhile, less informed spiritual minds will likely continue imaginative interpretations and lingering conflicts with more factual concepts of natural transcendent forces. However, it is not the casual opinion of contrary modern man but the underlying power of an unstoppable evolutionary process that will slowly change man from imaginative aspects of belief to more confidence in his own factual findings. The pace of change may be far more significant than man can yet imagine, and changes may even take place abruptly without much advanced notice. Although accurate temporal projections are increasingly difficult to make, Ray Kurzweil, a popular futurist, suggests that forward-looking predictions in technology are based on a process he calls the "law of accelerating returns" (50). He claims amazingly predictable trajectories and notes that the steepness of the learning curve is now almost mind-boggling. Creative and imaginative human thought is steadily advancing, accompanied by correspondingly advancing capability of reasoning.

A major forward-looking question with respect to the religious issue discussed here is how human progress will relate to natural brain enhancement and especially to brain changes that might be effected by the ingenuity of future man (40). Artificial "brain enhancement" is not necessarily expected to be accompanied by spiritual change but there is much current discussion about adding computational and other devices to modify or supplement human brain function in some manner that might substantially affect its overall ability and capacity. However, neuroscientist Vilayanur Rammachandran suggested that it would be difficult to "reverse-engineer" the human brain because of its haphazard evolution mentioned earlier. On the other hand, some technologists (41) insist that "super intelligent" networks have now arrived and that the human brain must be prepared to "collaborate" with artificial intelligence on a regular basis. Those who have the knowledge and courage to think ahead, envision an entirely new world of complex human thought with less humanlike emotion, emotion that Antonio Damasio claimed to be required at some level for effective decision making. When and if such a new world descends upon us, it will likely be a more godless world, compatible with the anticipated long-term natural direction envisioned here for man.

There is no question that effective brain changes will certainly encompass a modified view of natural stimuli and religious belief and reflect directly in the character of human life. The most likely changes are responses to new information and continued improvement in applicable reasoning capability. Even without the influence of unusual brain changes, traditional religious belief is already teetering on the fence in many areas and civilization may well be in for some unpredictable surprises. Exactly how religion will fare as a function of time is difficult to predict because of the current make-up and distribution of belief and its complex mix with culture, race, technology, nationalism and processes of government. These deepening mixtures only verify and support the concept of ongoing change. Today there are very large groups of believers that emotionally rely heavily on the existence of a benevolent god and on an ideal heavenly afterlife. For many narrow believers, life is very serene as church membership and goodness reign in their lives and in the lives of closely related society. Future generations will no doubt interpret and appreciate life quite differently.

**High-speed evolution:** Expanded "cultural evolution" and mind development are expected to make it possible to sufficiently separate spiritual emotion from reality, to equally serve the fundamental non-religious and non-imaginative interests of all man, interests that include freedom, happiness, and satisfaction, whatever life can offer, all based on advancement in human development. As control of behavior is disassociated from the fear of an imaginative god, man will tend to rely on modern human understanding, suitable monitoring, and enforcement of established rules and laws of ongoing society.

Mind evolution in this book means a combination of adaptive genetic mechanisms with those mechanisms called "cultural evolution", a descriptive term coined in earlier publications. Cultural evolution is based on much more rapid mind change by successive influences from generation to generation, not genetically, but through inter-operation of minds through senses and memory. Cultural evolution not only creates and propagates specific behavioral characteristics but also keeps them alive for long periods. Together, these two important mechanisms may provide substantial change in human character over time. The biological mechanism of traditional genetic evolution is quite well understood although there are still some differences in interpretation of their working details. Such evolutionary change is generally reflected in progeny, defined in its resulting genetic structure.

The mechanism of cultural evolution is quite different. Progeny and others may be influenced and molded by teaching, by directive, by observed behavior, and by example, with respect to persons who are part of the environment. Cultural influence is exclusively transmitted through the five senses and may remain effective over a long sequence of generations. Further, its effect may well result in a rapid influence on globalizing processes discussed in chapter nine. It is this form of evolution that most effectively applies to the historical transmission and to multiplication of spiritual belief and should equally apply to expectation of its future reversal.

While traditional biological evolution acts to create permanent changes in structural formation, both external and internal, the combination of the two forms of evolution can produce a fully modified man, given sufficient time. The recognized historian, Daniel Lord Smail recently wrote in his *On Deep History and the Brain* (42) that "the imperfect copying of past behavior and small, often unconscious preferences, can push a society in a new direction, even without anyone aiming toward a particular goal." "Cultural evolution can be rapid and it can help human beings adapt to their environment, but it needn't be intended or progressive." More descriptively, it is said to be "teletropic" and can work to affect the mood of others, stimulating a "wash of neurochemicals" at a distance. He agreed that cultural evolution in the past has progressed at rates that Darwinian processes cannot begin to approach. Although he speaks of the psychotropic mechanisms shared with animal ancestors resulting in a directionless series of ongoing behaviors through human history, this book suggests a specific spiritual direction, a selective direction in which belief and factual human knowledge must eventually converge to close the issue of spiritual uncertainty.

The June 22, 2006, New York Review of Books contained an excellent review by Professor Freeman Dyson of Dennett's *Breaking the Spell* in which he quoted a friend and famous scientist as saying, "Religion is a childhood disease from which we have recovered." His apt reference was a reminder, not only that religion is frequently seen to spread and function as a contagious disease, but also that it has pretty well run its most infectious course and is now experiencing constructive rehabilitation. Similar sentiments about contagious spirituality are contained in these chapters. It suggests that the world history of religion demonstrates how extremely contagious it can be, how easily it infects the relatively uninformed, the vulnerable, and the immature, how its promises deny proper diagnosis and relief, and how long it has variously tormented man. It has, in reality, functioned very much like a contagious pathogen for which perhaps, there is no more effective remedy than expanded factual education. After many millennia of ineffective effort, realistic education is only beginning to reach and permeate world minds. It is now only in "kindling" status but is expected to serve as the necessary fuel for the anticipated flare-up of the fires of new reality.

**The cognitive revolution:** A recent op/ed by David Brooks (43) succinctly updated the principal belief issue, largely as specified in this book. He cited new literature by Newberg, Siegel, Gazzaniga, Haidt, Damasio, and Hauser, suggesting that beliefs will soon become the subject of much "wider discussion", that we are in the middle of a scientific revolution and "there will be a big cultural effect." Although mysticism and imaginative visions will no doubt continue for a time, they will largely come from the deeply spiritual, uninformed, and older population segments. Contrary to the Brooks view, however, in many systematic respects, the advancing brain does indeed work rather like computer logic and will therefore eventually see man through the perennial "god problem." It now seems clear that the cognitive revolution, already well in process, will dominate man for at least the next half century and will gradually "undermine faith", both in the traditional god and in the Bible.

Although one myth has always been readily replaced by another more updated myth, the recent ages of man have clearly demonstrated a slow change away from imagination toward greater fact and reason. It is a fundamental long-term human direction most likely contained in the underlying adaptive mechanism of human evolution. Despite widespread current religious views, eventual secularism is therefore expected to materialize in some yet indefinable persistent manner. The so called search for truth that has been so much a part of past theology will certainly move beyond imaginative objectives with the continuing process of seeking and finding "greater reality."

**Invisible change:** Society, however accustomed to periodic challenges to tradition, currently seems to have little overt awareness or appreciation of the significance and intensity of such an ongoing contrary movement. But it is nevertheless destined to proceed forward, perhaps with a very bumpy ride, one that will endure major new challenges and one that will slowly modify many aspects of traditional human life. In all ages of developing man, slow change has allowed for proportional generational adjustment and is therefore, fully recognized only in subsequent history, and that paradigm is expected to apply here as well. The prospect of an abrupt, widespread radical rejection of belief would otherwise suggest untold social chaos. Nevertheless, mankind seems to have become so enwrapped, so confused and so misdirected in its spiritual diversity that it may have to endure some form of "spiritual shock therapy" to transcend conflicting imaginative beliefs and modify its direction.

Having investigated many of the mysterious and complex corners of knowledge, I feel that the emotional human mind has allowed man to generate a plethora of deceptive notions, cleverly hidden within languages of philosophy, religion, psychology and science. Some of those notions may be ensconced in the very same area of the working mind as imaginative beliefs, notions not always carefully based on acquired and stored factual knowledge. They tend to leave one with a false sense of mystery, not subject to observed laws. They all need to be brought into the modern open mind of reason and logic, reevaluated, and placed in full synchronism with recognized reality.

**What is delusional?** One can readily leave the world of reality in a "dream" by the use of drugs (44) and even by harsher means, but return to reality, the base of factual knowledge, is always the preferred way of life. Recent examination of deficits in brain growth and function among the world population has taught man much about reason and mind reliability. It simply demonstrates the chemically directed detachment effects of thought processes with respect to reality. The human mind is very sensitive and its response to many forms of stimulus must be carefully considered. Some minds may function with a lower reality threshold or react differently because of different chemistry or structure and may even feel or desire a sense of "mind control." Such minds may be prone to "delusional thinking" and may interpret belief quite differently (45). When such thinking and behavior reaches an abnormal condition the mind should be submitted to therapy. The Internet now contains websites through which delusional thoughts are thoroughly discussed and interchanged. The American Psychiatric Association, however, in its guidelines, says that if a belief is held by a person's "culture or subculture" such as for rituals of religious faith, it is not considered a delusion. This suggests a rather "free and different" interpretation of social behavior when religious belief is involved.

**Ideological unity and peaceful coexistence:** Strong emotions are part of all deeply religious processes. Imagination and reality can readily be intermixed, a process that can enhance acceptance of imagination. But there are now very few inaccessible mysteries of importance contained in cortical folds, or at the roots of long white beards, hidden within words, like phenomenology, prophesy, existentialism, or psychoneuroticism. Except in the area of spiritual belief, there are no longer serious concerns with spirits, devils, ghosts, or shadows of the darkness. There are simply things that advancing man does not yet know and has to learn largely through the pursuit of science in terms consistent with established working knowledge, reliable information that will benefit and interest man, factual knowledge that will lead to improved common understanding and therefore to greater satisfaction and peaceful coexistence.

**Satisfied without a god:** Christian societies like America have been "saturated" with the fear of living without a god and a potential afterlife. In contrast, those elements of societies that have advanced beyond imaginative belief are relatively successful, have low rates of crime and corruption, and are supported by strong modern educational programs. In Phil Zuckerman's *Society Without God* (4), he describes interviews with citizens of Sweden and Denmark, societies that do not worship any god, that do not pray or have an abnormal fear of natural death, and are generally happy and content in their lives of secularism. This book suggests that these are clear examples of trends and expectations for all society. The reality of secularism may not be as contagious as is the imagination of spirituality but it is based on an assured stable ideology.

**Just getting smarter:** In closing this chapter on the forces of change, mention should be made of the yet unexplained steady rise in IQ of children and some adults over the past century, called "the Flynn effect." This effect was postulated a few decades ago by James Flynn of the University of Otago in New Zealand. Nutrition has been temporarily ruled out as a factor, but scientists believe that it might well be the result of using minds in new ways and of dealing with more complex technologies. Whatever the specific causes of advances in mind function, this book claims that, as the result of natural forces, they will be a significant factor in future spirituality and in the behavior of man.

Continued newness and change are simply expressions of the unstoppable and all-encompassing natural forces that accompany the chemical function and evolution of life and will endlessly modify and guide the path of man on his planet. The term "forces" is used symbolically although effective physical forces do exist among all the particles of chemistry, among all the members of the solar system, and among elements of the galaxy and their galactic nucleus. Man has little control over fundamental mechanistic forces and is entirely subject to their dictates.

However, evolved man has a strong natural proclivity, indeed, a basic instinct, to learn and to know. New research has found that learning encourages neurogenesis, the growth of new neurons in the brain, and preserves the neurons that are thereby created. It is an essential process in human development favoring factual, as opposed to imaginative, knowledge. Therefore, with respect to human direction, the essential sequence of learning, reality, and secularism represents important steps of his long continuing journey into the future. Clearly, man must resolve the god problem before he can attain true understanding, deal effectively with his place in the universe, and achieve responsible global behavior.

# Chapter Two

## The Spiritual Mind

> The mind of man seems so far removed from anything known in other animals and the animal mind seems so inaccessible to us that those who approach the problem from this side seem prone to seek a way out through metaphysics or mysticism, though relief of this sort is obtained only at the expense of profound narcosis of critical and scientific method.
>
> C Judson Herrick—1924

Man is naturally equipped with several principal senses, each connected directly to his mind, through which he can communicate with others and can selectively detect the nature of his vast outside world. Religion has made claim to a presence felt by the mind that reaches beyond those senses, a presence imagined by believers to be a sacred or divine mystery, a mystery that cannot be fully understood. Although no fundamental difference has been claimed between the mind of a believer and that of an atheist, the spiritual mind has been led to believe that a humanlike god is responsible for universal creation and function. Such pervasive unverifiable understanding, perpetuated by believers and emotion-driven leaders, explains why so much of surrounding society appears to be religious. But the modern rational mind is beginning to see it differently. The sixth "sense of spirituality" is slowly being replaced by a more understandable "sense of reality." In all the history of man, none of his senses has ever detected any valid sign or actual existence of any humanlike divine being. The imaginative nature of emotional response, however, often goes into high gear spontaneously. Modern mind study increasingly explains how and why such imagination becomes inexorably mixed with reality.

**Mysticism is unreal:** This book fully recognizes the natural basis of human emotion and its widespread tendencies toward mystic interpretations of human interests such as love, beauty, environment, birth, suffering, and death, but without surrendering appreciation of any such aspects of life, it separates the residue of earlier uninformed, superstitious, and emotional responses from those of modern minds of enhanced understanding and reason, and presents an optimistic view of man's future of more rational thought. Therefore, this book does not acknowledge the existence of any unsupported creations of human imagination intended merely to satisfy spiritual emotions, for such creations are considered to be violations of accepted principles of reality (46). It recognizes only the biological senses of man, measurable known senses that alone can be confidently exercised in all communication and learning processes and reliably utilized for all judgments throughout life. It clarifies the fundamental spiritual issues with humility and courage and provides a pedagogical directive in keeping with the human goal, the realization of optimum life, not in the imaginative, unreachable, idealized heaven that man has constantly sought through emotional hope and desire, but in the realistic heaven available on his own planet during his lifetime.

**Life as discontinuous chemistry:** With that said, the reader should take a giant step backward to a sound platform of modern understanding. The biochemical process of human birth slowly adds population to the once bare planet Earth, population that is produced entirely from its vast chemical resources, only a microcosm of the endless universe of similar content. At death, human chemistry is again returned to the planet and the process goes on. When there are more births than deaths the population rises, as has been the case throughout most of human history. This inclination for continuity through reproduction is one of the natural propelling forces contained within human instinct as it was before man in his animal ancestry. The established sequence will fail only if mankind is subjected to some overwhelming cosmic event, a major debilitating disease, an unfavorable change in climate, or if man should become self destructive.

Human propagation is not a characteristic or an intention imparted by any god but a natural biological evolutionary process (47). It is consciously encouraged by the limited longevity of man. Although humanity is said to have first emerged from animal form on the continent of Africa, there are now billions of subsequent generations distributed over the entire globe and it is estimated that a few billion more will be added in the coming century. Surprisingly, the record shows that almost five million births took place in America alone for the first time last year. If factually educated, such new minds will represent part of the anticipated march of modern knowledge that will dominate the next age of human thought and behavior. The reader should also contemplate the vast universe of endless dimension and its billions and billions of uninhabited, unsuspended bodies roaming its immeasurable realm. But it is likely that it has little fundamental meaning for the human life we know. It is merely an infinite forest of orbs that man can admire and explore as he is capable of doing so. Man will eventually fix this expansive image in his mind without fear and emotional concern and learn to live a satisfying earthly life of relatively confined reality.

**Interacting minds:** It is indeed difficult to imagine the increasing complexity of so many new interacting, yet unsynchronized, minds on our own planet. They are all interconnected only through the senses. It is no wonder that there is so much difference and confusion in thought and behavior and the need for some sense of unity and order has never been greater. Today's difficult human problems, distributed throughout the continents, will be further multiplied in complex permutations and combinations. But among their many concerns intelligent humans have acquired an obsession with simple "origin" and "creation", concepts unknown and completely "foreign" to the underlying non-human forces of nature, seminal forces of which the mind of man should be aware though he may never fully understand them. The "god problem" is subject to logical resolution and must inevitably reach clear undivided agreement among man. Spiritual unity and peaceful coexistence based on realism appears to be a most essential step. Meanwhile, life on Earth is expected to go on well into the foreseeable future and eventually the planet may even approach saturation with living chemistry. Interpretations, such as the prediction of an apocalyptic end of the world on December 21st of 2012, as currently seen by "survival groups" (48) based on the Mayan calendar, must be regarded as just another imaginative exercise.

**The contagious influences:** A paper titled "The surprising power of neighborly advice" based on work at Harvard University that appeared in the March 20, 2009 issue of Science may be significant in connection with reactions to what others do or believe. It says that people tend to imagine the essential features and not the incidental features when they know how a neighbor reacted to an event. It further suggested that when mental simulations of events are inaccurate, the forecasts that are based on them tend to be inaccurate as well. Although the issue is much more complex, people evidently mispredict their reaction to future events because they imagine those events inaccurately. This new work is mentioned here because this book frequently refers to the strong influences of what others believe in the process of acquiring a belief. It is an important part of the process of cultural evolution discussed throughout these chapters.

**The forces of understanding:** Today, we find ourselves on the very cutting edge of spectacular human progress and at a defining moment of further advancement. It is not all prevailing technology. The forces of understanding also continue to accelerate and spirituality will gradually merge with reality to become a singular view. The movement toward reality is basic and not subject to the decision of man or of any of his gods. It is the natural direction of the expanding human mind impelled by forces far too unreachable and far too powerful to be fundamentally altered by any act of emotional man. It is not the power of any god that is at stake, but the unstoppable power of continuous natural development and inevitable change. The theme of this book, suggesting a more realistic vision of spiritual understanding and guidance in interpreting human behavior and direction, begins by examining the most controversial and significant characteristic of unique earthly life, namely, how the controlling mind reacts to its emotions and senses. It examines the relevance of belief and reality, still unresolved even at this advanced point in human development.

**Man doesn't feel the change:** Carefully tracing the known history of constantly adapting life, science has recognized the early reflexive behavior of ancestral fish and reptiles, the affective awareness of early mammals, the cognitive awareness of later primates, the self-awareness of great apes (49) (also see list of supplementary references) and finally, the amazing "awareness of awareness" in humans (34), awareness that engendered the initial emotions of spirituality. To simply attribute such development to some "creator" in the light of modern knowledge is certainly an unreasonable cop out belonging to the distant past. We now know that evolutionary advances have involved major changes in physical form but more importantly, they have produced much increased capacity of developing brains. This is especially significant for those who understand brain function. But most certainly, that cannot be the end of human change as the long chain of future lives continues to "begin and end" at its connecting links. Based on modern understanding, how could anyone think of biological change, indeed, constantly reacting living chemistry, ever coming to a fixed condition? By every indication and measure, even the foreseeable near-term change in man promises to be nothing but revolutionary; change is continuous (50). But how will change specifically influence human belief processes? Clearly, it will be through the function of the expanded capability of a larger, older, more fully interconnected, and increasingly knowledgeable human brain. Will inherited emotion and expanded human "awareness of awareness" allow more modern spiritual understanding? Will new reason eventually come to the rescue of earlier misguided emotions? As it envisions a spiritually modified future, this book covers an important, yet widely obscured trend and offers predictions that are largely explained throughout its textual commentary.

Because of the intrinsic genetic code that functions like the seeds of a plant in human origin, development, and growth, most people currently have reasonably like characteristics. However, the form and behavior of man may well change radically in the course of long time. Natural mutations in the gene structure, such as by constantly showering high energy cosmic radiation, produce corresponding changes in progeny. Some genetic differences, however small, routinely occur from generation to generation and the details of new life are always unpredictably subject to further change. But naturally adaptive mutations are effectively compounded with respect to both body and mind, which means that some minds will always have more advanced capability than others, including greater capability of knowledge, reason and judgment (51). These are often minds of new vision, vision that always leads the processes of major social change. Some members of society suggest that current Asian and Jewish minds demonstrate superiority in such vision.

**Developing brains:** Although all minds are of similar structure, they are of widely different content, but they are all now becoming increasingly interconnected because of expanding and migrating global population and rapidly accelerating means of communication as never before experienced in the past. These processes represent unprecedented factors with respect to change and consequently, the proportion of minds of more advanced understanding and reasoning is expected to increase. But not all minds develop uniformly or easily adapt to advancement. Many inopportunely still fall into the comfort zone of simplicity or are guided largely by tradition, satisfied with their lesser role among the masses. These minds will gradually be replaced, obscured, or supplemented in expanding education and advancing understanding of new generations. In popular terminology, the developing mind of man becomes "wired" not only in accord with his basic functional needs but also in accord with learning, experience, and emotional response. The introductory comments reminded us that the mind of a deep believer clearly gets wired in a manner somewhat differently from that of a firm non-believer. The resultant behavioral "rationale" is therefore correspondingly different. The mind must, however, always function in accord with its own wiring despite apparent similarities and wiring is not easily modified. This chapter is intended to explain how it becomes different and the effects of such difference. It also suggests an increasing preference for wiring in accord with accumulated "factual knowledge", referred to as "reality" in this book, as opposed to knowledge established by presumptive a priori.

A new mind is effectively "blank" at birth but, as it interacts with the outside world in various ways, its influx of information tends to become mixed, as though in a great boiling cauldron of human knowledge, giving and taking throughout its functional life (52). Some minds take, reconsider, reconstitute, and contribute new information to the central cauldron of knowledge before death, while others just live and die with a select portion of knowledge that gives them most immediate fulfillment and comfort. When minds have similar exposure to the outside world they generally harbor similar thoughts and, therefore, similar beliefs. Minds subject to substantial religious teaching readily develop the ability to imagine a complex spiritual framework, one that may continually seek even further imagination to gain "fullness." Educated minds, not subject to religious influences, may absorb greater reality but as man continues to develop in experience and knowledge over time, all minds are expected to rise in vision and capability.

**The sanctity of factual knowledge:** Because of limited longevity, new information to be contributed to the central bank of human knowledge must be deposited before death. But the master bank of knowledge, built on factual and well tested observations, is sometimes contaminated with questionable information derived only from emotional interpretations, feelings, guesses, hearsay, exaggeration, misinterpretations, and imagination. It is a common human trait, in ignorance, to resort to imagination in order to neurally satisfy needs and desires imposed by strong natural emotions. The world is replete with past examples of such influences and misjudgment. In that way, imagined information is carelessly assumed, disseminated, and recycled among man. Detailed concepts of related brain function, as to memory and emotion, are much more complex than can be discussed here. The reader is referred to modern books on mind function. As an example, a brief list of books on the subject was shown in a review titled *How the Mind Works* by Rosenfield and Ziff in the New York Review of Books of June 26, 2008. Steven Pinker (53) and Jaak Panksepp (34) are also among many other capable authors on cognitive and biological mind function some of which can be found in the reference section.

The most significant single factor in forming and maintaining traditional beliefs is the universal practice of influencing new minds to think in terms of an imaginary humanlike god from birth and encouraging its continuity throughout much of early life. Despite its purely imaginary nature there is currently a great deal of such information in circulation, not unlike that of civilizations of past ages whose visions man has long abandoned as unreal. Some live by, and revel in, its promises and its appeal to satisfaction, but intellectuals now tend to view such information as an "educational detour." They insist that it is the responsibility of every member of society to preserve the sanctity and purity of accepted factual knowledge in the best interest of those who depend on it, of those who work to advance it, and of all future members who seek true understanding of their world. Those who tend to violate the principles of known reality without adequate disclaimers of uncertainty and doubt must be regarded as self-serving agents of potential deceit and of regressive human progress. This book speaks strongly to the detrimental effects of contaminated factual knowledge in advancing civilization. It is particularly concerned with the contamination promoted by example in high positions of leadership, especially in governments and educational institutions of developed nations.

**Life is learning and memory:** Following birth, every human being undergoes a learning process lasting throughout most of life, one that, like a computer, involves human data acquisition by coordinated operation of the five senses in conjunction with their centrally connected brain (55). Knowledge and experience are there biologically recorded as memory for possible later access, and recording in the form of protein and associated circuitry is as physical as in modern electronic equipment (54). Memory, the very seat of human consciousness (56), is an electro-biochemical process, typical of most species that have developed a central neural structure (57). Although seemingly complex to the present human mind, learning and memory are natural processes of dynamic electrochemistry, the basis of life that, in humans, begins to function even before birth, probably sometime late in the second trimester and ends at death. Vladimir Nabokov in *Speak, Memory* tells us that existence is but a brief crack of light between two eternities of darkness. Modern neuroscience suggests that memories are not stored in specific points in the brain but involve small neural regions in a dynamic process (58). This book questions the quality and effectiveness of some of the unsupported spiritual information selected for retention on which society indiscriminatingly bases its thoughts, behavior, and the teaching of others.

**Planted religious seeds:** Religious belief in particular is the result of a human learning process of special significance and consequence. It is not a natural consequence of individual life, for most human brains are regularly subjected to religious inputs from parents, teachers, and spiritual advisors, starting from the very beginning of life, inputs that are fraught with ideal promises and spiritual expectations but accompanied by relatively few questions of doubt or references to reality. In their formative years, children become innocently reliant on information taught by their respected elders. Religious instruction is most often taken very seriously and becomes a deep conviction. As suggested earlier, it can easily become neurally locked by its promoted intensity as does an electrical circuit with positive feedback.

It is the resultant dogmatic nature of such widespread teaching without the expression of reasonable doubts or disclaimers that deflects freedom of judgment, a basic human right, indeed, an important inalienable right that should be afforded and guaranteed every man, woman, and child. Such restrictions are not properly observed and society is replete with students, teachers and philosophers obsessed with some form of imaginative spiritual vision. Major responsible and authoritative elements of society may even recognize this condition but they are apt to be ineffectively moved because of their own spiritual leanings. Because of the pervasive spread of imaginative belief, society as a whole reflects a conflicting bias and is yet unlikely to take the matter sufficiently seriously to consider or require suitable protection.

The "natural" spiritual inclination of man is generally overplayed. Religions of varying form and depth are most often acquired involuntarily, a concept that forever more is induced in the naked mind, perhaps even foisted on others as though they were nourishment, but nevertheless, they produce prominent mental images associated with common emotions relating to fear, awe, security, survival, death, and the like. They are most significantly associated with the end of life, with the strong desire and expectation of some form of ideal life beyond death. Once induced in the eager mind, these mental images are regenerated and recycled entirely within the mind and prominently interspersed with or hidden in many other normal visions, for religion is based only on concepts of faith and belief without the benefit of conscious reference to comparative reality. Once they are introduced as central concerns in the mind complex, they remain a prominent basis of belief, often laced with uncertainty forever more.

**Consciousness a series of pulses:** Long-term memory is processed in the brain through an organ called the hippocampus (70). Although it is not widely recognized as such, signaling from the sensory areas, as well as within the brain, consist of varying series of ionic or electrochemical pulses that would be totally unintelligible if presented to the senses in raw working form (57). Consciousness might consist of a series of groups of many pulses that vary in timing and in time spacing within the groups. These pulses correspond to the firing and relaxation of inter-stimulated neurons. It is simply reacting biochemistry but undergoes an amazing natural interpretive process. Consciousness and intelligence, therefore, consists of a sensation of complex pulse signaling of which one is not consciously aware, and religious belief comes down to a particular arrangement of such dynamic neural electrochemistry (58). Physiologically, belief may be thought of as an electrobiological judgmental condition that most easily relates to an elecrobiological emotional condition, providing a quiescent neural condition of relative equilibrium.

**Imagination is unlimited:** In this manner, the capricious mind can very easily overstep the bounds of realism in its willingness to entertain a simultaneous imaginative framework of desired images. The emotional mind exercises no rational restraint. A certain false sense of reality is often inadvertently created by frequent reference to lingering ancient biblical lore claimed to be the word of God, and is strongly reinforced by conscious recognition of the vast number of surrounding religious adherents. Imaginative visions, often inspired and encouraged by exaggerated or contrived ancient history of similar origin, consequently gain easy acceptance by the mind despite the contrasting reality of the surrounding world and accumulated modern knowledge. Virgin birth, resurrection, transubstantiation and similar miracles are but a few examples. The mind may even be induced to feel that the ancient religious world was spiritually quite different and incomparable relative to the present world of reality. Once desirable imaginative visions become settled as assumed reality in memory regions, they appear to be somewhat intractable and preferentially accessible.

Some searching minds become very "knowledgeable" about details of promoted biblical history and are little concerned with contradictory aspects relative to the factual modern world. Interest is often spurred on subconsciously by a strong underlying feeling said to be "fulfilling." Religious study is not only prevalent in society but also tends to be academically intense with respect to imaginative details without serious interest in verification. With full understanding and appreciation of the natural human sensitivity to spiritual influence, this book is openly tolerant of consequential imaginative inducements but admonishes current society to exercise reason and minimize reliance on ideology and claimed spiritual events of the past that were humanly derived and clearly inconsistent with current understanding of reality.

**Keep telling them:** Early religious learning is generally associated with virgin memory and often undergoes considerable repetition. It is not only usually established involuntarily, but it is liberally supplemented over a long period of time. There is probably no teaching more intensely directed in the minds of society than that of religion with repetition. As in any forceful teaching process, it is the most powerful key to spiritual penetration, figuratively known as "drumming it in." Because of its imaginative nature, religious thought cannot be easily integrated with other accumulating memories of reality as the brain develops and changes throughout life. In effect, it may even occupy a favored neural "compartment." Natural biological parental relationships may in time convert subconsciously to a god relationship.

Worldwide, the specifics of one's resultant belief are largely determined by chance, depending upon family circumstances and environment, most often acquired directly or indirectly from a variety of uncertain outside sources (59). All widely accepted traditional beliefs tend to follow a singular idealized pattern, a vision of an all powerful humanlike god that one can look to in times of need, and a promise of a heaven of perfection with universally sought immortality. However "too good to be true", it is understandably a promise that most every non-discriminating mind finds difficult to refuse.

**To learn is to know:** Decision-making in belief is an extremely complex process when considered from the viewpoint of neuroscience (60). It involves the executive system of the prefrontal cortex, a process of neuronal interchanges based on stored and real-time information. Decision-making may very well be subconscious but, like all other mental activity, biologically involves active ionic chemistry. Whether imaginative or factual, it is always referenced to what one has been told, what one has learned and knows at the moment of decision, and it is invariably influenced by one's emotions. Unfortunately, judgment by the religious mind is not based on reality but on the strong "pseudo reality" established by the well-rehearsed and universally accepted imaginative frameworks.

Other than memory and the senses, there are no known contributory influences, spiritual or otherwise, in belief decisions. There are no known contributory signals in the form of extrasensory perception or mysterious invisible commands from any outside sources such as from a benevolent spirit, or from any electromagnetic-like signaling systems known to man that would otherwise create or influence judgmental function. If there were any direct contributions to thought processes in reality they would have to be injected in the form of the previously mentioned neural pulses of specific format, subject to natural system compatibility, compatibility with the neural system that underlies consciousness. Despite frequent implications to the contrary, such decision-making is entirely intracranial.

**The notion of a god:** Neuroscientists and other scientists from time to time have attempted to establish a god/brain relationship. In some instances MRI and similar scanning systems have recorded response by particular areas of the brain when the mind is stimulated in various ways. Andrew Newberg, a scientist at the University of Pennsylvania, is among those doing research on the relationship between religious experience and brain function (61). In other research by Michael Persinger the brain is modified by transcranial magnetic stimulation (62) in order to attempt to induce spiritual feeling. This is done by a magnetic device called "the god helmet" that contains solenoids producing a weak but complex magnetic field in the vicinity of the right hemisphere parietal and temporal lobes.

This study has concluded that whatever neuroscientific data may be derived from these experiments it will probably serve good purpose but it is expected to lead only to continued speculation with respect to the results sought. Based on other information and experience there is no reliable evidence or indication that any humanlike god exists to influence brain performance. The evolved human brain is indeed a complex array of sensitive neurons the interconnections which can produce a large variety of interpretations under different conditions of stimulation whether it be through the senses, from memory, or by some direct modification of brain elements. The judgment of a normal brain is neurally determined as described in these chapters and not by divine influence through some imagined communication path.

**When religion is something else:** The mind, within itself, usually settles on some particular religious interpretation that is self-judged and rationalized to be compatible and consistent with other learned information to which it may be consciously or subconsciously referenced. But despite the newly discovered plasticity of the human brain, this process most often affords comforting spirituality the benefit of the doubt. The prefrontal lobe of reason does not entirely mature for two or more decades, therefore in early life, there may be limited judgmental filtering. In the course of life, religion is often mixed with other emotions generated in associated neural structures. In the highly interactive modern world, pure monastic belief is now quite rare, for it is generally found woefully insufficient for full satisfaction. Therefore, religious emotions may exhibit an indefinable and unpredictable relationship with varying aspects of race, culture, music, social status, nationalism, and the like, and may even extend to more extreme militaristic or political ambitions. The personal and private nature of religion is thereby frequently compromised and misinterpreted, even distorted by many factors, by organizational requirements, by conversion efforts, by public base-building, by financial opportunity, by other more worldly ambitions, and, in extremes, by politics and mass activism, even armed conflict (63).

Because of the impossibility of settling on a singularly defined world belief based on reality, an infinite variety of religious interpretations has arisen and the choice of belief is therefore sometimes subject to abrupt change, especially if new information is judged sufficiently different to merit it or if new circumstances provide more opportunity to take advantage of, or expand, specific objectives. Recent surveys suggest that almost half of the United States citizens end up with a different concept of belief than in their early lives (22). Once comfortably acquired, however, beliefs are generally retained, continually rationalized and defended.

**Belief established by chance:** By its very nature, religious information maintains a relatively high position in the hierarchy of learning and memory. It is therefore easily recorded but difficult to modify or erase. There is absolutely no basis to think that spirituality is evenly distributed directly to man in some divine extrasensory fashion.

Indeed, religion is emotionally distributed among many other aspects of mind function that arise quite randomly in earthly man. A particular mind under other circumstances might respond totally differently. For example, the most militant Muslim might have become the most peaceful Buddhist monk or vice versa, but for the chance of birth, location, and time. This randomness may be a significant factor in the serious analysis, appraisal and evaluation of acquired spirituality.

Scattered and splintered world beliefs are many and the more populous, rapidly globalizing planet now brings them all into much closer contact and consequently into greater question and conflict than at any time ever in the past. The larger religious factions become most prominent in influence and competition and tend to attract most new members (64). Claims and promises are given a sense of reality by the shear numbers of adherents. But there are also many small religious groups, some called cults, that have arisen from a different interpretation by local leaders and operate in various parts of the world. Participation is not only for deep spiritual reasons; there is a great deal of joining for power, novelty or excitement as well as for social and family reasons. Limited education and knowledge is also a large factor. There is often an element of pressure due to cultural traditions independent of religious tenets. In times of religious wars, of which there have been many, association, loyalty, security, possession, territory, and nationality also play significant roles in religious participation. In earlier times, many vanquished civilizations were simply converted to a belief system by governmental decree or military force. The many sordid human events in religious history tend to undermine the meaning of lofty spiritual claims.

But religions are the product of the natural emotions of man and clearly not subject to organized overall control by any singular wise and benevolent divine master, except as imagined. Nor are they all tied to any single well-focused world center of peace and serenity. On the contrary, in the world of reality, they remain largely separated in constant turmoil and in competition, and often in disorganized conflict, sometimes unified in large groups under principles of mass psychology (65). They are typical of creations by the undisciplined emotional minds of mixed human cultures derived from separately evolved living circumstances and traditions in different geographic areas of the planet.

**The power of coherent minds:** Irrespective of the quality and nature of underlying spirituality, its power through mass psychology may be used to advantage whenever the opportunity arises. Influence, largely a local phenomenon in the past, has now become a worldwide matter of more universal concern. Civilization is increasingly challenged with fundamental religions of expanding size, distorted by mixtures of strong nationalism and sometimes by militaristic leanings. This book speaks to the increasing dangers in mass conversion to man-made imaginative notions, unmoved by reason or by accepted norms of justice, leading to inconsiderate and disruptive behavior that may adversely affect the interests and rights of other members of society. This book recognizes a natural central force as the basis of human development that may tend toward long-term benefit, perhaps guided by the fundamental objectives of man. It suggests that spiritual needs may gradually be weakening and building toward a unified, more worldly path consistent with this study. The forces of change are not apparent in one lifetime but over a long sequence of many lives, each being a significant link in the chain of advancement. The unity of secularism with increasing globalization contributes to the realization that long-term human improvement must be based on natural principles that maximize peaceful and cooperative coexistence. Like minds without the enigma or stigma of belief tend to strengthen all mankind.

The days of teaching unverified idealism to eager minds that cannot resist absorbing it like a sponge, are now beginning to fade in many regions. The Earth is shrinking rapidly, both physically and electronically. Education is becoming more factual and rational. Effective missionary efforts to new and undeveloped areas of the planet are almost ended. The globe is becoming increasingly finite for its occupants. Whereas highly religious individuals and groups were instrumental in founding universities and other institutions of higher learning, modern education is producing new and different independent minds, many of which ironically oppose the very religious beliefs that were associated with their founding. This is a harbinger of further spiritual inversion to come.

**Eternal doubt:** Faith in an idealistic religious framework is frequently accompanied by the same doubts that troubled Mother Teresa. Lingering doubts in religious circles are universal but yet little recognized and rarely publicly expressed. They are hidden amidst declared, but loosely practiced, membership in organized religions. Casual conversations with members of religious organizations often reflect little knowledge or interest in core tenets. In a recent survey in Catholic Italy, only one in four respondents indicated that religion was important (66). Dennett tells us that many people merely believe in "belief" itself (24). Indeed, religion is ripe for an eventual remake and the bubble is deflating just where it originated, in the highly emotional human mind.

For some, the ancient biblical world may be associated with a promised heaven. The natural sensitivity to spiritual emotion is thereby easily triggered. A subconscious sense of fear of being, of uncertainty, combined with a desire for survival, is universal. This is being offset by expanding knowledge of evolutionary history. Although biblical glitches constantly appear over time (67), the widespread power of

Christianity, as an example, is assumed to have arisen from visions and promises of Christ, a bold expansion of earlier Jewish philosophy, initiated around his teachings of the first century and continuously evolved, and expanded through many ups and downs, constantly modified and promoted by a sequence of inspired followers (68). It is difficult for people today to see Jesus as merely an "ancient teacher and prophet" as were many other "preachers" in biblical times. After he was baptized he evidently preached as the son of God and promised the "coming of God's kingdom" and other unusual events. He was condemned to death, largely for such views, and was crucified by the Romans of more rational view. The followers of Christianity believe that Christ arose from the dead and ascended into heaven as the "son of God." Despite such impossibility based on factual knowledge, its interpretation is somehow hidden in catch-all "faith." An interpretive account of his life and teachings was accordingly written in four gospels, thirty or more years later.

Europe, representing a relatively mature society, currently tends to lead the modern world reversion to secularism. But in America, Christianity is still the most popular game in town, where about three-fourths of the adult population regard themselves as Christian, though with widely different depths of participation and private belief (22). This book views religions, worldwide, as having passed their peak, resting in relative confusion and jockeying for more modern positioning and resolution in the face of increasing human rationality. From this changing process, there seems to be emerging a definite inverse relationship between depth of "broad" factual education and depth of imaginative belief.

**Sins are OK for some:** Despite widely varying degrees of acceptance, religion is kept alive along with political, social, economic and other interests that simultaneously occupy human life, lacking in appropriate emotional alternatives for many. Those who subscribe to non-belief, tend to shun or suppress publicity and do not readily promote equivalent non-religious organizations. Realism must therefore become the natural product of factual human education and, in that manner, must eventually reach a stable form of general secularism. It has often been said that religion is only for sinners. In a sense, that may be true, for it is behavior that frequently first comes to mind when one attends church or thinks spiritually. Many still believe that their sins can, in some manner, be divinely forgiven. Such forgiveness, combined with perceived availability of heavenly assistance and immortality of the soul, constitutes an extremely powerful competitive package.

It is time for all modern thinkers and, indeed, all civilized and educated men, to exercise reason and logic with respect to the basis of morality. Moral behavior is not the sole province of religious minds (69). The exclusive relationship that has long been promoted between religion and righteousness is clearly a myth. The notion of a judgmental and guiding spirit to control human behavior is slowly vanishing. The behavior of man, based on complex psychological factors, is now known to be about the same with or without religious membership, although it is said that religion sometimes affords rationalization to engage in sin with an eye toward forgiveness. It now seems appropriate, even essential, for some great learned central "voice", indeed, a new "earthly Christ", to rise up among men on earth, to explain human existence to the entire world in terms of modern logic and reason, and set the stage to usher in a new and more fitting era of realism in the best interest of modern man. Such an accomplishment must be the product of a superior respected mind, one based on factual modern knowledge and understanding, a sensitive mind, but one not easily turned by emotional reactions to the vicissitudes of life.

**Life with an imaginative base:** Spiritual belief, destined to follow a unidirectional path of change, must not only come to rest in factual reality, the end of the imaginative line, but its associated troublesome stigma and its ubiquitous conflict will also become diluted and begin to vanish among the earthly ways of man. The current modern era of man is surely much different from any major period of human development of the distant past, and future eras may be successively different. Many of the accepted current patterns of society may not even be recognizable upon review by societies of the more distant future. Long projections and critical analyses of future behavior are simply beyond the ken of existing minds. The average mind is primarily occupied with what it knows and is accustomed to, what it has learned over time, what is contained in its memory, and what it observes in the normal intercourse with other minds of society (72). Developing man is still largely a product of inherited instincts, specific forms of education, and natural emotions that will continue to change slowly with increased understanding, generation after generation. There has never yet been any successful program instituted by man to radically change his world independent of the natural forces of change.

**A spiritual tightrope:** The spiritual emotions of man require new filtering and greater separation from otherworldly emotions and social issues in order to provide better understanding of their meaning, interpretation that may pave the road to spiritual unity. This can only be accomplished if human belief is removed from its highly imaginative realm and is placed on an ideological basis more consistent with ongoing factual knowledge. This may already be happening at the roots of the world educational systems, but its progress is still severely limited and inconsistent. Its conversion is a matter of some urgency, for man can no longer afford to base his entire existence on an imaginative vision of false hopes and desires and on the mixed consequences that it is bound to produce.

Persistence of emotional memory: Although religious thought is always associated with strong fundamental emotions, it differs somewhat from most other subjects of learning, however overt or subconscious. Emotional learning is known to particularly reinforce and solidify the biological memory process (71). This is because traditional religion relates to deep personal questions such as the meaning of life, the appraisal of good and evil, fundamental personal behavior, inner perceptions of loneliness, hardship, fear, abandonment, security, and expectations about eventual death.

Located deep in the brain, the amygdala interprets emotional content and amplifies memories that are both pleasant and fearful. Current religious teaching can therefore reflect both extremes. It is a human phenomenon that has always played out in accord with the observed functional pattern of the evolved spiritual mind described earlier. One can assume that questions relating to meaning and to the emotions of spiritual fear, among others, will gradually decline and disappear as the established facts of human origin and man's place in the universe are properly taught, understood, and accepted as given. Even the strange atmosphere and forces experienced by modern astronauts are becoming routine for many. The humanlike god, created by early man to serve the emotional interests and to control the behavior of mankind, has probably all but completed its earthly task in most regions and an alternative modern view is in order. Will advanced minds of distant future man accept a faith-based notion for which there has never been a lick of evidence of any sort? The answer most certainly must be "no."

**Fear plays a prominent role:** The inherent emotion of fear is perhaps the most effective of all emotions in determining behavior. The book by Lars Svendsen, A Philosophy of Fear, explains why the powerful emotion of fear still enters every aspect of modern life (73). Religious belief in a protective god and in life after death tends to be a substantial ameliorating agent. There are many examples of similar effects of fear in normal life. One that has become significant in recent years is the fear of nuclear power. People appear to relate all matters that sound nuclear in nature to the devastating nuclear bomb and fiery death. The establishment of essential nuclear power in America until a better substitute is perfected has not yet been possible despite its utility, high safety standards, and desperate need. As in the case of religion, much thinking has been irrational. Several other countries are not only safely using nuclear power plants more extensively for expanding the increased human need for power, but are in the process of making it more available and safer. The industry tells us that no member of the public has ever been killed, seriously injured, or unduly exposed to health threats in a modern nuclear power plant. Theodore Rockwell, a well-known nuclear engineer and assistant to the late Admiral Rickover in the development of the first nuclear ship propulsion, considers nuclear energy to be, not a "Faustian bargain", but a "near-perfect providential gift" (74). However, beneficial usage will remain limited until the nation's citizens and particularly their leaders "learn" to overcome unjustified fear.

**Imagination with impunity:** In the case of organized and developed religions, their well-structured frameworks of spiritual images and related rites, created and defined by man in the name of their established gods, are frequently adopted as an important part of current daily living. Is it not a scary reminder of the mythical days of Greece and Rome? Why do current believers feel that their god is any different? In a worldwide sense, however, organized beliefs today carry a great deal of additional political and cultural baggage and therefore significantly conflict with one another well beyond their intended spiritual base. Because large portions of their differences remain in the imaginary realm, successful efforts toward unity cannot be expected to make effective progress without more serious analytical concern and factual teaching applied to underlying root factors.

Neuroscientists also tell us that, for emotional memories, the biology of the brain may be modified in a manner that can well affect further learning, thought, and behavior. Therefore, the physical presence of accumulated religious memory elements in the brain becomes a significant influence throughout life, an influence that may be consciously recognized or may even remain largely in the subconscious realm as a subtle behavioral factor (71). For many people, religious belief may actually be regarded as a "significant event" in life and for such significant events, mind function may spell behavior somewhat different from the norm. Accordingly, it readily becomes a subconscious part of one's every thought and action. We can, therefore, generalize by saying that belief is not only a special form of knowledge, an ethereal ideology retained by a believer, but it also becomes an influential physical part of the functional biological brain. The marked functional influence of its physical presence in the guiding brain must therefore be acknowledged. The effects of its many neural associations remain part of the thought and behavioral system forevermore in life. Only new life and new generations can reflect significant neural differences.

**When one learns how it really is:** This book speaks foremost to that presence, to the emotional effects of that powerful retained subconscious influence, one that may range unpredictably from spiritual quiescence to violence. It, more generally, speaks to the ongoing influence of imaginative belief on human behavior amid the stark reality of the interacting modern world. When man understands fully how he came into being as a human entity, and how the Earth and solar system before him were formed, as we now know many similar worlds regularly form in his universe today (221), when man learns that he can find no humanlike god anywhere that responds to his needs, when man knows that biological death is final, then his mind will respond quite differently to his most advanced "awareness of awareness." It will respond in a manner similar to the minds around him, all of which must eventually subscribe to the same factual human knowledge. Metaphysical interests will move from an imagined spirit to a higher level of concern, to the natural mechanistic character of the vast universe and circumstances of its origin and function. Man must learn that the universe was never specifically intended to involve man. Life developed spontaneously in the course of natural chemical function and its structure and evolution are now well understood. But for explanation to ongoing man, origin as such is not of practical importance. Human longevity is limited and highly problematical. Man must hasten to organize to mutual benefit and welfare and must find means with which to endure major natural events to which he is exposed. He is rapidly learning that he is not in the hands of a benevolent spirit but in the hands of the medical profession. He knows that prediction and explanation of destructive cosmic concerns now rest in the hands of our very capable cosmologists, astronomers, meteorologists, theoretical physicists and their increasingly capable technology.

Religion, largely a fear-based emotion (75), is considered a natural and somewhat involuntary human reaction to the state of being in the chanced circumstances of human existence, especially the fear of certain eventual death. Modern neuroscience is bordering on the cutting edge of research in learning and memory. A rapid advance of memory understanding would substantially modify human capability to deal with memory traces. As suggested elsewhere in this study, the entire matter of memory and memory modification as it relates to fear and religious belief may change abruptly. Recent work has now penetrated the very core of memory (76), research which could advance and perhaps modify man's understanding of spiritual response and resultant behavior.

**Neural fixation as a virus:** One should not underestimate the computer-like structure of the brain system. As mentioned elsewhere it consists of billions of electrochemical elements with sensory input circuitry, conscious and subconscious "processing" capability, and large memory capacity. It contains a complex mixture of reactive and inhibitory elements. Information injected into the human computer elements creates appropriate circuitry, finds paths that are not easily traced or fully understood. Some interneural functions including subconscious emotional responses might be akin to a computer virus that shows up only in output expression and behavior. It can go into "neural oscillation" with positive feedback. Chapter seven for example, mentions an uncontrolled mental condition described in a book titled *When God becomes a Drug* (77) authored by Father Booth. These extreme properties of the functional mind system should be given serious consideration when teaching imaginative religions, when suggesting ideal promises to highly sensitive minds. This book frequently mentions the need and indeed, the responsibility for an appropriate disclaimer when input information is based, not on rational certainty, but only on "faith and feeling."

**The blockage of rationality:** However, the religious mind has historically been encouraged to feel that life, as it plays out in reality, is incomplete and that one can look beyond it with spiritual expectation. In a sense, it is the creative imagination that remains to be penetrated by the restless mind, preferably after it surveys and settles the landscape of reality. Meanwhile, life is a function of cellular growth and is ended when its reacting chemistry expires. That seemingly harmless notion often becomes the target of a powerful and unreliable mix of conflicting explanations and interpretations, a mix that clearly continues to generate disastrous conditions of human conflict and irreconcilability. The "god problem" is also a very significant factor in the resolution of human interests and differences with respect to social issues such as homosexuality, birth control, abortion, and the more liberal use of stem cells. But, in the world of reality, they should be non-issues, primarily subject to medical considerations and to human equality. The god problem must first find popular resolution and the resolution of many other issues will closely follow.

Religion might also be considered an "end product" of underlying reactive emotions in a neural structure, largely inherited from animal ancestors from which the human formative structure itself was derived (78). It is an emotion which should be found increasingly less significant for future man in the light of advancing development and knowledge. However, for most people, such emotion still tends to exaggerate the fundamental question of existence, perceived needs, and dependence, and as a result, full satisfaction within one's mind still requires a liberal application of imaginative thought for which the presently evolved form of mind stands well prepared and is highly capable. Because it involves a purely mental process, religious acceptance and satisfaction is not considered to require specific factual proof or acknowledgement of hard reality.

**Saints among sinners:** Libraries of the world are replete with representations and publications about "holy persons" whose lives reflect a special relationship with their god. We know from the work of William James, referenced at the very outset of the introduction, that some spiritual minds tend to pursue religious belief with unusual depth of feeling and emotion. His *Varieties of Religious Experience* is no doubt the most profound compilation of religious practices ever published. As a doctor and psychologist, James studied the experiences of many saints and mystics of earlier times and investigated evidence of the religious worlds they considered to be quite real, perhaps heavenly states with little concern about fact or reality. To James, the purely religious mind seemed to generate and contemplate vast arrays of imaginative detail of a close relationship with God. His lectures made a great contribution to the world in defining genuine religious experiences but could provide no explanation as to its causal basis and had to await later understanding and reasoning.

Visions of the searching religious mind are formulated by strong emotions, of hope, desire, and the perceived need for an all-powerful superior being. They are fabricated like dreams, from bits and pieces of learning and memory, and stem from electrochemical interplay that remains entirely within the neural system, without physical definition and without evidence of reality. They have significance only in terms of what transpires in the mind of the beholder, all in accord with the principles discussed at length in these chapters. The incidences of new "holy men" are expected to diminish rapidly with time and have already virtually disappeared in their earlier form as a result of greater communication with the modern world of reality.

**Life is fragile chemistry:** Prior to life on our planet, we now know that the Earth, upon violent formation and cooling, consisted of very diverse chemistry, chemistry that included the heavier elements produced in its cosmic reaction, essential for forming life. We also now know that it was a primordial mix of such substances that first resulted in simple biological life forms and that, through evolutionary change and development, eventually became the more complex human life we now know, indeed, a very long natural process (79). Little consideration is given the fragile nature of life on earth. For example, the motion of the molten earth's core produces a strong protective magnetic field, without which the earth's atmosphere would be wiped away by the solar wind. Also the atmospheric ash from major volcanoes has at times wiped out plants and animals around the world to a level of near extinction. Other natural conditions render biological life marginally possible. It is said that simple lifeless viruses, assemblies of various molecules that seriously interfere with biological life, existed at the time life was in formation. Although still somewhat incomprehensible by the average mind, man remains the highest and most important form of that initial life-forming process, one that he has increasingly studied and now reasonably understands. Up until recently, human life was considered to be organic chemistry uniquely imbued with a special spiritual quality. It is now considered to be complex biochemistry with appropriate dynamic properties as with all such life, and science tells us that basic life may soon even be created in the laboratory from similar lifeless chemistry (80).

Relatively few minds, including those of many learned theologians, may yet be equipped with sufficient knowledge about the detailed structure and function of man and about his greater material world to make firm reasoned judgments with any degree of conviction with respect to critical aspects of reality and belief. Wide exposure to all these important realities is still a limitation in our educational systems, a limitation that even the most educated and learned believer may be unwilling to recognize. On the other hand, philosophers and scientists, especially trained in logic and reason, generally have the advantage of greater universal exposure, allowing more rational explanation and conviction.

Informed new generations will no doubt substantially advance this condition within the next century. Meanwhile, many minds remain exceedingly vulnerable and capable of easy justification of desired beliefs through overwhelming emotional forces that tend to allow indifference to underlying factual judgment. Therefore, once a desired religious concept is fully envisioned, casually filtered, and accepted by the still limited judgmental mind, it tends to remain fixed as an intractable belief. It may even become permanently lodged in an affective neural region as is often found in troublesome post-traumatic memories (82). It may actually become indelible as neurally formed, a solid bedrock of "realistic" imaginative belief, a subconscious misinterpretation of meaning and purpose. But even the most powerful religious mind with all its accumulated vision is forever erased from the universe at the instant of death.

**Inevitable transitions:** The accepted reality of man's universe and his familiar home planet are becoming quite impersonal and should no longer induce any special emotions of fear and loneliness, but the mind does not yet easily accept the reality of inevitable death. In that respect, human weakness and courage still have some distance to go. Religion was largely created by the human mind to stem the fear of death, and all the soothing and consoling words of man can only play again and again on the powers of imagination for those that still believe, to "conjure up" another life in the vast unknown for those departed. Steven Weinberg (81) aptly expressed his views by quoting Philip Larkin, saying that "this is what we fear, no sight, no sound, no touch or taste or smell, nothing to think with, nothing to love or link with, the anesthetic from which none come round."

You often miss what you've had: Julian Barnes, a respected current author, wrote "I don't believe in God, but I miss him" (83). Barnes was also concerned with the "god problem" and frequently wrote and meditated on the subject of mortality. The reviews of his book, *Nothing to Be Frightened Of* praised the mature way he talked about death. However, religious belief prepares man very early for his time of death by purposeful indirection, teaching immortality of the soul. But both believers and nonbelievers experience exactly the same ending of life, a time when life just disappears for all time, as though it did not ever exist in the first place. Logically, there can be no organized remnants. However, the present mind tends to benefit by such belief in reducing the chemistry of fear and of depression, and suicide-bombers may thereby escape the fear of death in their determined destruction of others, but in the course of future time, all minds will be guided by the facts of realism as discussed in these chapters. The fabricated chemistry of man is increasingly used to modify or replace natural emotional chemistry. But, however it reflects in consciousness, those who have known him shall also miss God.

# Chapter Three

## Imaginism Versus Realism

> Indeed, no scientist, by virtue of his science, has the right to pass judgment on the faiths by which men live and die. We can only set out the data about the brain, and present the physiological hypotheses that are relevant to what the mind does.
>
> —Wilder Penfield, *The Mystery of the Mind,* Princeton University Press, 1975

**Only two concepts of belief:** There is little purpose in discussing secondary details of religious beliefs without first addressing the more fundamental question, the primary "god problem" of Belinsky and Turgenev. To streamline that issue for purposes of this book, all spiritual beliefs are reduced to just two basic concepts. Each of them may involve many diverse, secondary details defined by prevailing belief systems but only one interpretation of the world and human existence can be the correct one. This book suggests that the most realistic answer has already been reasonably determined by confirmed scientific observations of modern men of vision, and that their more rational views must be seriously considered for they could well lead to eventual spiritual unity. In brief, they are:

1) Belief in the existence of a superior humanlike being, the creator and controller of all things, a being who resides somewhere in ideal surroundings, one that benevolently relates to man, monitors and judges man's behavior, and offers him some form of immortality. Considered a natural response to human existence, it is a belief system that the emotion of man strongly desires, one that has been proposed by earlier human minds and traditionally taught in various forms over the past millennia. It is based not on observation or evidence but only on faith. It is what the average believer has been told or imagines to be the case. It is therefore here called IMAGINISM.
2) Belief in the existence of an all powerful governing natural force, the underlying origin and purpose of which is not yet fully understood and may never be understood by the human mind. It is considered a natural transcendent mechanistic force responsible for the universe and its function, one that follows observed physical laws and measurements but has no humanlike character and no personal relationship with more recently evolved parasitic man. It sees man as a reacting biochemical organism, recognizing the inevitability and finality of human death. It is what man calls nature, what is believed to be reality by most informed scientists, intellectuals, philosophers, and men of reason. This book therefore calls it REALISM.

Challenges of modernity: Imaginism, in some form, is currently the most prevalent universal belief in most civilized areas of the world for it best fits the deep emotional needs and desires of current man. It claims that a humanlike god created the universe as well as man himself, and assumes that man can therefore not only beneficially communicate with his god through thought processes at any time and at any place, but can also maintain an immortal spiritual relationship. It is a belief system derived from early Jewish philosophy that initiated the principles of Christianity in the first century and, a few centuries later, led to Islamic belief. Many skeptical minds believe that if a humanlike god truly existed, it would involve a singular notion that would be made known to man by this time. Although the notion of imaginism is challenged repeatedly by modern knowledge for its lack of evidence and substance, it is difficult to argue against a "promise" that serves man's strongest emotional needs, especially one that so easily escapes effective inquiry and serious verification.

The most recent challenges to imaginative religion by select modern man, earlier called the "new atheism", are based on factors relating to continuing human development and his increased knowledge about himself and his surrounding world. The new atheism not only sees a lack of any meaningful evidence to support the existence of a humanlike god, but is increasingly concerned with the conflicting state of imaginative religions undergoing expansion in the complex globalizing world of unprecedented human interaction. Traditional religions are found to be incompatible with the educated and informed rational modern mind. Conflicting religious laws and intractable fundamentalism have stirred modern global society as never before. Conservative elements of society are not only finding conflicts with respect to ideology and social behavior, but a disturbing sense of militarism and terrorism has also crept into the unsettled religious sphere. Further, some scientists feel that continued imaginative belief will act to deter the ongoing evolutionary process of greater rational mind development. The principle of "blind faith" is not only being increasingly challenged but is slowly losing rational acceptance, first in the most educated areas.

**The survival of reality:** The first chapter discussed widespread human response to spirituality, the role of natural human emotion, the imaginative basis of resulting beliefs, contrasting views of modern reality, and an indicated gradual movement toward universal secularism. The second chapter discussed the functional role of learning and memory relative to the stimulated spiritual mind, providing deeper comprehension of the root sources of belief and behavior. This third chapter offers commentary on the two principal underlying systems of beliefs, one based on a "humanlike god", and the other on a non-human "mechanistic" transcendent force, commonly called "nature." It discusses how these concepts permeate and influence changing and developing society and where they might be headed. The division of primary belief into two bare concepts emphasizes the need to look at fundamentals before considering commentary on specific traditional religions and practices. This book suggests that, in terms of contrast and logic, modern man carefully consider both basic views before accepting any belief to be rational and appropriate. It further, suggests that only one of the two, favoring realism in some form, is likely to survive on its relative merits. Imaginative concepts tend to become more imaginative and concepts based on reality tend to become more firmly established and real.

**Emotion requires little intelligence:** Although the twenty-first century brought man to a relatively abrupt new level of human knowledge and to a more rational understanding of how his world functions, fundamental origins are still in question. Among its extensive lack of definition, imaginism leaves open many practical questions, such as the origin of matter of the universe and the origin of the humanlike creator itself. Most traditional notions of belief that still persist just fail to be concerned with practical details implying that an all powerful god can do anything. Up to now, the simplicity of imaginism has been sufficient to provide much of mankind with a sense of human purpose and emotional satisfaction, albeit without reason, proof or sound philosophical basis. Cardinal Christoph Schönborn (84) indicated that the *Catechism of the Catholic Church* contains the concept that "Human intelligence is surely already capable of finding a response to the question of origins" and "that the existence of God the Creator can be known with certainty through his works, by the light of reason."

On the contrary, this book has indicated that human intelligence has now reached a level at which such ideas must be rejected as irrational. The human brain is expected to continue to grow in capacity and capability. However, the concept of realism in its godless mechanistic concept does not yet pretend to fully define the origin of nature or provide full explanation of the underlying basis of observed functions. It simply fits current measurements and factual understanding more fully and more realistically. Modern theorists tend to trace the origin of the expanding universe back to an initial "singularity" with a beginning known as the "big bang." Despite its simplistic assigned name, it is a concept now generally accepted and widely discussed in scientific and philosophical research (85). There is no other competing rational explanation. The bodies that occupy the universe appear to be moving away from each other in a process known as "inflation" and, under certain conditions, could experience a reversing process called "deflation." But such cosmic movement at breakneck speed is relatively "slow" in terms of human time. The concept tends to fit physical theory, especially in mathematical representation, which separately would be completely unintelligible to the untrained and unscientific mind. Unfortunately, the concept of realism does not offer man divine communication or the prospect of life after death.

**The endlessly inquisitive mind:** Scientists are constantly in search of additional explanatory clues for better understanding of matter and origins. The fundamental make-up of matter is pursued relentlessly.

Many secrets of matter have already been uncovered through studies such as radioactivity and nuclear reactions, and Albert Einstein showed the equivalence of mass and energy by his famous formula, $E=mc^2$. With its principles established, there will be continued scientific interest in mass-energy exchanges in future science. However, average man may never understand or conclude its comprehensive significance.

Religious thought is commonly associated with both the origin and function of the universe as well as with man and his detailed structure. Modern man now knows a great deal about the nature and mechanics of the macroscopic universe. The new Hubble space telescope has suggested a new understanding of the details of our structure. In the microscopic world, some of the remaining secrets of matter are believed to be still hidden in sub-particles that make up the core of the atom. Scientists are currently in the process of testing a huge new high-energy particle "collider" in Geneva (86) to look for the possible release of one or more new subatomic particles. These are particles of "symmetry" called "sparticles" (each older particle name is preceded by an s) that could more satisfactorily explain mass, the mysterious dark energy of space, and other phenomena of interest. Because of its perceived significance with respect to defining the fundamental character of the universe, they talk about a "god particle" sought in these experiments.

In addition to ongoing experiments on matter, there is also theoretical work directed to finding a common mathematical description that would be compatible with all fundamental forces of the universe. Modern theoretical physicists have come up with a mathematical model called "string theory" suggested as a common description of the structure and behavior of all matter in both its atomic and macroscopic roles. Although existing particle theories explain phenomena quite well, physicists since Einstein have looked to develop a unified "mathematical" theory that would include both relativity and quantum mechanics. The string theory would replace the concept of particles but strings require space to have more than the assumed four dimensions. This idea was combined with the "super symmetry" concept in which every particle has a partner, leading to "superstring" theory. Further variations led to the M theory or the "theory of everything." These forward-looking theories are principally mathematical, still looking for empirical definition. However, it is important to note that neither the results of the collider experiment, nor the new mathematical formulation of string theory, is expected to modify or contribute significantly to the commentary of this book on spiritual belief.

The term "realism" is used here only to contrast with the term "imaginism." However, realism has always had an accepted meaning in philosophy, the doctrine that things exist in and of themselves, independently of the mind that knows them. Use of the term realism is defined here to be more like naturalism of philosophical theory, the doctrine that "everything is derived from nature and there is nothing beyond nature." Naturalism rejects the supernatural and believes that all things are subject to scientific laws. It is the essential basis of what has been called natural or cosmic religion by Albert Einstein as more fully discussed in chapter six. The terms used in this chapter to describe the two contrasting interpretations of existence are not intended to conflict with formal terms of philosophy, but used only for purposes of simplicity and clarity of description.

**Belief in disbelief:** Both views of human existence defined in this chapter are referred to as "spiritual beliefs" for simplicity, for they represent the two principal forms of human responses to the natural quest for understanding existence. Realism, the more modern ideology, may sometimes be called disbelief, non-belief, atheism, anti-theism, agnosticism, humanism, or secularism, for it principally rejects the notion of a traditional imagined humanlike god and immortality of an assumed human soul. The term agnosticism is more correctly a belief system in which man is non-committal, not having sufficient information on which to make a firm decision. The term secularism used throughout this book is considered to be the lack of or indifference to religion. Unlike imaginism, realism assumes that man cannot communicate with any transcendent mechanistic force or spirit and that every human entity must eventually face final and permanent death (87). Such assumptions correspond to observed reality. This study concludes that there are no transcendent beings or forces of any sort that can respond to human needs or wishes and that man therefore should not expect any response to prayer or supplication. Contrary to traditional teaching, man is assumed to be effectively "on his own" and is expected to work out a satisfactory relationship with both nature and his fellow man. Over-dependence on a god in heaven has seriously distorted world responsibility. Eventual world secularization will therefore represent an advanced stage of human maturity, the beginning of man's "age of responsibility" It will constitute his equivalent of "coming of age." Man is entirely dependent on the forces inherent in nature as reflected in his current existence. The resultant direction of man and his world is considered more fully in chapter nine on globalization.

**Powerless prayer:** Despite frequent prayer-like adoration of an imagined god offered today, and by so many believers over past centuries, no specific response by any god has ever been detected and confirmed. Believers generally rationalize reasons why the existence of God would not be detected by man and some even interpret natural occurrences as god responses. As much as we all would like it to be so, the concept of a benevolent humanlike god is clearly the product of emotion and imagination. The rational mind must be the final and indisputable judge. Also imagined cures of human sickness and disease (38) through prayer have never been verified and therefore are assumed not to be possible.

As a consequence, none of the many religious sub-beliefs derived from concepts of a humanlike god should be expected to have any effective meaning for man except in his own mind. But nevertheless, human belief in such concepts, with supporting biblical lore, will continue to be taken very seriously for some time into the future. The association of a humanlike "God Father" with the biblical Jesus, the claimed "son of God", a claim that tied heaven to Earth for man, is accepted without question by the most emotional spiritual minds. Although it seemed to fit the imaginative framework very well for early man, its impact is gradually diminishing in the light of advancing human knowledge, rationality, and experience as more fully explained in previous chapters. Will any distant future man bring himself to seriously believe in such a claimed relationship? Overt claims of theology with respect to heaven, hell, angels, and devils, together with numerous other imaginative biblical connections, are gradually disappearing from serious public concern in most developed civilizations. In his letter to the New York Review of Books of November 20, 2008, concerning an earlier critique of his own book titled, *Without God*, physicist Steven Weinberg noted, "it's just a guess, that in the long run there will be enough people who care about what is real that the decline of religious belief will lead to a decline in other trappings of religion."

Evidence of increased lack of seriousness about religion suggests its slow movement to a lesser position of neural hierarchy of the mind. Religion has often been trivialized in recent public conversation. It has become the subject of stand-up comics, moving picture productions and other events intended to create humor and entertainment (9). A recent news article (88) described "spiritual services" as the subject of a humorous automatic telephone response. Lampooning modern telephone answering practices, it featured instructions to press particular dialing numbers for prayers of repentance, supplication, forgiveness, or serenity, leading to typical caller exasperation and to eventual call termination.

**The endless search:** The only godlike power observed, and known empirically by man to actually exist, is simply expressed as the "god of nature" in this book, an undefined mechanistic transcendent force assumed to be responsible for the material make-up and function of the universe as well as for the function of its underlying chemistry, including the biochemistry of life itself. It is the force assumed to underlie the structure and function of every part of the universe, a non-human force that follows the physical laws and principles observed and measured by man. Accordingly, it is assumed to be the force responsible for man's solar system with all its familiar planets known to have been gravitationally formed from gases and space material. It is the force responsible for the chemical formation of life, yet known only on our own planet, and for the biological evolution of modern man over substantial time through a chain of biological ancestors of simpler form.

Within the broad terms of everyday experience, human existence on a huge spinning globe floating in space among billions of other globes constitutes reality. However, in human terms, it makes a strange story, but one that has been well researched and documented over recent centuries. It took many progressive ages of man to understand the origin and makeup of life, the evolution of human form and the underlying origin of the vast surrounding universe to the extent it is known today. It is this verified understanding that finally allows man to replace his earlier imaginative ideas about God and related religious frameworks with reality. It is truly unfortunate for man that his neurally based tendency to desire and worship a god preceded his capability to acquire such knowledge and understanding. The full findings of man constitute a long and complex story, far too long to include here even in simplified form, but the story is yet incomplete. Some of the many references contained in this book may be helpful for those seeking further understanding. Man can look forward to accelerated search for appropriate extensions of his knowledge of reality but it is not likely that he will find any in significant violation of these predictions.

**Man is now beginning to know:** Based on accepted scientific data, man has determined that the universe formed about fourteen billion years ago and that his solar system formed about five billion years ago in one of the arms of our great spiral Milky Way Galaxy, one of many billions of galaxies of the universe. Because of the unique environmental conditions of planet Earth, simple microscopic life is believed to have spontaneously formed in its chemistry sometime within the first billion years after cooling and, through its reacting genetic "chemical" structure, gradually evolved, resulting with man as the most intelligent of known species up until now (19). This type of information is the result of intensive observation, research, and measurement by the most advanced scientific minds. It is therefore considered to be part of the continuously expanding bank of factual human knowledge. It reflects the continuous changes that are known to have taken place before and following the formation of the planet, changes that are basic to current human understanding. This is what man knows and what he must accept while establishing his proper place both conceptually and physically, while conducting his life in a manner most appropriate for peaceful coexistence.

But why did life form on our little planet located among billions of orbs in our galaxy and why did life form in our galaxy among billions of other galaxies in the universe? In time, science and particularly astroscience will no doubt determine some of these answers, but they are not expected to bring us to any better explanation of religion than we have already discussed based on what we now know. Nevertheless, it is important for man to have knowledge of his larger world and to know where he stands in the whole of his universe at least to the extent it can be determined by observation and reason. Consideration of what we know about the near part of our universe is amazing enough to man. It is like a child's dream, a bedtime story, and it should become well known before there is any consideration of the "god problem." It is best considered with some pre-understanding of the principles of astronomy and physics.

In realism, man is regarded as a form of terrestrial "parasite" made up of part of the same basic chemistry as one expects to find in the rest of the universe. With knowledge "closing in" so effectively, there is no longer a need for an imaginative humanlike god or a fabricated surrounding religious framework. Man's "religious sixth sense" is now modified by doubt and understanding. Although realism rejects a god of humanlikeness and notions of immortality, there may be some who prefer to interpret nature as a "god of nature," an underlying transcendent force completely out of range of human communication and understanding. Some even believe that the mechanistic universe was fashioned by a humanlike god who thereafter sat back and simply monitored it, an early notion called deism.

**Religious dependency:** Needless to say, there are very important social issues that depend on imaginative belief. Some have plagued society for many years. Issues such as abortion, gay marriage, and contraception are mentioned elsewhere. Another issue mentioned earlier relates to the fundamental questions of voter rationality, frequently raised whenever there is a major election for high office in America. It is ironic that spiritual expression, based solely on faith and imagination, contributes so specifically to the perceived qualification of a candidate for high office, whereas a well-informed, non-religious candidate, who relies on factual knowledge, cannot today be democratically elected. Some intellectuals even suggest that imaginative belief may be associated with limited education and comprehension, even poor reasoning ability, perhaps a fundamental neural weakness unbecoming of a responsible candidate for national leadership.

The playing field for imaginism and realism is certainly not even. The thousands of religious institutions, distributed liberally throughout the world, simply take their religions for granted and function as though there were no question about the credibility of their narrow beliefs or about the availability of advertised divine benefits. There are very few institutions that explain or promote realism other than those of science, but science has other important objectives and is still far too limited with respect to spiritual recognition and promotion by average man. Therefore, the world finds itself virtually inundated in imaginism and effectively immune to the remedial challenges of reality. Despite scientific observations and their implications directed toward the accumulation of new factual knowledge, there remains an underlying human penchant for the excitement of the mystic and the mysterious, and reality is frequently compelled to take a temporary back seat. This may be acceptable for entertainment and casual stimulation but not for a system of belief that seriously affects all society. When it comes to the interpretation of existence and the determination of the basis for human behavior, factual understanding and reality must indeed become the first and most important criteria of man. There are a few organized groups of nonbelievers but none that have yet reached the promotional proportions of organized believers.

**The richness of atheism:** In *Philosophers Without Gods*, a study of what philosophers have determined about atheists, edited by Louise Antony (89), one learns about the "richness" of atheistic belief. Its essays suggest that atheism is not only an acceptable and valid alternative to traditional religion but also a profoundly "fulfilling and moral way of life." This differs from some of the observations presented in the following chapter on stigma. But atheists are often demonized as arrogant intellectuals and considered antagonistic toward religious belief. The new study finds that there exists "marvelous diversity" among atheists and that few dismiss religions as being "stupid or primitive." This book also recognizes natural spiritual stimuli and vulnerability and therefore does not refer to any religious belief as "stupid." However, it suggests that spiritual response and neural wiring corresponding to faith and strong belief, may be the result of early involuntary learning and should be subjected to question and rational analysis at the age of reason. Although it retains great respect for deep religious study and sincere belief and applauds the righteousness and good works of its teachings it warns of the development of a narrow and biased faith-based society.

**Theological bias:** There do not appear to be any theological seminaries that study or promote secularism, atheism, or the "god of nature". Some interested groups such as those already mentioned have been only partially successful (90) in communicating such principles to the general public. Imaginative religions, on the other hand, have had a huge following, endless publicity, and the aggressive ministries of thousands of churches. If the matter of atheism is raised by them at all, references would likely be on the side of disdain and negativism in an unfriendly religious environment, no doubt in fierce defense of the traditional god. Little room is left for question in most organized beliefs and any form of disclaimer would tend to weaken confidence in their teachings. This book favors a requirement for appropriate disclaimers in all public promotions that lack absolute certainty.

**The study of nonbelief:** One might expect that organizations such as the Center for Theological Inquiry (91) and the Center for Religious Study (92) in their academic postures, might be expected to have an interest in the inquiry and study, not only of a humanlike god, but also of non-humanlike transcendent forces described earlier. However, they do not seem to have any effective programs that include both. Non-religious universities that strongly influence the choice of one form of belief are particularly subject to criticism (93). Except for implications in the normal teachings of their departments of science, it seems that spiritually oriented academic groups simply fail to acknowledge the possibility that a humanlike god might not actually exist. This might also be the position assumed by typical trustees and alumni bodies around which a university effectively functions.

The Princeton Center for Religious Study claims "generous" corporate support for its programs, aimed at the "intellectual basis for Christian thought and practice in the United States." Activities include postdoctoral fellowships for scholars engaged in research on historical or contemporary aspects of Christianity, congregations, or clergy; internships for graduate students to gain experience teaching in a seminary or divinity school context; and a public lecture series on topics concerned with the public understanding of religion. It claims to support the "scholarly mission" of Princeton University while also serving the interests of people of "faith" who wish to think deeply about the historical resources within the Christian tradition, making such information accessible to a wider audience. To my knowledge, there are no "offsetting" programs of realism that complement imaginative interpretations of man to serve or encourage those who have "no faith."

**Conflicts in education:** In the context of this book the "god problem" is an appropriate university level issue and eventually becomes a concern for alumni and, indeed, for all society. Many universities are fundamentally based on the rational search for factual knowledge in one area, and overtly support the pursuit of imaginative knowledge in another without adequate reconciliation or explanation. Few universities, if any, become deeply concerned with appropriate public clarification. They ignore adequate disclaimers and are clearly not prepared to take a firm position as dictated by their own factual sciences. It seems that man has just not yet reached far enough in academic maturity to face reconciliation and recognize educational responsibility with respect to this still sensitive matter.

Although biblical research and history of religion are generally included in most university curricula at appropriate department levels, there is yet little appetite to recognize and support concepts of realism in any formal course material or to publicly field related questions. It seems that the academic world believes it is not necessary in consideration of the overwhelming popularity of the established traditional god and that overt recognition of imaginism better meets university community "standards." The significance of spiritual conflict as presented in this study should be academically recognized. For the long-term solution of world differences and unrest, there should be as much instruction about the reality and roots of atheism as there is about traditional theism, and differences should be kept on a high academic plane. If the problem is not recognized at the university level, it will remain "hidden" as an important issue, never to be properly aired, and leaders in education will have been remiss in molding and controlling a better informed world.

**Promoting factual knowledge:** Science should be more academically tied to religious belief as has been done in the early explanatory chapters of this book. Teaching all views as formal university disciplines in an appropriate manner with appropriate understanding would tend to encourage awareness and reduce conflict based on festering belief differences. Is there a fear that it would unleash the spiritual restraints on human behavior? Is it similar to the fear that the study of sex may induce undesirable sexual behavior? A university pledged to the advancement of factual knowledge has an implied commitment to encourage all important disciplines for defining and guiding social relationships vis-á-vis world behavior.

Failure to recognize and consider the existence of a natural mechanistic god in education constitutes failure to prepare society for the future age of spiritual unity. It is a matter of importance to discuss and compare views at the university level where mature rational thought is established before students are released to the world. All stages of education leading up to that level must also assume part of the total responsibility to stimulate modern thought. Universities, which normally lead the processes of "advanced" learning, cannot be indifferent in this matter lest they fall behind in their intended role.

**Eccentric theology:** Not long ago, the Princeton Center for the Study of Religion conducted an interdisciplinary program in public theology. In the third year of the project, it focused on an "Era of Scientific Challenges." It addressed questions on the role of theology in a scientific culture, specifically on topics such as bioethics, the mapping of the human genome, advances in molecular biology, astrophysics, the farthest reaches of the universe, microprocessing, artificial intelligence. These are all normal creations of a rational mind. However, it was stated that theological attention must be paid to these issues and to questions about human life and its place in the universe, questions that increasingly needed to be answered for they have "far reaching implications for theology itself." A "center" should not become central to any one view. Programs slanted toward imaginism should be fairly balanced with peripheral alternative views of realism to meet the expectations of full academic responsibility. A university should be capable of determining how that should be properly done for current society. Education at all levels must recognize, not cloud, the advancement of the secular trend.

This issue should be raised to encourage better understanding between theology and science and it should involve the most learned scientific scholars and theologians. The bias instilled by tradition must be considered. Although teaching may appear to touch on realism, it seems to inevitably revert to the comfort of imaginism, emphasizing depth and meaning in its own special terms. It is as though a university must assume the existence of a humanlike god for its most acceptable and efficient function. From an outside viewpoint, it seems that there should be only "one department of religion" rather than the creation of a "supplemental ministry."

The Princeton Center also introduced a Faith and Work Initiative suggesting that most students, workers, marketplace professionals, and leaders wish to live a "holistic life" that integrates, among other things, faith and work, but have few resources to help them do that. Its purpose is to "generate intellectual frameworks and practical resources for the issues and opportunities surrounding faith and work." It is intended to investigate the ways in which the resources of various religious traditions and spiritual identities shape and inform engagement with such workplace issues as ethics, values, vocation, meaning, purpose, and how people "live out their faith in an increasingly diverse and pluralistic world." It is further intended to "explore market topics, global competition and its ramifications, wealth creation and poverty, ethics, diversity and inclusion, conflicting stakeholder's interests and social responsibility."

Although its meaning is highly commendable on an absolute basis, such an initiative is based on the assumption that participants all subscribe to imaginism and it tends to "separate" those who, without "faith," have exactly the same objectives. It is also a program that encourages participation in imaginative belief. There should be no educational bias or weakness in the pursuit of the highest university principles and objectives on which the academic world has placed its reliance. This book questions the path of such initiatives in terms of its emphasis on forces of spiritual change acting on the postmodern world. The course of recognized and responsible institutions of learning requires careful new consideration of this issue. The proper approach might be to clearly separate "divinity school" affairs from other academic affairs with the same fundamental transparency afforded the realms of its scientists and researchers.

**Humanism at universities:** Harvard University (and perhaps other Ivy League universities, such as Cornell and Columbia), maintains active humanist programs of secularism associated with the Harvard Chaplaincy (94). It is said to have been active for over thirty years and sponsors well organized community programs for students, alumni and friends at least once a week. A community of college students, known as the Harvard Secular Society, aims to create an opportunity for nonreligious people in the area and is responsible for the selection of the annual Lifetime Achievement Award in Cultural Humanism. The chaplaincy seems to have overcome much of the common stigma of atheism and qualifies as an academic nucleus for explaining the logic of secularism in the rational world.

The March 19, 2009 bulletin of the Harvard Chaplaincy spoke of a rash of new publicity in major media recognizing "all this attention because, with the non-religious rising in all fifty states in the new *American Religious Identity Survey*, the Chaplaincy seems to have become a symbol of an idea that can help the tens of millions of American Humanists, atheists and agnostics bridge the gap between numbers and strength." "We are working toward the day when people at Harvard and across the country will think of ethical service to one's neighbors and to the planet as an act of humanism as much as it is now associated with churches and other houses of worship. We are being good without a god."

An article in the *Princeton Alumni Weekly* (95) described the Chapel tradition at Princeton's fall 2008 opening exercises. It was called "an awe-inspiring building, big and beautiful enough to make even a godless heathen wonder if it's not time to start hedging his bets." The ceremony has become increasingly multi-denominational as incoming students include Judaism, Islam, Buddhism, Hinduism and other faiths. The program included a prayer for Princeton, hymns and music in addition to recitation of appropriate religious texts. It also pointed out that there was nonbeliever participation in which a female student read a passage from *The Catcher in the Rye*, suggesting the passage to be in the "secular humanist tradition." However, in line with discussions of stigma in this book, she admitted that she really called herself an atheist, but thought that "describing it as an atheist tradition" would not sound good. Is this an appropriate concern for a student of any university, especially one that claims the highest ratings among similar world institutions?

**Not religious but spiritual:** Students arriving at universities look for some continuity of rites and traditions and should be given ample opportunity to reconsider them. Diverse denominations under such circumstances should coexist without conflict or rancor and should contribute to student comfort in adjustment and, indeed, to cooperative mutual understanding. The Princeton Office of Religious Life (ORL) suggested that "Religion is happening on Campus". Contrary to the theme of this book, it expressed the view that religion has been growing, not just on campuses, but also in the wider world. By way of clarification, however, it stated that it would not necessarily use the word "religion" but would say "spirituality" adding that so many people say, "I am not religious, but I am really spiritual." Spiritual is defined "as relating to, consisting of, or affecting the spirit." In turn, the spirit is defined as "an animating or vital principle held to give life to physical organisms." Concerned about the aura of existence, man is naturally spiritual, but is increasingly enjoying "secular spirituality" as atheists, agnostics, and humanists in various satisfying interpretations of reality. A university experience must inevitably rely on reality and factual knowledge presented without ambiguity.

**The growth of humanism:** Harvard, by its formal humanist chaplaincy, may have assumed a position of leadership in inevitable growth of the movement. Higher education can no longer be interpreted as preparation for "service to God" as was assumed by the highly respected Presbyterian founders of Princeton University. But a pronounced "missionary atmosphere" at the university level in the twenty-first century seems both unnecessary and inappropriate. The encouragement, promotion, or emphasis of any belief system, or even the impression given to the public, should be without bias. Factual information and its influence on society and world affairs are not only primary areas of concern but require constant monitoring as changes occur. Resolution of world problems, complicated by conflicting imaginative beliefs, must not be relegated to a secondary or "narrow" position of educational preparation.

The activity of humanist organizations is now more widespread and much information is now readily available on the Internet. Typical examples are the Center for Inquiry in Los Angeles and the Free-thought Society of Greater Philadelphia. Most of them have well-organized programs and lectures featuring alternatives to traditional religions. Participants appear to be just as enthusiastic as those who are driven by the religious spirit. They not only sponsor forums and newsletters but also have secular book clubs and links to similar activities. They claim to have hundreds of members. The American Humanist Association of Washington, DC even paid to place billboards on city buses asking, "Why believe in a god? Be good for goodness sake," and the Philadelphia free-thought group placed a large trimmed Christmas Tree of Knowledge in a public park alongside faith-based holiday displays. The tree ornaments were full size color copies of the covers of atheist and humanist publications.

**New freedom not to believe:** At the University of North Carolina, executives elected not to put up the usual Christmas tree last year out of concern for the sensibilities of non-Christian and non-religious students. In accord with this writing these organizations are increasingly expressing their feelings in the face of strong religious expression of traditional denominations. Up to now, such feelings were largely individual and unexpressed, having no organized medium in which to advertise their opposition to dominant imaginative religions. This book has claimed that mankind is only at the very beginning of an age in which a substantial portion of the developed civilizations will follow the course of secularism but it will take a great deal of honest education, indeed, over several generations to even begin to effectively challenge the entrenched position of imaginative belief. Reorganization of school programs is a critical factor in the timing of its clear "break-out." As suggested earlier, deep believers in every area of civil organizations will tend to oppose and slow the eventual establishment of unifying secularism for all of mankind despite the hope for spiritual unity and eventual peaceful human coexistence.

Increasing numbers of atheistic organizations are also appearing in other areas throughout the country. In South Carolina one is called the Secular Coalition for America and another, United Coalition of Reason. The Secular Student Alliance is said to represent a proliferation of college groups with 146 chapters. In a reflection of stigma discussed elsewhere, they liken their strategy to that of the gay-rights movement, through which minority groups acquired the courage to go public. Reluctant to make their beliefs known in the past, some members now say that "the most important thing is coming out of the closet." These are all further indications in support of the claims of this book with respect to long-term universal secularism.

**Jewish secularism:** The Center for Cultural Judaism (7), associated with the Posen Foundation, has a major interest in international support of secular Jewish education and educational initiatives on Jewish culture in the modern period, as it had in Jewish secularization over the past three centuries. The foundation sees modernity "deeply associated with secularization" and rejection of the supernatural and religious institutions based on the "natural world." Despite the persistence of religious traditions, Jews are said to be "the most secularized of modern people." Thinkers from Spinoza to Mendelssohn, Heine, Freud, Einstein, Herzl, Ahad Ha-Am, Berdichevsky, and Bialik charted out alternatives to traditional Judaism. They all posited "being Jewish" without adherence to traditional religious laws and beliefs. As mentioned in the introduction, the Posen Foundation suggests that half of American Jews consider themselves secular and the percentage is even higher in Israel.

The Posen Foundation, based in Switzerland, is an organization fully committed to secularism backed by strong intellectualism and sponsors many high quality programs of education worldwide. Their courses include history, sociology, anthropology, and other important modern disciplines in well organized interdisciplinary programs. The foundation is committed to the study of Jewish secularism in higher education, secondary education, and adult education. It not only examines secularism over the past three centuries but focuses explicitly on secular traditions in modernity. The early secularization of intellectual Jews is most interesting as it relates to the theme of this book. Raphael Patai's *The Jewish Mind* (51) reminds us of the unusual vision, foresight, and capabilities exhibited by Jewish intellect over the centuries. It is generally recognized that Jewish mental acuity is dominant in fields such as physics, mathematics, philosophy, medicine, legal, financial, bridge, chess, and in many other complex disciplines and activities. Jews have made unique contributions to human knowledge, as in the case of Einstein described in a subsequent chapter. This firm commitment to secularism among intellectual Jews represents further evidence of the validity of the claims made in these pages.

**Teaching teachers first:** Many public school teachers and administrators in America have not experienced comprehensive education in the factual sciences and tend toward imaginative beliefs. They eventually parrot just what they have heard about religion, whether knowingly or subconsciously. Their training is normally focused, not on human reality, but on excellence in techniques of teaching and in administration of teaching. They therefore easily fall victim to the emotional forces described in earlier chapters. Subject to the views of parents and school boards that lean heavily toward traditional imaginism, their fundamental teaching effectiveness is somewhat slanted.

Popular justification for imaginism hinges on the perception of evolution as though that issue were the most vital factor in religious belief. Interpretation of evolutionary processes has therefore been widely questioned. With a large contingent of Evangelicals in America, there has been much discussion in recent years about "school prayer," "creationism" and "heavenly design" (96). In 2005, the teaching of "intelligent design" was considered rather thoroughly by the courts in Pennsylvania and was finally banned from the public school curriculum by a federal judge.

Unfortunately, the god problem persists without resolution, especially in America. The battle continues relentlessly in states where the basis of the Pennsylvania case is still beyond full understanding and acceptance. Florida has been in the midst of conflict (97). Legislators in six or more states have been influenced to require classrooms to be "open" to views about the scientific "strengths and weaknesses" of the Darwin theory. The state board of education in Texas has recently built a majority that subscribes to the notion of intelligent design with the blessing of the governor. Its action to modify textbooks in such a large and influential state is considered an educational threat to the entire country (98). These anomalies of rational thought in education represent an important reason for writing this book.

**Interpreting science with emotion:** The principle of evolution is currently considered clear-cut good science (99), reality fully accepted throughout the well-educated rational world. Certain, less progressive regions of America seem to be influenced by an expanded wave of distorted cultural evolution as defined in the preceding chapter. A number of strong believers, some of whom seriously think that the formation of planet Earth is only a few thousand years not a few billion years old, will no doubt continue to attempt to promote their beliefs in creationism by including "creation science" and the teaching of "strengths and weaknesses of evolution" in public school classrooms. Their personal beliefs may even encourage some distortion of the facts. This is simply another way of expressing intelligent design by indirection. There has even been pressure to open the issue to "academic freedom," that is, teaching "both systems of science," another clear subterfuge for including faith-based creationism.

The public does not seem to be aware of an implied conflict in education. At some point, advancing society must make a considered decision to pursue the direction based on accepted factual science for the issue is already effectively resolved, subject only to recognition and acceptance. It often comes down to the education and understanding of specific persons of leadership and authority (100). Meanwhile, it should be possible to openly discuss imaginism anywhere in the educational system without stigma or backlash but it must be properly explained. America seems to be without strong central educational authority with an open view of spiritual reality. Scientific fact must be recognized as fact without ambiguity or emotion and formally separated from the mish-mash of feelings and opinions expressed by uninformed society, opinions that continue to jam the vital gears of scientific expression and human progress.

**America a theocracy:** In the long course of human development, the past two centuries have seen spiritual change everywhere among civilized nations, but no change has been more eventful than that in America. At the time of its founding, religion was still considered the means for explaining man and his universe (101). Discoveries and explanations of science were still in their infancy, even at that relatively late time in history. Colonists, free from the yoke of European oppression, established their own religious practices in the new world and many required close monitoring and tight control, often at the risk of punishment or even death. The more fundamental religious groups even eliminated dissent by hanging or by exile. Government itself was obliged to support religion in regulating morals, punishing blasphemers, and assuring church attendance. By virtue of widespread citizen belief and behavior it even became the responsibility of the state, a virtual theocracy, to support "religious truth." The accumulation of a large number of different faith-based religions in the colonies provided unusual diversity. However, growth of the nation saw the political need for a major effort to separate church and state, an effort only minimally successful. Despite the encouragement toward secularization in the affairs of man by intellectual and scientific leaders and indeed, by established law, effective "separation" in America today remains highly problematical.

America remains a very religious nation in which the "glory of God," virgin birth, and the resurrection of Jesus are widely recognized. However, recognition is of far more limited depth than ever before. The twenty-first century, following a century of great advances in knowledge, has resulted in a nation in which religion, politics and science are now significantly dividing the country. As stated at the outset, new discoveries by advancing minds have included most disciplines but nothing new about the god of man. Advancement is sometimes slowed because life processes are biological in function and new understanding must always await appropriate growth, development, and assimilation. This book, largely concerned with the world struggle to rethink religion, finds America lagging in its adaptation of rational spiritual thought and means to clear the way to improved social unity. Many citizens have undergone mixed religious experiences as described in earlier chapters. Therefore, one must expect these conflicts to continue, gradually finding adjustment through appropriate analytical processes, most likely only over a series of generations of improved science and education. Differences must eventually be settled by the reason and logic underlying factual knowledge as opposed to determination through emotional feeling and the claims of early scripture. Despite the widespread promotion of imaginative religion and its support by various faith-based organizations, this book promises a gradual change from generation to generation in the form of greater realism through cultural evolution as it becomes the principal product of a more comprehensive and expansive worldwide education system. In Christian America alone, the number of people who describe themselves as atheists or agnostic has increased about fourfold from 1990 to about double the number of Episcopalians in the country. But this is only the beginning. This book sees the gradual development of a sound base of secularism as worldwide unity in rational understanding.

The famous words of Thomas Jefferson, a recognized founder of America, originally spoken to the principle of "belief in freedom," might now well be applied to "belief in the freedom of realism": "May it be to the world, what I believe it will be, (to some parts sooner, to others later, but finally to all) the signal of arousing men to burst the chains, under which monkish ignorance and superstition had persuaded them to bind themselves, and to assume the blessings and security of self-government." In this case, it will not be bursting the binding chains with respect to self-government, but the binding chains of narrow and unreal belief systems, to gain the blessings of emotional freedom that will contribute to the nation's strength through spiritual unity.

In the imaginative world of beliefs, one can clearly select exactly whatever one's mind desires to have. One can also make changes and adjustment at will, for there is never hard established reality to recognize or to violate. One can even adjust the tenets of a religion to match advances in human progress. That seems to be what has happened over later years without specific recognition. Christian religion of the first few centuries was far different from that of the twentieth. An omnipotent, omniscient and omnipresent god, a benevolent and controlling god, is always preferred by the imaginative mind and therefore, given all the highest known traits of man. Some may settle for less, but others may want much more. Some may privately modify the tenets of their religion in "thought exercises" to further provide exactly what they feel they would like to have. This "adjustment" process continues both privately and publicly in many areas of the present religious world.

**God characterization:** For purposes of "ease of understanding," man has been "smart" to retain a god with a human type mind and with human type emotion. It is only because he knows his own nature. But one would expect a genuine god to be quite different and make himself known to man in some decisive manner to settle the pervasive enigma that tends to destroy man or renders him hostile. There has never been a reliable impression of an image of a claimed god and God has never been verifiably seen or heard. For millennia, literature and art have therefore carried physical impressions of an envisioned likeness in the many different forms imagined by man, and popular religions commonly speak to them, and of them, in prayer and in simple human conversation. The religious world and public museums are replete with paintings and sculptures of visualizations of God. Xenophanes, an ancient Greek, once observed that, if animals were religious and could draw, they would picture their god as an animal. Characterization of the god concept is subject to change while the concept itself is gradually replaced through increased factual understanding and greater human reasoning capability.

**A human soul:** Religion introduced the notion of a separate ghostly form in every human body called the "soul." Despite close genetic structure, animals were not included. It was a bold attempt to explain immortality without retention of the deceased body. Whenever or wherever derived, it became an important element in teaching and in propagating religious concepts. The idea seemed attractive to those of earlier centuries. Only recently there was reported an ancient monument to the soul (102) dated in the eighth century before Christ. It was found in the mountainous region of southeastern Turkey. The inscription instructed mourners to commemorate the afterlife of a royal official named Kuttamuwa, with the words, "for my soul that is in this stele." Archeologists suggested that it gave indications that the people in the borderlands where Indo-European and Semitic people interacted in the Iron Age, practiced cremation. By contrast, Israelites and related contemporaries believe that the body and soul remain inseparable which, for them, made cremation unthinkable, as suggested in the Bible.

**Fluidity of belief:** A survey on religion and public life (22) produced some interesting data that support the thesis of this book. First, it emphasized the surprising current "fluidity" of religion compared with the past and secondly, suggested that nearly half the population had switched from, or quit a formal religion. It claimed that no religious category had grown more than the category of "no affiliation at all." As suggested in earlier chapters, it stated that young Americans were the most likely to undergo change by shedding religious affiliation. Many people now seem to just "shop around" when it seems socially appropriate to find the most suitable church association for themselves and their families. Clearly, it tends to be a social convenience, losing much of the traditional depth of fundamental belief.

The large drop in "born Catholics" is said to constitute the principal purpose of a recent papal visit to America (103). It is claimed that the nation as a whole remains largely Christian but that there is substantially less response to all religious doctrines. Despite lagging interest, it seems that few want to be considered "churchless", an important concern with respect to considerations of stigma. The underlying theme of this and other religious discussions suggests a clear pattern of religious change, especially in young America, just as we had witnessed in older developed nations. It is significant that many of the surveys on religion are initiated by concerned religious institutions looking for spiritually weakened areas and means for increasing their support.

A timely news article (104) recently described how the Internet was being used to attract new recruits. It said that "enticing young people to become regular members of a Christian congregation is a major challenge." It pointed out that the Barna Research Group, which tracks religious trends, in 2006 showed that only a third of people born after 1984 attended church on a typical weekend compared with half of the baby boomers. Every modern technique seems to be employed to encourage religion for others including profiles on Facebook, MySpace, MyChurch, and endless other usages of modern electronic information technology. It spoke of the beginning of a new website series called "Tongue-Pierced", showing a photo of a man's tongue with a cross pierced through the middle of it. The article also implied that biblical texts seem to belong to "another time and culture", and are difficult to translate to modern life. Using language and tools with which young people are familiar was said to be an effective way to connect them "to the sacred." These activities confirm the claim of earlier chapters regarding the relentless effort, an assumed religious imperative, to "market" imaginative religion.

Students of theology have arisen from the most educated and scholarly depths of society to fill the very best seminaries and universities of the world. The very desire for advanced seminary study suggests a natural human desire for deeper inquiry and understanding, often the desire for a potential life of righteousness in the care and guidance of a god in which they believe. Usually advanced pursuit of theology is rooted in an already established belief. It may also suggest a strong interest in determining the meaning and underlying "fabric" of religion. The word "inquiry" is often used in describing such studies but inquiry usually involves no option other than pursuit of the inquiring belief itself.

**Religious sites:** Continuing studies define, modify and perfect religious history. The records are constantly examined and reinterpreted to provide more modern and more acceptable understanding of earlier concepts. This book is intended to contribute to that process, an updating document, a document of spiritual review and consideration based on factual recent knowledge. On the other hand, archeologists stimulated by ancient accounts of religious events, have frequently sought remains and relics that relate to established biblical events. The searches for such documents and religious sites have occasionally raised the specter of earthly appearances of the god or related characters in which man believes and they are often given overwhelming significance, well beyond reasonable terms of rationality.

The Holy Land is well known for its biblical relics for it constitutes much of early "biblical geography." But most anything that can contribute "reality" to imaginism gains special notice. Claimed sightings of the Virgin Mary, validity questions surrounding the Turin Shroud, the endless search for the Holy Grail, remnants of Noah's ark, and numerous other objects and evidence of people and events tend to bring excitement and discussion to the world of belief. Books and films of wide popularity are often created around them. Despite the high quality and intellectual depth associated with such pursuits, the imaginative aspects remain fixed and intractable even without a suggestion of positive verification. To rational modern man, most of these pursuits seem highly imaginative and unnecessarily dramatic. Except for mention in biblical writing, there has never been any credible verification of the existence of any transcendent personal god or evidence that one might have been represented on Earth in any manner. It is truly a primitive concept suggesting wishes and hopes that still permeate many generations of vulnerable human minds. Dawkins would say that the concept of "faith" covers all errors and omissions.

In a special issue on "sacred places," *US News and World Report* (105) stated, "They are as varied as the human sense of the sacred and as the world's many spiritual traditions." It suggests that sacred places range from entire cities to a special room in one's home, and can be man-made or part of nature. Often associated with saints or holy figures, "they are places that draw pilgrims, sanctified, in some cases, by great churches, mosques, temples, or shrines." "They are outward and visible signs of an invisible, numinous order. All play a part in the common effort to define the cosmos, to name the divinity or higher principle behind it, and to locate the self and the community within it. They also partake in the ambiguities of religious life." It reminds us that it is Jerusalem where, over the centuries, the peoples of three different great faiths came to fight almost as much as to pray. For better or for worse, sacred places are part of the complex web of our collective spiritual heritage. However, many modern minds of reason decry the concept of spiritual heritage and stop abruptly at the point of reason and good judgment. The element of time will gradually dilute its importance to changing and globalizing man.

The Holy Land has undergone many stages of change but it is still a magnet for religious people. Israel has become a very modern and technically capable nation, however small and militarily strong. In a recent Israeli election, a secular candidate defeated an ultra-Orthodox member of Parliament in a hotly contested race for mayor of Jerusalem. Voting in the divided capital city is said to be significant, since it had not had a secular mayor in several years. As the largest city in Israel, it is the most complex politically and religiously, and its central government will ultimately decide the city's fate with respect to the Palestinians. The troubled city is highly symbolic and holy to Muslims, Christians and Jews, but its livelihood depends largely on tourism. The area remains the center of conflict among principal religions and will continue in that role until secularism becomes reality. Its biblical significance will no doubt fade in a future secular world.

**Immiscible cultures:** A recent news story (106) confirmed how difficult it is to separate Palestinian interests from Jewish interests because of strong religious belief combining with immiscible cultures. The Palestinians seek Israeli withdrawal from the West Bank and control over cities like Nablus, one of their largest. Almost a thousand Jewish pilgrims recently descended on its "holy place" known as Joseph's Tomb, a stone compound in the heart of the city believed to be the final resting place of the son of Jacob. It is said to be a parcel of ground that Jacob bought for a hundred pieces of silver according to Joshua 24:32, suggesting that its ownership is not in doubt. The religious conviction that this land is the Jews' birthright was said "not to be up for grabs." The pilgrims were said to dwell in a hastily built sukkah as men danced in circles to the tune of a lone clarinet, feasting on sweet and spicy kugel and orange squash. They celebrated the "redemption" of Joseph, betrayed by his jealous brothers and sold into slavery in Egypt. These strong emotional feelings relative to beliefs and the distant past seem to discourage positive projections with respect to modern man. It confirms a need for accelerating the processes of fresh ongoing cultural evolution. Must Middle East settlement await world secularization?

It is clear that sacred places are sacred because the literature, lore and continued religious practices promote and maintain their sacredness. They are emotionally built into the traditions, culture and the ongoing religious teachings of man. Granted, such physical mementos of the historical past can bring strong emotions to bear but, absent those frequent historical triggers, they would have little lasting meaning. The vulnerable and greedy spiritual mind requires memory and excitement and assembles whatever it desires for emotional stimulation. In accord with arguments already made, anticipated changes in spiritual understanding will gradually normalize the significance of sacred places. Changes will include rethinking the past, more desire for earthly realism, expanded education, improvement of human reason, a greater diluting mixture of cultures, improved forward vision, and increasing separation of generations. Archeological and cultural interests may be preserved but the underlying religious meaning will gradually go the way of imagined

Greek and Roman gods. It seems far better for society to apply equivalent effort to the help and support and indeed the advancement of current man to assure his opportunity for a modern experience.

**Religion misused:** Some forms of imaginative belief have attempted to take religion a step farther. Evangelicals in the US have redefined dependence on a benevolent humanlike god (107). The Islamic world has also reinterpreted its fundamental beliefs (108). Unlike Christian and other continuing moderate faiths, the beginning of Islam is said to have been fraught with military might. Desert tribes were often conquered and converted to the establishment and practice of its newly introduced belief system. In the modern world of Islamic belief, some elements again seem to emphasize dominating force (109). Militant and terror issues have raised an important question of religious meaning and purpose. Even in this late period of man there is no meaningful evidence of effective ideological unity among principal religions. Widely different religious definitions and claims under the same assumed god tend to degrade all religious beliefs. To the rational, intellectual mind this becomes further evidence of emotional human origin rather than divine.

**Factual principles:** In the rational and more factually informed scientific camp of realism, many highly recognized scholars have carefully studied the nature of the physical world, the structure and function of life, and the surrounding material world in a consistent and reproducible manner. Their mission has been to seek reality, utilizing imaginative powers only within the confines of the established scientific method. Their factual studies have been directed to building a bank of reliable knowledge, adding new dimensions to reality. Science does not directly oppose religious responses but cannot accept violations of firmly established principles, the defining and guiding principles of the observed and tested laws of nature. Failure of underlying principles would violate all human sense of reality and reason, indeed, all the vital factual knowledge established by the human mind with so much effort and care to date.

As discussed more fully in chapter eight, science is known to be a method of determining realistic and verifiable knowledge and understanding of the world and its living creatures, a method that advancing man must preserve and follow above all else. Man's efforts are constantly directed to interpret both man's universe and his thought processes. The methods of science can, and do, apply to the study of religion itself. This book has indicated that many realists believe that the era of imaginism should gradually be coming to a close as modern knowledge increasingly supplants the long period of human ignorance and misinterpretation in past understanding. The greatest stumbling block consists of mustering sufficient human courage to face up to established reality in view of deflecting ideal spiritual promises. In this respect, the communication bottle-neck rests largely with leaders and "teachers," themselves. Indeed, the current period of man is still fraught with the powerful, long "nourished" human emotion mixed with reason and it is not yet known how long it might take for the evolving judgmental mind to equal or overtake the neural strength of the emotional mind where idealism and promise are so effectively lodged.

Traditional believers are often critical of new and different liberal minds associated with persons called "free thinkers." Bertrand Russell, an English philosopher and mathematician, spent much of his life writing and thinking about mathematical logic, meaning, truth, and important social issues. He was a determined pacifist during World War I and, at one time, was briefly imprisoned for writing an article contrary to public views of the times. He wrote on "marriage and morals" and on "new hope for a changing world." With respect to views on religion, he reflected the common interpretation held by many philosophers and scientists. He argued that Christianity was an "enemy of moral progress" and wrote *Why I Am Not a Christian* (110) in which he set forth the case for atheism. In it, he refuted attempts to prove the existence of God using reason. He lived through the period of the hydrogen bomb development and became very active in nuclear disarmament issues. He taught at Trinity College, Cambridge and periodically at schools in America, contributing much to modern mathematical philosophy and logic. He was indeed one of the early promoters of the new atheism that has since gained considerable momentum. His unique courage to raise contrary religious issues at the university level in an academic world, largely of imaginative belief, was admired and well respected among his discriminating learned colleagues. Russell is typical of how serious long-time study of reason and logic frequently leads to views of atheism.

**The born again syndrome:** There are occasional instances of an abrupt perceived need for a new mental outlook that requires the "immediate assistance" of a powerful emotional force. In some cases, following a very difficult experience, subnormal behavior, or feelings of guilt, when "down on luck", depressed, at a "dead-end", or when experiencing weakness in belief or in non-belief, a person might undergo a sudden "epiphany" (111). It is said that he has now "found Jesus" or has "discovered God" in a transformation known as a "born again" experience. It is a natural property of the plastic human mind, a desire for relaxation of its mental imprisonment or recognition of rebellion against social norms in some manner. Although it is somewhat meaningless from a religious viewpoint it is an effective resolution of a personal dilemma. It is merely an exaggerated resolution to make a firm, decided change, utilizing the special strength of "religious type" emotion. It is a sudden enlargement of retained imaginative belief to achieve an immediate end, but given religious interpretation. On the other hand, as explained earlier, many believers who have become more knowledgeable and informed undergo an opposite transition and just as suddenly reject all imaginative ideas of a humanlike god and, in effect, become atheistic realists or fully secular.

At times of great trauma, man becomes frustrated and panicky, and looks to something or someone that may offer immediate and effective help. Tradition immediately suggests invoking his all-knowing and all-powerful god. It is quite automatic and universal, despite some level of insincerity. I recall an account of a plea by an American soldier, pinned down at the difficult Anzio beachhead in the Italian Campaign of WWII. Desperate and facing impending disaster, he finally called for the highest authority he could muster, saying, "God come help me. Please come yourself, Don't just send Jesus."

**This side of death:** An article about the Vatican and globalization (66) suggested that the core benefits of religions, unlike other worldly institutions, most often relate to the afterlife, the instinct for survival. Social scientists, however, argue that many benefits of church membership are to be had "this side of death." Religions were likened to the advantages of a club of like-minded people. They provide rules to live by, solace in times of trouble, and a sense of community that can promote higher levels of education and income, more marriage and less divorce. The article goes on to say that such a "club" needs strong, believable rules and, like marriage, membership becomes more valuable the more committed participants are to the common cause. Demanding rules, like celibacy, or avoiding meat during Lent, are said to "enhance" the level of commitment. Although some of these points may be psychologically valid, this book discourages any principle of promoting a framework of rules based on a false premise and imaginative concepts. It seems to be like offering the gifts of an imagined Santa Claus for good behavior. Many informed modern minds are now moving beyond that notion and recognize the complex of unresolved world problems hidden among the "strange mixture" of imaginative beliefs. Society would do well to promote effective world programs of improved human interaction, emphasizing pleasant and satisfying coexistence. Default dependence on imaginative religious references does little to advance the position of society in a rapidly developing global world.

Does one have to maintain an imaginative belief in a framework centered around a humanlike god to be emotionally happy and satisfied? The modern rational mind doesn't think so. Satisfaction and happiness in secularism and atheism were discussed earlier. Mere belief in immortality does not save one from ultimate biological death and baseless imagination in the modern world of understanding brings only tentative spiritual relief followed by periods of disappointment. On the other hand, belief in the supreme power of nature, in cosmic or natural religion, provides deep and permanent unvarying appreciation of the full beauty and harmony of nature as described by Einstein and many others who claimed to be deeply "religious", even more religious than many traditional theologians. Their satisfaction was based on the fact that such belief was without uncertainty, knowingly based on invariant reality, and consistent with advancing civilization and factually determined human knowledge.

As stated in earlier chapters, one can rely on major long-term changes in society, changes that are well hidden among those that cause them. These are changes not often visible as they take place, for they can be recognized only by later generations. For traditional religions to reflect some higher degree of solidity and acceptance they must first find a common spiritual basis for integration and merger into one "true" religion. In so doing, their inherent weaknesses would be revealed. Based on the length and depth of past conflict, any such rational attempt would most likely result in the formative stages of a mass transition to secularism as proposed in this writing.

**No divine guidance:** Careful examination of traditional belief systems today reflects a lack of firm agreement to any one principle. Differences and "reformations" have appeared throughout religious history. Leaders simply want to maintain independent recognition and control of their specific interpretations, typical of human business and government development, but quite contrary to ideal spiritual principles. New leaders often produce new detailed religious interpretations. Such division always reflects uncertainty and weakness and the unlikely future achievement of unity. It is a human characteristic spread throughout the entire religious world, suggesting no divine guiding core at all in a matter that is claimed to concern but one god and all man.

**Organized uncertainty:** Surveys of American religion (22) demonstrate the extent of the changing mix of detailed beliefs. Protestants who were, by far, the largest percentage of believers in America, now claim to constitute only about one half of the population. Although about a quarter of the population is Catholic, its organization has also undergone a great deal of change. It has lost more members to other faiths and to "no faith at all" than any of the other prominent organized religions. The referenced survey further suggests that immigrants may be serving as replacements and about a third of Catholics are therefore now Latinos. The survey also suggests that many practicing Catholics are "illiterate" about the principles of their faith. This may mean that simple "belief in belief," as suggested earlier, may be the intermediate course of average current man.

Adults under the age of thirty years, considered to have the more "modern" minds of the current population, were the most likely to drop from religious affiliation. This is in line with earlier commentary. One in four indicated that they are not now affiliated with any organized religion. Overall, less than eighty percent of the population claimed they were Christian and more than fifteen percent said they had no affiliation. Jews, Buddhists, Muslims, and Hindus now make up less than the remaining few percent, all inclusive. Despite the large percentage claimed as Christian in America, this book suggests that most are becoming increasingly indifferent to the principles of their faith. It again suggests a declining interest in knowing or caring about the underlying tenets of their religion reflecting a marked decline in church attendance. It further suggests that the young Latinos, in the role of "replacements," are likely to participate in greater education of "reality" and will, in time, tend to follow the route of indifference and then secularism, influenced by the power of factual modern knowledge. The change appears to be related to immigration from the southern regions of the continent, much of which had been overly influenced by earlier missionary efforts.

**Secular recognition:** The Catholic Church clearly recognizes the trend toward secularism as claimed in the theme of this book. Despite the large number of Catholics and Ecumenicals addressed in his recent visit to America, Pope Benedict XVI warned of serious "secular challenges." In a speech to American bishops, he also warned of the "subtle influence of secularism" that can "co-opt religious people" and even lead Catholics to accept abortion, divorce, and cohabitation outside of marriage. The reality is that a large fraction of Catholics already tend to disregard one or more of those prohibitions and the trend is unlikely to reverse. Unfortunately it may leave a sense of unnecessary guilt. They simply do not make it a public issue and it is rarely disclosed in any survey. The Pope also admonished that "Any tendency to treat religion as a private matter must be resisted." At a later meeting of the United States Conference of Catholic Bishops, the Pope read responses to three questions, submitted in advance, that reflected the bishop's prime concerns: 1) secularism, 2) Catholics abandoning their faith, and 3) a shortage of vocations to the priesthood. The Pope seemed to be acknowledging that the Church is fighting a slowly losing battle, the result of the unstoppable forces of change that continuously affect the direction of human life on the planet. At the same time, its members are confused by the uncertainty building between its teaching and reality. It is for that reason this book is intended to provide realistic information on current trends and a steadying rudder with which to direct human lives without fear or guilt.

**Fading interests:** Prior to his last visit, the Pope made clear the realistic condition of the Church (29). It was reported that hundreds of parishes were being closed or consolidated. There was not only a severe shortage of priests, there were also insufficient funds to maintain the aging churches. The demographic changes and "not enough people attending Mass" could not justify keeping some parishes open. Dioceses were said to be closing parochial elementary, junior, and high schools that had provided a rigorous education for many generations. Of some eighteen thousand parishes in 2007, over three thousand were said to be without pastors. More than eight hundred parishes had been closed in just one recent decade and additional closings were expected. There were many fewer new priests, "way below replacement level."

A member of the Life Cycle Institute at the Catholic University of America considered it a "crisis." Specific mention was made of a "devoted" lawyer in Camden, where the number of parishes had been cut in half, a lawyer who spent every Thursday, from eleven PM to midnight, praying before the blessed sacrament, keeping the Eucharistic adoration for forty-eight uninterrupted hours every week. The article did not indicate if that effort constituted an appeal to their god to intercede. Some church administrators suggested that consideration be given to opening the priesthood to women and to married men to help solve the problem. Indeed, the quiet movement of society toward "no religion" and secularism has been continually taking its toll.

**Immigrant workers:** An article datelined Owensboro, Kentucky (112) discussed the desperate importation of foreign priests to work in America. It pointed out that a third of those studying in American seminaries are foreign born. In the Roman Catholic diocese of Owensboro, priests were recruited from Nigeria, Uganda, India, and Kenya. Mexican priests were preferred because of the need for Spanish-speaking capability. But it turned out that Mexico had a priest shortage of its own, with only one priest for seven thousand Catholics. There are many problems, not only in recruiting, but also in adaptation of foreign priests to the American communities. Problems included refusal to learn to drive, hostility to other nationalities, inability to adapt to cold temperatures, and involvement in romantic entanglements. The pool of priests was shrinking not only because of retirement and deaths, but also because of removal from the ministry following accusation of sexual abuse of young people. Religions are indeed in a difficult state of transformation.

**Extreme measures:** Another news article (113), the headline of which included a "mix of sacred and silly," suggested that an effort is being made to build interest in the Church through the use of a channel on a satellite radio system called the Catholic Channel. The new satellite radio system is a subscription-only network, popularly known for its immense drawing power based on its relatively profane and pornographic programming. David Gibson, author of *The Coming Catholic Church*, suggested that the church's foray into talk radio may reflect official acknowledgement of the need for a new, more interactive relationship with believers.

**Legislating behavior:** The most recent indication of the earthly nature of organized traditional religion is demonstrated by yet another "reformation," the break-up of the Episcopal Church (114). The Episcopal Church is the American branch of Anglicanism, a Christian religion that traces its roots to the theocratic formation of the Church of England with the Archbishop of Canterbury as its spiritual leader. The new schism was caused by the ordination of an openly gay man who lives with his gay partner in New Hampshire. About fourteen dioceses took offence and four of them broke away. In order to remain Anglican they joined the authority of bishops in Africa and Latin America. There has since been concern about ordaining women as priest or bishops. They seek organization of a new American province. There are pending complications, such as questionable freedom to take church property with them. Such reorganization and reformation would also face complex organizational problems. It comes at a time when religion is losing ground and when there are so many other churches available to the less discriminating worshiper. Sometimes the deep purpose and meaning of a church is lost in the lack of agreement or rehash of unimportant details of ideology. A victim of changing morals and modern social behavior, most theologians are pessimistic about the reestablishment of a long-term successful program that can match the narrow solidity of the original Episcopalian Church. Although public behavior is best guided by the established mores, laws, and rules of overall society, the control of private behavior must be guided by conscious education and the best interest of man. It cannot be effectively legislated by religions or governments unless it adversely affects the rights of other persons.

**When fellowship is salesmanship:** A local Presbyterian community church recently sent a mailing in the form of a poster size ad that included commentaries reflecting the times. It said, "Let's face it. Many people do not attend church these days," that people would say "Christianity is out of touch with modern life." "Sermons are boring and irrelevant," and "I can find spiritual fulfillment in other ways." With respect to summer camp, it also said, "Your children will have a blast . . . playing with their peers and learning Christian truths." These are indeed wonderful social programs of fellowship and music for both children and adults, but they fail in their underlying dogmatic pursuit of "God-centered lives" of prayer for each other as "members of the family of God" and in their definition of "religious truths." Public "faith" and "belief" seem to be hanging by a thread, and the guiding principles of such formal religions must receive basic reconsideration.

But we also know that imaginative belief has not represented reality and that its views are continually in conflict with modern reason and understanding, undergoing uncertainty, and change together with advancing underlying life, all redirected by natural forces and by newly acquired human knowledge. Despite all the religious promotion and missionary work, despite the pressure applied to alliances with major gods, despite the special benefits they appear to bring to man, the increasing conflict with reality in such effort only delays the time scale of the eventual secular forces. Man is now living in a fully "real" world, a world that will become increasingly more real with each generation. But his amazing cognitive powers of imagination, that begin and end in reality, must never allow major "detachment."

**Theological alternatives:** The question of what can replace fading religious emotion has been the subject of frequent discussion and has been variously answered in these chapters. There must first be full understanding of the mechanisms of religious stimulation, the imaginative human response and resultant behavior. The reality of modern knowledge must be acknowledged and must become integral in the education process. Ordinary man is not going to understand his position in these matters without substantial help and it therefore must be made an important part of the worldwide educational process. Education is vital and cannot be allowed to be spiritually based in isolation. It must be made uniform throughout society. Traditional promises without explanation, such as noted in these chapters, are dangerous and misleading. Once one becomes convinced that the notion of a humanlike god is beyond reality, it must be abandoned forthwith and some alternative line of thought adopted if true origin and meaning is of interest and desired to be pursued. The observed natural mechanistic force of the universe has been offered as a more factual explanation for modern minds.

There are many alternative choices for contribution to, and for participation in, the lives of fellow man, indeed, in the process of encouraging factual understanding and peaceful coexistence. These include serving the material needs of fellow citizens, the effective organization of people, the study and resolution of remaining scientific mysteries of nature and of its people, and the care of the environment of the planet, and many others too numerous to mention in detail. The most important service to fellow man is the resolution of conflict, the elimination of stigma and useless differences as frequently suggested. There is no need to create a huge sphere of "empty theology" that continues to have no reality except in the psychological and neuroscientific sense. The fact that it has been created, and that it has existed without satisfactory explanation for millennia and only temporarily contributes to man's satisfaction, is no reason for encouraging its continuity or even looking for an alternative other than reality itself. This book claims that, once man is properly educated about himself and his world, the human response to spiritual emotion will be decidedly secular and will lead to a sense of human satisfaction centered about realism. Freedom from fear-based imagination, and increased service to fellow man, as promoted by humanist organizations, provide a real sense of purpose in place of a sense of "imagination."

**Does belief matter?** Some people feel that it makes little practical difference as to which of the two principal modes of belief referenced at the outset of this chapter is more acceptable as the correct interpretation, saying, "What is IS, and it will all work out as it is meant to be in the end, independent of the human viewpoint, that man has no say in the matter anyway." Feeling that all life begins and ends in the same manner, they are openly willing to accept and enjoy life just as it comes. Concerned with the god problem, Blaise Pascal, a seventeenth century philosopher, mathematician, and physicist, once wrote an essay on "the necessity of the wager", suggesting that if one simply bets on belief in God he might have nothing to lose. If there is a beneficial god, it will probably serve man equitably. If there is none, nothing is then lost.

But modern man feels it is not that simple, that man is passing through the stage of imaginative speculation, that he now knows much more and has a major responsibility to himself and to fellow man to resolve "the god problem" if he can, at least to the extent that available facts can reasonably provide an answer. At the very least, advancing man is expected to rule out irrational and unproductive notions, notions that not only waste much human time and effort and encourage misguided reliance, all without benefit, but also notions that lead to dangerous conflict among highly emotional, inflexible, and demanding fundamental believers. Abnormal concern with belief often engenders abnormally high emotion. There is no rightful place or justification for dogmatic and obsessive imaginative believers in a peace-seeking society. Although serious believers might feel that they have certain responsibilities to their assumed god that must be fulfilled, some concerned theorists feel that continued irrational belief might unfavorably deter the process of ongoing human improvement with respect to natural evolution of greater reasoning, that the character of future man might then be in greater question.

Following earlier considerations and explanations of the mechanisms of the spiritual mind in initial chapters, this third explanatory chapter is intended to define the two most basic contrasting views of human origin and belief. It is also intended to suggest the more probable rationale for educated modern man based on present knowledge and anticipated direction of human change. It is further intended to introduce commentary on observed effects that such changes may have on the life and behavior of globalizing society. Further effects on society are included in subsequent chapters in order to emphasize the high degree of continuing religious influence and undefined uncertainty. Believers, who followed and understood the introductory chapters of this book and managed to read the most important references, might now view traditional religion a little differently, or might at least be willing to offer a more receptive mind with respect to consideration of realism. Few persons who have long lived with the satisfying and pleasing concepts of imaginism want to readily replace its ideal promises with hard reality no matter how rational and factual they know it may be. Both defined notions of belief represent only states of mind, conditions of thought and possible convictions, one based on desire and hope, and the other on observation and reason. However, neither ideology is directly subject to acute personal consequences other than possible evolutionary deterrents and social stigma.

**Waiting for the rational mind:** As with my earlier publications, this book is written in the interest of recognition of, and continued pursuit of, realism. It is neither religion-bashing nor is it intended to produce wholesale conversions of imaginative believers. It not only defines reality as determined by current factual knowledge, but also envisions a trend toward a new atheism movement that is expected to accelerate until it results in a unified world. It is a trend that is expected to move well beyond existing human thought, but only on its own time scale. Its direction will not depend upon any god or man, or on any religious ministry, but will depend only on the powerful natural forces of change as defined in the first chapter. These forces underlie the further growth of the rational human mind and the strong continuing natural interest to pursue factual understanding, both important factors in the anticipated advancement of ongoing man. Whatever the current position of the highly imaginative emotional mind, it must be looked upon as the most powerful progressive machine on the planet and its inherent long-term instincts will eventually prevail. The mind will naturally find its way to understanding in reality without any salesmanship, without the need for any proof, without promises, threats or torture. It will survive the powerful grasp of numerous longtime traditional beliefs no matter how extensive or deeply entrenched, and will eventually find its own path to factual and intractable reality. Since it is based on the most powerful fundamental transcendent forces of independent nature, forces that have produced the universe, life, mankind, learning, knowledge, it is expected to remain on an irreversible path toward reality and secularism. It will certainly continue its advance with or without the contribution of this book.

The recognition and pursuit of reality of this study clearly suggests a viewpoint contrary to much of present society, but it is accompanied by a clear and valid explanation. Its purpose is not to belittle any positive benefits of imaginative religious beliefs, but to understand them and why they exist, and to transform them to terms of accepted factual knowledge and modern thought. Imaginative religions have long served man with a form of emotional satisfaction and have brought him pseudo-freedom from the fear of death. Religions have also given man temporary purpose and meaning and have encouraged morality and ethical behavior of the masses, albeit under ignorance and the threat of an assumed god. By their nature, they have also variously served the interest of the poor and suffering in society. But society will continue its advance and will survive all the suggested changes.

**Distorting the developing mind:** In closing this third chapter, I again remind the reader of certain secondary dangers inherent in continued imagination and distortion of factual knowledge. Failure to recognize man's position of reality and to move in concert with the natural forces of change constitutes failure to encourage the processes of human growth and natural advancement. Man requires unprecedented "unity of belief," without which his emotions can find no long-term satisfaction and peaceful coexistence. Extreme emotion surrounding imaginative religions can run amok in the present unbridled mind of imagination and can easily spill over into dangerous and destructive conflict as it has recently and, even more seriously, at various past times in history. The promise of a glorious heaven following death not only distorts fact and reason, but also induces men to act against current worldly society in a futile attempt to anticipate its advertised benefits. Lastly, there is an underlying concern that extended imaginative interpretations of existence may negatively interfere with ongoing natural evolution of reason and judgment in still developing and changing man.

**Truths begin as blasphemy:** The very suggestion of realism in these pages may indeed appear as "blasphemy" to those who were taught to accept traditional belief. Its concept is very different from the imaginative views that have pleased so many palates and have held sway in so much of the world over so much of the past millennia. Man has long grown accustomed to the claimed personal God who he believes can interact with man. He has become locked in "spiritual loyalty" from which he cannot escape without sacrificing all the promised benefits.

Many have reached a point at which they have become addictive with little alternative. Neural wiring has become semi-fixed and can only be modified effectively through progeny. In the first chapters, references were made to scientists who had developed early religious beliefs and in later life were unable to neurally shed a religious bias in their work. It was implied that a recent nominee for directorship of the National Institute of Health, might well fall into that category. Sam Harris, who retains websites known as the Reason Project and the Scripture Project, recently published an op/ed titled "Science is in the Details" (115) that questioned Dr. Francis Collins' suitability despite his impeccable credentials. In general, the Harris philosophy has been in line with the principal theme of this and my earlier publications and his concern in this matter should be given a fair hearing. This constitutes a most important position with respect to neuroscientific research and future direction as emphasized in earlier discussions. Applicable scientific programs and religious interpretations in this field throughout the next decade are absolutely critical.

Rational minds of increasing numbers of intellectual and scientific leaders in modern society have largely concluded that the concept of a humanlike god and the framework built around it constitute understandable wishful thoughts, products of early human desire. But modern citizens of the world have also come to realize that denial of a God in heaven may no longer be considered blasphemous. George Bernard Shaw was once quoted as saying, "All great truths begin as blasphemies." His meaning might well be applied to realism as defined here, a reasoned view, contrary to tradition and emphasized in this book as "truth."

**Truths begin with ridicule:** Schopenhauer's concept of the "three steps to truth," namely initial ridicule, violent opposition, and admission as self-evident, may indeed also be applicable here.

Disbelief, in one form or another, based on modern minds of reason, has already suffered its share of "ridicule" and has experienced violent "opposition" by theologians and believers. Many persons, along with informed intellectuals, scientists and philosophers, are now advancing the new movement to the third stage of the Schopenhauer principle, claiming "self-evidence."

This book assumes that the "truth," represented by fact and reality, is firmly in the process of establishment, backed by unstoppable natural forces, and will ultimately work its way through all three of the Schopenhauer stages in all nations, throughout all levels of society, to find an eventual sense of unity that will ultimately benefit future generations.

# Chapter Four

## Stigma in Belief

(References from "Crossing the Divide," about rejecting creationism, Science, February 22, 2008)

The inconsistencies he found led step by step . . . to a staunch acceptance of evolution. With this shift came rejection from his religious community, estrangement from his parents, and perhaps most difficult of all, a crisis of faith that endures.

—About Stephen Godfrey, Curator, Calvert Marine Museum, Maryland

Nothing else I have done in my life has made me such an outsider I still have childhood friends and relatives who won't speak to me.

—About Brian Alters, Director, Evolution, Education Research Centre, McGill University

Cantor argues with a relish that the Hebrew Bible is a tissue of unverifiable myth and historical fantasy, and is racist to boot.

—Review of Norman Cantor's *The Sacred Chain* by Mark Silk, New York Times, February 5, 1995

**Freedom from inequality:** Stigma, based on differences in religious belief, is one of the most divisive of human evils. It is especially inappropriate and without purpose especially when considered in terms of imaginative beliefs. It not only exists between believers and nonbelievers but can also exist among the many of the thousands of competing religious systems and sects. It may well be a subconscious function of the uninformed, undisciplined, immature, or prejudiced mind. It involves the same or similar forms of emotional prejudices that are found among racial and cultural differences, but often with greater subtlety. However, it can easily become intermixed with other stigmatic emotions leading to significant unrelated incidents of conflict, indifference, separation, isolation, divorce, imprisonment, violence, warfare, and even destruction of national entities. World history is replete with major incidents of religion-related anguish, torment, and persecution, not to mention conflict and disappointments among members of society in ordinary daily living. The references listed in the eppendix contain numerous examples of stigma and their effects on society, but only a few of the most interesting representative cases have been selected for discussion in this chapter.

**Stigma in the Military:** Religious stigma has reached deeply into the U.S. Army. For example, an article in the *New York Times* (116) spoke of threats and lawsuits specifically related to religious stigma. Specialist Jeremy Hall and the Military Freedom Foundation filed suit in federal court in Kansas alleging that Hall's right to be free from state endorsement of religion under the First Amendment had been violated and that he had faced retaliation for his views. In November 2007, he was allegedly sent home early from Iraq because of threats from fellow soldiers. He had held a meeting in the previous July for "atheists and freethinkers" at Camp Speicher in Iraq. An officer attended and severely berated Hall and others about atheism.

The lawsuit raised new questions about the military's religious guidelines. In 2005, the Air Force issued modified regulations in response to complaints from cadets at the Air Force Academy. They charged that Evangelical Christian officers used their positions to proselytize. The armed forces presumably have regulations covering rights to their own religious beliefs, including the right to hold no belief.

Specialist Hall said, "They don't trust you because they think you are unreliable and might break, since you don't have God to rely on. The message is that it's a Christian nation and you need to recognize that." He had organized a chapter of the Military Association of Atheists and Free Thinkers near Tikrit to support others with similar views about religion. An officer evidently charged him with "disgracing those who had died for the constitution and, one day, you will see the truth and know what I mean." Complaints included prayers "in Jesus' name" at mandatory functions, which violated military regulations, and officers proselytizing subordinates to be "born again." A retired Air Force judge said, "Religion is inextricably intertwined with their jobs. You're promoted by who you pray with."

Hall was raised as a Baptist and his grandmother read the Bible to him every night. In the summer of 2005, after his deployment to Iraq, thinking he was going to Iraq because "we had God on our side," he became friends with servicemen of atheist leanings. Their questions about faith prompted him to read the Bible more carefully, reading that bred even more doubts, deepening over time. He said, "There are so many religions in the world. Everyone thinks he is right. Who is right? Even people who are Christians think other Christians are wrong."

Within a different unit, Hall said the backlash continued. He had a "no-contact order" with a sergeant who, without provocation, threatened to "bust him in the mouth." Another sergeant allegedly told him that, as an atheist, he was not entitled to religious freedom because he had no religion. But in very religious America, there are new stigmatic problems everywhere. It is most apparent in large public groups. Religion and its role are said to be especially in dispute at the service academies (117). Following the scandal at the Air Force Academy a few years ago, there were frequent complaints about pushing religion by faculty, students and staff at both West Point and the Naval Academy. The simultaneous rise of Evangelicalism and persons of "no faith" in America led to a collision course. The ACLU threatened legal action if daily prayer at lunch was not abolished. Commander Austin of the Naval Academy responded with the comment that "the Academy does not intend to change its practice of offering midshipmen an opportunity for prayer or devotional thought during noon meal announcements."

**Is early religious teaching fair?** At West Point, mandatory banquets began with prayer. Cadets and staff claimed that the Commandant routinely "brought up God" in speeches and events. In his farewell presentation, he told his cadets to "draw your strength in the days ahead from your faith in God. Let it be the moral compass that guides you in the decisions you make." Cadets who did not attend religious services during basic training were sometimes referred to as "heathens." The message was constantly being sent that, to be considered a successful officer, you must believe in God. The military field manuals on leadership emphasize "spiritual values" in development of duty, honor, and country. At his commencement speech at West Point, the Secretary of the Army started and ended with biblical quotations and represented the Iraq and Afghanistan wars as "a clash between American and radical Islamic religious differences." It is interesting that most of the complaints by servicemen appeared to have come from members who were brought up in religious organizations and later became agnostic or atheistic.

**Stigma everywhere:** Religious conflicts are also replete internationally. In present-day life in Egypt, the clash of Moslem with Coptic Christians (118) exemplifies the extent to which imaginative religions have differed and how this continues to negatively affect the relationships among man. A monastery was ransacked and monks forced to spit on the cross. Do they not look to the same god? There is constant Coptic fear that a son or daughter may fall in love with a Muslim. There are frequent disputes among farmers and among students. Religious identity is paramount, even more important than common citizenship. Coptic priests are accused of wearing black to mourn the Arab invasion of Egypt in the seventh century, and many are affected by the belief that the Koran ordered Muslims to kill all Christians. There is hatred and fanaticism on both sides, and the Egyptian security force has grown to twice the size of the country's army. Officially there is liberty of speech and religion but it is, in fact, kept under tight control.

Another vignette of social stigma and conflict is evident in the current social upheaval in remote India. It is largely a stigma of equality and only indirectly related to religion. Once considered to be "untouchable" in the ancient Hindu caste order, many citizens are now escaping hunger and humiliation, if not social prejudice, largely because of the last several years of new capitalism and education. This represents the changing character of advancing civilization through cultural evolution and the steady mechanistic transcendent forces frequently emphasized in these chapters. India's new knowledge-based economy now rewards the well educated and highly skilled, rewards that were previously preserved for the upper castes.

The first "freed" female of her city is said to have persuaded her family to allow her to defer marriage so she could finish college and get a stable salaried government job. She said, "With education comes change. You learn how to talk. You learn how to work and you get more respect." One active provocateur said that economic expansion will neutralize but may not entirely end the caste system in fifty years. However, in fifty years much more will happen in India with respect to social freedom and freedom of thought. Factual human knowledge will reach out to even the most backward of human society, and greater logic and reason will begin to influence, and indeed control, all human thought and behavior. Social and geographical borders are expected to succumb gradually to new understanding. The cultural and religious stigmas will tend to disappear with true globalization, awaiting only sufficiently improved education, as suggested throughout this book.

**Stigma in Politics:** At this time, no high elective office can be readily filled by a nonbeliever in democratic America because the electorate is largely made up of believers and they typically insist that a candidate must not only be "blessed and guided" by their imagined god, but also must outwardly subscribe to his goodness and righteousness (119). In other words, the believing electorate majority feels that the candidate must first be "one of them" religiously as well as politically and sometimes racially, and not necessarily a choice of the most able, wise, knowledgeable, righteous and rational. This concept is also discussed in connection with other issues considered in this book (120).

Whether honest or deceitful within his own political strategy, a candidate is generally well aware of the influence of his spiritual position for favorable electability. This process restricts most all elected leadership to a single emotional or "narrow" class of citizenry. That may become a limitation or problem in the successful association of nations looking forward to expanded globalization in a very complex and difficult world. However, in line with its underlying theme, this book predicts that presidential candidates in America will eventually include atheists, just as in recent years political slates have expanded to include Catholics and African Americans and as the Episcopal Church has recently included gay bishops. Nevertheless, the forces of change described in chapter one will continue to act effectively on both man and his world.

**Removing religion from polities:** Karl Kaiser, a German political scientist (3), recently pointed out that "Religiosity now seems at least as important for public office as leadership qualities. The entrance condition for the American presidential race is 'being religious'." The first words of the Bill of Rights read, "Congress shall make no law respecting an establishment of religion, or prohibiting the free exercise thereof." It is assumed that freedom to practice "no-religion" is implied in this mandate. Constitutional freedom must be appraised with care for, under the guise of a fabricated pseudo-religious establishment, it seems possible in America to devise any sort of imaginative spiritual cult with impunity, promising substantial benefits to vulnerable minds in order to gain loyalty, membership, support, power and perhaps even considerable monetary benefits, without issuance of a word of doubt in the form of disclaimer. It indeed suggests open season on the unprotected sensitive emotions of vulnerable common man. It also suggests the possibility of the germination of highly unstable fundamentalist organizations with military or terrorist inclinations. Historically, there have been numerous dictatorial rogue-leaders around the world who have established unchallenged control of citizens through promotion of an appealing ideology while conducting a regime of oppression, of greed, of sin and pleasure, exercising ruthlessness, protected through a position of military dominance and control.

Religious belief in competition for elective office is a very important social issue. It suggests that a successful presidential candidate in America has to be a "preacher" of sorts, implying that he must exhibit all the qualities of a religious leader. He must not only believe in the traditional god, but also must create the impression that "all persons of faith are good, righteous and moral." All candidates for high office have to attract the most faithful majority without upsetting the secular minority. If this book is correct in its view, and there are few valid arguments to the contrary, the religion factor will continue strongly in elective processes in America for some time and only very gradually disappear. Perhaps there will be no religion-related stigma in politics in the next century.

**When man misuses power he created:** Meanwhile candidates will avidly seek the backing of the so-called moral majority or equivalent existing groups of organized religions. If they follow the published disclosures of recent elections, they will hire religious consultants to hold periodic sessions among potential voters and relate current issues to appropriate portions of the Bible to assure them that they are doing the "Lord's bidding." This clearly creates stigma. However, there is no greater hypocrisy than to build, exaggerate and purposely take advantage of spiritual vulnerability to assure maximum elective support. The very best of candidates, with respect to leadership and reform, might well be atheistic but, by implication, he or she might be effectively charged with lack of "Christian qualities" and would not stand a chance of election. Despite this "high" principle, many religious persons have taken advantage of elective public office to engage in fraud and misuse of funds, often in abject violation of the religious principles that were responsible for their election.

Social issues surrounding elections frequently include questions relating to abortion, environment, immigration, same sex marriage, stem cell research, cloning, contraception, and the like. These are all issues that strong believers specifically relate to the assumed interests of their god and to biblical interpretations. It is implied that they somehow claim to know the mind of God. Rather than to look at each issue on a factual, medical, moral, and scientific basis in the best interest of advancing society, it is only the imagined god relationship that counts most for them. Some simply publicly choose the most advantageous interpretation for the situation at hand. It seems that the principle of separation of church and state is constantly compromised or blatantly violated. Only in pure secularism can these influences be properly and rationally dealt with.

**Values are universal:** There is a great deal of talk about "religious values" in American elections, and their importance may well vary depending upon each individual's overt expression of belief. The cry is frequently "faith, family and values," suggesting that family and values are not possible without religious faith in a humanlike god, without an imaginative vision. This book suggests that such teaching, no matter how sincere, is basically wrong and misleading and creates a strong unjustified bias, indeed a damaging basis of stigma that is unfavorable for those who do not share that belief. It is a tool of conflict and separatism. In a perfect election, the truth should be factually explained and considered by the electorate. The electorate should ideally be unbiased. Meanwhile, it continues as a decided glitch in the conduct of American government and must eventually find correction. To a rational intellectual, the whole process of elections seems fundamentally flawed, subject only to improvement with further educational development and spiritual maturity. Experience has shown that leadership has left much to be desired and that a large fraction of time in office is directed to preparation for the next election rather than to current service. Once in office the candidate's position and supporting funds can variously influence such continuity of office. How can such behavior serve all Americans alike?

In 2008, a Mormon presidential candidate had to make a special public "appeal" to prospective voters in order to insure recognition as "only an American" who would make decisions unrelated to his faith (120). This suggested that even the expression of natural or cosmic belief as described in this book, the belief of Albert Einstein, would be looked upon as unacceptable in America. Recognizing the unique power of his superior mind, Einstein was offered the unsolicited presidency of Israel in the period of its early formation and development as an independent state without public opposition, an offer that he respectfully declined. As earlier stated, even if a nonreligious candidate were to have all the required administrative qualities, an honest and able diplomat with considerable political experience, his chances of successful election in America would be about zero. This speaks strongly to the problems of inherent religious stigma emphasized in this chapter. It also speaks to possible problems of a truly successful democratic government. Where man imagines the existence of a benevolent humanlike god that mandates good behavior and overt human response for obtaining divine guidance and eternal reward, there have been unnecessary added complications (69). Good behavior and guidance have been associated with imaginative belief and have raised the ante with respect to widespread stigma (122). Perhaps the feeling that guidance of a humanlike god supersedes that of the state represents a decided weakness in government and politics.

**Whose word really was it?** There is a clear element of naivete in a mind that regards the dictates of biblical expression to apply to current man with exactitude. Such misinterpretation may even go beyond naive understanding. It may be centered in emotional interference with proper intellectual reasoning. Most Evangelical Christians appear to believe that the Bible is the very word of God. Deep belief and overt religious recognition in high office could well lead to governance problems if given too much literal biblical emphasis (32). Several candidates for president running in the 2008 election stated that they believed the Bible to be the "word of God," and other candidates expressed belief in creationism as opposed to the accepted scientific principle of evolution. Could they all be serious and sincere? Has their education been found wanting? Can any such candidate gain the respect and confidence of all citizens, especially the confidence of the country's important intellectual base? Do imaginative beliefs fully "control" the national electorate? Indeed, what does that say about the political function of the country as a whole? What does it say about the claimed separation principle?

**Expressing values is good politics:** A poll taken in July 2007 (123) confirmed that America is a Christian nation and therefore that "religious values" should serve as a guide in choosing candidates. However, it found that agreement with this principle was entirely dependent on one's own spiritual belief. Among nonbelievers, agreement on this question was less than ten percent. Among Jews, it was nearly twenty-five percent, and for Catholics, a little over fifty percent. Among Protestants, the number increased to over sixty percent and to nearly ninety percent for Fundamentalists. It is no wonder that some candidates must consider distorting their inner beliefs if they want to be assured of successful election in America. It is further indication of the importance of this book, to encourage recognition and the advancement of normalizing reality in functional society. It is particularly interesting that, among people who earned an adequate wage, almost fifty percent were in agreement on religious values, while for those at the poverty level the percentage was nearly sixty percent. Dependence of belief on social and educational level must be increasingly recognized and aired in public discourse.

It is interesting to look more closely at the stigma issue with respect to lack of support of a particular candidate. For a Catholic or Jewish candidate, about ten percent of voters said that they were not particularly supportive. For Mormons and Fundamentalist Christians about thirty percent were not particularly supportive and for nonbelievers this figure rose to about sixty percent. It is clear that a believer running for office might expect twice the support a non-believing candidate would receive. A candidate, who was said to have become a Christian following his college experience, said that if we remove religious language from our conversation we forfeit the "imagery and terminology through which Americans understand personal morality and social justice," adding that secularists are wrong "to leave their religions at the door." This demonstrates a significant level of religious bias in current American society. But religious influence has always entered the election process in some manner. In a much earlier campaign, Catholic bishops denounced one of the Catholic candidates for his position on abortion, suggesting that he should be "denied Communion." Indeed, the serious need and growing prospects of unifying secularism become increasingly significant.

**Good reasons to believe:** Candidates themselves are keenly aware that voters respond well to claims of faith and belief. They actually design their campaigns to use people connected with the ministry to seek out those voters sensitive to religious belief. Some people may even be unknowingly "used" in this manner. Faith-based political consulting firms are hired to act as matchmakers between candidates and clergy in order to help navigate the Christian media markets and to develop messages that cross party and religious lines. Other consultants are hired to shape a candidate's religious identity. Nuns are sometimes "inspired" to man phone banks. Serious campaigns run a weekly news bulletin and hold regular "faith forums." They all talk about the god that guides and sustains them. Religion is made a significant issue. Perhaps for some, it becomes more important for election than at any other time. How can a traditional election be fair and equitable and, in the end, effective until the factor of belief is better isolated? An atheist, commenting about the issue of religion surrounding elections, said that it is indeed hard to "find a reason to believe at all" (124).

Can an atheist win a popular election in Catholic Italy? It is interesting that in recent competition in the Italian national elections, one of the fringe candidates was Ferrara, considered Italy's most mercurial provocateur (36). He was running on a single slate devoted to "pro-life." He had been a newspaper editor and former government minister, best known as a talk-show host. He claimed, "I'm not a converted Catholic, I'm still a nonbeliever, even though my idea of reason is the idea of a reason which is open to mystery." His crusade was said to be as much about the "power vacuum in Italy as the power itself." The critic Chiaromonte observed in the late 1940s that, "In Italy, the church offers, not heaven so much as protection from the sheer impact of history." In the end, Berlusconi won over Ferrara, serving his third stint as prime minister.

**Stigma defined:** Social stigma is defined as "a mark of shame or discredit, an identifying brand or characteristic." If two minds cannot eventually adjust to find compatibility, they should, at the very least, adjust to achieve a more reasonable level of tolerance and understanding. In today's world, most serious believers still tend to look down on atheists and nonbelievers with overt disdain for the reasons outlined in earlier chapters. This book refers to hidden stigma associated with the difference in views of believers and disbelievers. Stigma involved in such difference may be as illogical and inappropriate as it has always been between two established competitive beliefs. This book demonstrates the folly of such stigma as an example of "the pot calling the kettle black." One imaginative belief simply looks critically at another imaginative belief. It is the result of long accumulated, well established and perpetuated imaginative visions. These chapters have suggested that man can only work toward improvement through accelerated education and factual knowledge in new generations. Defective spiritual stigma must be corrected in national policy as it has been improved in racial stigma.

The stigma in question is not merely an outward reflection of the difference between acceptance of the principles of realism or those of imaginism. Believers also tend to consider nonbelievers to be void of spiritual influence, righteousness, and protection by their all-knowing god, a god who is considered to be ever-present, representing the ultimate in goodness and morality. Believers may even feel a position of "preference" or the advantage of their "chosen" spiritual position. In some instances, they may even feel "tainted" by mere association with atheists. On the other hand, nonbelievers tend to consider believers to be without adequate knowledge and void of a reliable rational basis of thought and judgment. In an elite, well educated social group, stigma tends to be less prevalent and of less concern. But at moderate and lower intellectual levels, and in less sophisticated society, contrasting differences tend to clash more readily and with greater effect. Society may eventually learn that all its members may not actually be "created equal" but that any resulting differences must be understood and afforded equal treatment and respect. It may also learn that differences in belief are not fundamental but with causal influences as described in the first two chapters.

**Appreciation for rationality:** Stigma concerning atheism constitutes effective rejection of, or disregard for, an appropriate rational judgment of important contributory members of society, particularly those who have devoted a lifetime to discovery, furthering the development of human knowledge and essential methods of factual science, efforts through which all human understanding has been derived. Such persons are frequently agnostic or atheistic in contrast to traditional spiritual belief for the reasons earlier described, but represent some of the most advanced minds of learned man, those that indeed provide leadership, clarification of man's position on Earth and in the universe, those who teach understanding of the body and of the mind, and those who have sufficiently studied the world to distinguish and appreciate the difference between imaginative belief and a sense of reality.

Many such persons are broadly intellectual and philosophical, with extensive knowledge of nonscientific disciplines in addition to their knowledge of the deep principles of science. In most instances, such persons do not overtly advertise or reveal their innermost thoughts about belief, which if revealed, might come as a great surprise. As implied earlier, they describe the world exactly as they see it but they are not as aggressive as those associated with religious ministries. This may be the result of their greater awareness and acumen, more extensive study and the exercise of greater thought and reason.

**It is so because I want it:** Why are believers not called dreamers, imaginers, or promoters of "pie in the sky"? Why are they not considered to be "irrational" and unable to reason independently with adequate sensibility? Why are they not considered uninformed and ignorant? Why are they not considered inferior? Why are Christians targeted by Muslims as infidels to be summarily eliminated from the Earth (125)? Why were Christians forced to seek refuge in the catacombs of Rome? Why are modern humans still so much influenced by ancient writings? What level of dependence and reliability can be attached to the position of "faith only"? How long will these conflicting questions exist among man? Although they represent emotional differences based only on different learning or different interpretations of the very same thing, man's very existence, they have far reaching influence on current society. Why is factual knowledge not accepted? Why are established scientific data inundated by contrary imaginative promise at this late period of human enlightenment? Why does modern man not take more positive steps to understand and eliminate his most troublesome enigma and stigma? Why are processes of belief and stigma not more specifically included in mandated educational processes?

**Emotional sensitivity:** The very mention of words like god, religion, Bible, devil, sin or atheism in ordinary conversation immediately sharpens the senses and activates the emotions. It alerts emotions that may respond to acceptance or disapproval. It may suggest a prior argument or a distasteful experience, or a comment concerning religion, but it is more often a contrary reaction to information biologically implanted and retained in the mind, as explained in chapter two. Because of neural "overlap," differences in belief can suddenly be converted to "stirred-up" emotional feeling relating to race, culture, or politics. Within the mind's processes, religion might react quite differently from other memory contents. It may react somewhat like emotions of sex which generally involves high emotional levels and a sense of strong inherent desire and relates to personal and sometimes difficult, conflicting feelings. Few people want to precisely reveal their private, most profound inner feelings, but man can generally sense the feelings of others in these matters and responds accordingly, even without overt explanation.

**All differences can be resolved:** Pervasive imaginary spiritual thought is exceedingly strong in specific areas of the world and is directly responsible for instances of local conflict and friction, both on a personal level, among groups, and even between nations. It is rare that these instances heal themselves in a single generation, even through detente, diplomacy, or through sincere and careful application of logic and reason. A paper titled *Sacred Barriers to Conflict Resolution* (187) is an interesting reference relating to advice in settling differences. With respect to the defense of belief, Richard Dawkins once suggested that the word "faith" alone is a defense that obviates the need for any rational analysis or proof. Unfortunately, this simplified view has become the position of much of surrounding society. The inevitable mixing of spirituality with cultural and other emotions increases its complexity, modifies its meaning, and enhances interference with judgment. Time and generational differences, indeed, are significant balancing factors.

Clearly, early religious interpretations, the basis for much continuing stigma, were formed well before man developed adequate factual knowledge. Much early understanding was based on imagination and speculation and then molded into various forms of more deeply seated spiritual beliefs. Stigma, in earlier years, was particularly open and intense and often very destructive. Religious development of all forms has since involved a "backing and building" process over many millennia without interruption. Biblical literature and scriptures remain replete with suggestions of human ignorance and misunderstanding (126) and have been surprisingly little subjected to serious modern reinterpretation. Although their spiritual interpretations sometimes tend to transcend the dictates of established secular rules and laws, none can be reliably claimed as the word of some superior humanlike power. Acquired factual knowledge about man and his world developed only gradually, over many centuries, but the most significant and reliable of all the information known today is based on relatively recent determination, largely acquired in only the past half century, well beyond biblical times. However, despite all that has been learned by advancing civilization, including new psychology of the functioning mind, stigma remains as a significant social thorn in the side of man. With the continuing help of favorable unidirectional forces of change described in the first chapter, mankind may well find gradual improvement. Perhaps the only times that were completely free of stigma were those rare instances when all citizens in question were part of the very same belief and the very same culture. This is a goal for secular humanity.

**Knowledge will become more global:** This book anticipates a substantially increased flow of factual learning in the next stage of human development. Even the most modern knowledge of man has not yet fully penetrated planetary civilization. Full and effective distribution is unfortunately slow and problematical. It has not even reached a satisfactory level in the most developed regions of society. Much important accumulated data on details of man and his universe are not yet processed and evaluated. Important areas of teaching are often not included with respect to the issue described in this book. However, the laboratories and ascribed records involve mounds of data and ongoing experiments that will eventually contribute to answering existing questions of importance. This leaves the currently expanding world overly ripe for significant continued stigma, differences, and conflicts, conflicts that, in the past, have been limited to only relatively small areas of the globe, often with self-correcting results. New conflicts are hidden everywhere within the shadows of globalization. With expanding communication and increasing population, conflict and difference may even become far more acute and widespread before man reaches an adequate corrective turning point of more universal rational understanding.

Differences in appearance such as racial characteristics, black, white, brown, and yellow, in many areas also tend to remain a significant hidden basis for prejudice and stigma. They seem to involve some of the same or similar neural mechanisms. However, when in combination with religious differences, they can produce deeper and more immediate conflict. Emotions can also neurally flip-flop, somewhat like a solid state binary computer element, reducing the effectiveness of any singular therapy. Difference in education and inability to express knowledge also tend to set the level of inter-human respect. However, a man of reason finds it almost incredible that one can dislike or violently hate another by virtue of a differently acquired view of spirituality, a product only of the mind, one that might have no bearing on significant personal traits. Although contrary views may serve as the basis of conflict, it is usually the resultant behavior that causes reactive clash. The imaginary factor, as introduced in this book, simply adds to the degree of futility of ongoing spiritual stigma. But the stigma problem may extend well beyond religious considerations, for one form of prejudice promotes another, often ending in an indefinable subconscious barrier to resolution. As history will confirm, small initial differences and prejudices have not only led to conflict but also to long-term friction and occasionally to major unresolved stand-offs, even to warfare.

**Thwarting religious freedom:** Because the tradition of religions and their practices developed in accord with random processes discussed in the first two chapters, religious freedom has not improved very much with modernity but, in some ways, has actually gotten more out of hand. Worldwide, it is in a mixed state of chaos and ignorance. It has been in a perpetual state of transition. In early America, where many pilgrims, immigrants, and other believers had sought refuge from oppression, the new nation was founded on strong principles, including principles of religious freedom (127). However, real and full freedom has never really taken place as implied in these chapters. Under the rubric of freedom, it has been possible to offer and even "sell" spiritual promises to naive and unsuspecting citizens using strong sales techniques and funding to assemble religious groups, including universities and other institutions (128). Religions have been given the freedom to promise most anything, to collect money, to spread fear, and to enjoy tax-free organization, and on and on. It is all done under the influence of widely popular imaginative concepts without any notion of a disclaimer that might suggest uncertainty or lack of proof. Yet, in America it is illegal to publicly sell financial instruments without a disclaimer of some sort applicable to its claims. Most citizens are in favor of religion of some form, and the Dennett "spell" has indeed consumed the land. In fact, its recent past administration worked hard on faith-based promotion allegedly because of personal belief (129). The Constitution has been broadly interpreted but never fully clarified to provide all the rights and protection that every citizen should expect concerning religion. The court must, in fact, interpret it somewhat differently for different cases, as described in other chapters.

The preceding chapter suggested that religious leadership involves a large number of highly educated and well-respected persons of all denominations who fill universities and theological seminaries. These intellectuals must be treated with great admiration and respect for their sincere beliefs and their selfless efforts to contribute to society. They are spurred onward by their "different" neural wiring which induces them to promote their views of proper contribution to society. A number of respected scientists and intellectuals with similar motives have also made great efforts to relate religion to theoretical physics and other aspects of science (130). They tend to keep the religious processes alive, increasingly facing the enigma and stigma described here. But this book looks to newer generations for better understanding and interpretation. Indeed, the forces of change are many and varied on the growing mix of society.

Modern religions also have many extensions. Some enterprising businessmen even use religion as a basis for profit, while others are concerned with the financing and management of religious interests. There are several well written modern books on religion and economics, and there are lending institutions that particularly cater to Christians. There are now a number of visionary organizations that will match

Christian singles (131) on the Internet, while another, displaying a family and a church front on its website, specializes in Christian auto financing. Use of the word Christian in America today represents good advertising.

**Conflicting law:** Stigma is sometimes emphasized by differences between religious and civil laws of an effectively theocratic nation. There is no way that such religious differences can be easily resolved in a multi-faith society without compromise and realistic considerations. Something must give, even if it has to involve changes in culture or tradition. A recent article (132) appeared in the New Jersey newspapers emphasizing conditions of incompatibility. It spoke of a Muslim entering a nursing home and requesting a diet consistent with his faith. It spoke of a Christian whose boss asked her to work on Sundays and of a Jew rushed to the hospital, balking when asked to sign admission papers on the Sabbath. The State Assembly passed a bill requiring employees, health care facilities, colleges, universities, and public agencies to make reasonable accommodation for religious observations. An employer that fails to accommodate a worker's sincere desire to observe a religious holiday could be sued or administratively prosecuted under the discrimination law. These issues continue to present challenges for religious differences. Many employers do not appreciate that for observant Jews, the Sabbath begins at sundown which, in winter months, can fall before five PM on a Friday, requiring them to leave work early. A law required institutions of higher education and agencies that administer standardized tests to make alternate arrangements for students who are unable to take a test because of a religious observance. All religions should "modernize compatible timing" for the good of all society. Certainly, any reasonable god would want it that way.

These are further examples of the wide scope of conflict among religious groups that cause difference. People of one faith just cannot always understand the commitments of another. The faithless cannot understand why there should be such non-uniformity in the religious rules and regulations. In America, polls show that atheists are ranked lower than any other minority or religious group when asked whether they would vote for a member of such a group or approve of their child marrying a member. One woman, a nonbeliever, said that her husband was afraid to allow her to go public because employers might refuse to hire him (133). Another atheist was invited to give the invocation at a City Council, but half the council members were said to walk out in protest. In still another case, a chapter of Habitat for Humanity refused to allow volunteers to build houses while wearing shirts that said "Non Prophet Organization." This level of stigma is unacceptable in a rational society and is now beginning to receive appropriate attention. It will no doubt be corrected very quickly in currently changing society.

**When mixing is over:** Stigma has not only resulted from the laws set forth by the different religions but has also derived from outside lack of appreciation of differences within cultures with long traditions. In nations where large sections of the population follow a single religion, there is minimum conflict in this regard. However, in the modern world in which nationality and religious beliefs are constantly being mixed and broken into smaller groups, the problems of accommodation are enormous. Loyalty to an imagined god is so strong it often overwhelms rational thought. In many cases it precludes tolerance of any other viewpoint. This book has therefore been highly critical of any educational system that does not adequately review the basis of both principal forms of belief set forth in chapter three. Emphasis on imaginative belief at any educational level without appropriate disclaimers is an especially damaging influence on advancing society and should be subject to protective measures.

Religious views advanced in community administration also frequently create friction. In the Bible belt of Tennessee, a proposed Bible park unexpectedly ran into fierce opposition (134). The project was proposed by Safe Harbor Holding, a New York company that was developing Bible Park, USA. The park was described as having a Galilean village as its centerpiece, one side of the park was to represent the Old Testament stories and the other side, New Testament stories. Displays would contain excerpts from scripture and parts of the park would be reserved for Bible study. The detractors were ninety-nine percent against it in contrast to a theme park for an Olde South park. The only comparable Bible theme destination in the US is the fifteen-acre Holy Land Experience in Orlando, according to the president of Faith Based Amusement Association, though he claimed that there were dozens of traditional theme parks incorporating faith-based themes. One of the local citizens indicated that it was an "outsider" pulling the wool over the eyes of the Tennessee hillbillies.

Stigma develops, not only because of differences between beliefs, but also between different beliefs that may have widely different tenets and practices. Some religions reflect major contrast in practices and the "god differences" are difficult to justify. A young college student recently reported her experiences with dating in Saudi Arabia (135). Dating in the Western sense was not at all acceptable in that venue, either socially or religiously. Some dare to meet in coffee shops or restaurants that have separating partitions. Most young women prefer to get to know men through "accepted channels" like the Internet, friends, family, or the phone. The practice has been for a man to hand out telephone numbers to most any female, but the introduction of the cell phone has considerably improved the process. Songs and novels show how affectionate and passionate Saudi men and women can be, but many believe real love is that warm feeling a couple develops after their parents have arranged a match and the marriage contract has been signed. The author indicated that what matters to all is to find love somewhere around the corner, hidden in that mall or behind the tinted window of a car. This story is intended to emphasize the considerable difference and lack of realism, even in this modern age between societies not far distant from one another, knowing that they will eventually have to merge in belief and behavior in the ongoing process of globalization. It does not require a prophet or a futurist to see the handwriting on the wall. The forces of change, about which this book is constantly concerned, will act to produce eventual social uniformity.

**Government interference:** Stigma in belief is sometimes multiplied by government interference with religion that has penetrated a nation. The current situation in Russia is a good example. Although many observers have doubts about its reality, Vladimir Putin and his associates have evidently taken a strong position against multiple beliefs in the new government and have placed singular emphasis, not on earlier atheism, but on the Russian Orthodox Church (5). Putin now wears a cross and speaks freely about the importance of unity of religion. After the fall of Communism, Protestantism evidently flourished, at least in southwest Russia, but now it is experiencing strong suppression as are many others referred to as "sects." The government has, in effect, tightened control over faith as well as political life. The Russian Orthodox Church has, for the moment, become the de facto official religion of the country. Proselytization by Protestants has been banned by various indirect means. The Orthodox Church represents the restoration of nationalistic order following social disarray. Anti-western sentiment has been directed toward Protestant and other groups presumed to be associated with the United States. Protestant missions were accused of spreading "American propaganda directed at the Russian population." If stigma does not result as a natural consequence, it is indeed purposely created.

The number of religious organizations currently registered in Russia is said to be more than twenty thousand, many of which are Protestant. There are Evangelicals, Baptists, and Lutherans and many others, most of whom complain about being denied permission and opportunity to conduct normal services. It is a field day for the germination of stigma. An Orthodox priest was quoted as saying, "We deplore those who are led astray, those Jehovah's Witnesses, Baptists, Evangelicals, Pentecostals, and many others who cut Christ's robes like bandits, who are like the soldiers who crucified Christ, who ripped apart Christ's holy coat."

Although the Russian constitution guarantees "freedom of religion" on the surface, the government now suggests that their Orthodox religion is "inviolable and it should not be shaken." Russia has more Islamic residents than most of the other residents of formal religions, but they are not viewed as suspicious or dangerous. The Russians feel that they must live through the "second Baptism of Russia" referring to the Russian adoption of Christianity in 988. Behind this new mixture of government interference and religious change, the power holders are regarded as atheists who are merely adopting the Orthodox as a much needed move toward social unity. This current religious upheaval in a major developed country is indicative of the condition of the spiritual enigma of civilization and how intense levels of stigma can develop from variations of imaginative belief. In a sense the problem seems to stem from the ruling party that desires all organized entities to be firmly tied to the government. The adoption of orthodoxy may be considered the best basis to accomplish this for the moment. Outsiders must stay tuned and see what changes develop in a huge modern globalizing nation with much diversity. The new atheism will no doubt eventually catch up and clash with the old atheism but with the rest of the globe will certainly tend to overwhelm and "devour" it.

**Supreme Court dilemma:** In a town in Utah, the Summum Church proposed to erect a monument with the Seven Aphorisms in the city park (136). The Summum Church was founded in 1975 around elements of Egyptian faiths and Gnostic Christianity. Its followers believe that there were two sets of tablets given Moses on Mount Sinai, that the Aphorisms were the first and the Ten Commandments the second. The church founder claimed that he learned about the aphorisms during a series of "telepathic encounters" with divine beings called Sumna Individuals. The town mayor rejected the concept of such a park monument and the case, representing questions of government ban on establishment of religion and free speech, reached the Supreme Court of the United States. The case is mentioned here to demonstrate the wide scope of human emotion in imaginative religious belief and the cost and complexity of dealing with relatively small related issues in a democratic society. Again, this book suggests that education and the further natural development of rational judgment will tend toward secularism and the elimination of such conflicts. The governments may not yet be ready to accept the more decisive position of secularism in judging current issues that involve imaginative religious belief.

A recent work by Martha Nussbaum (101) expressed the problem very well in her treatment of the High Court views about applying the American Constitution to believers and nonbelievers in a fair and equitable manner. The first Amendment, mentioned earlier, was always considered a guarantee of freedom to exercise religion. Again, it reads, "Congress shall make no law respecting an establishment of religion, or prohibiting the free exercise thereof." On the surface, this seems clear enough. But with modern American involvement with thousands of different religious beliefs and practices in our globalizing world and their rapidly changing population and lifestyles, it seems quite impossible to find a "one scheme fits all" answer. The Nussbaum book makes a valiant attempt to set down some suggested principles, but falls somewhat short because the objective at hand just cannot be achieved without a great deal of "gymnastics." Her real objective is claimed to be satisfaction of the so called religious conscience of man.

Assuming that sincere "religious conscience" is to be protected fairly under all circumstances, one encounters some serious limitations. "Conscience" is a property of the mind and can be widely and rapidly varied to suit most any given condition. There is just no way of determining if a religious conscience is serious, valid, or false for purposes of applying the intentions of the constitution. Even its intentions, based on the minds of origin, may be in some question. Therefore, any accommodation afforded one believer must be afforded all other believers and non-believers as well.

This book has considered and claimed that traditional religious belief is derived from purely imaginative thought processes. If it were derived in any singularly divine manner or on the basis of pure reason, it would not likely result in such stigmatic quandary. The judiciary must therefore deal with widely differing situations and consider each case with respect to legitimacy and conformity. Should tax exemptions be extended to the nonreligious? Should nonbelievers be afforded the same type of financial support as has been offered to faith-based groups? Should conscientious objectors, based on their religious conscience, be given different treatment than nonbelievers? It is said that there are enough questions implied in these considerations to keep the Supreme Court busy for years to come.

**A higher personal power might be great:** There are other questions raised in connection with religious conscience that cannot be fairly and reasonably answered. Should government leaders and judges rely on their own religious principles or exercise religious ideas in fulfillment of their secular duties? Should beliefs be involved in decisions on animal consumption or treatment in research laboratories? What about prayer in schools or teaching intelligent design or exhibiting religious objects? Should labor laws take into account the variation in religious holidays? Are the words, "under God" in the Pledge Allegiance permissible under the Constitution? What about "God bless America" or the words "in God we trust" on coins of the realm? What about changing "an act of God" in legal language? Do they really make sense in the big picture? It is already assumed that the "under God" phrase does not endorse religion for it has faded into a "ceremonial deism." But why is it treated differently from other similar phrases? Would not a real god act to settle the pervasive enigma and stigma that so widely separates man? This book is intended to contribute to a firm resolution.

The attempt to protect expressed religious conscience seems to be totally without merit, for religious conscience cannot be determined with any degree of validity and in fact is no different from any other conscience. This book has shown that the creation of religious conscience, very much like religion itself, is a thought process that effectively excludes the human senses and relies on imaginative formulation within the emotional mind. There then can be no real basis for religious accommodation or privilege, except to selectively mollify the strong emotions that are involved. Individual reluctance to butcher an animal, to engage in abortion, or to kill as a soldier, is a legitimate emotional issue that must be carefully considered by modern society.

**The majority must be right:** A democracy is ruled by the people. In America most of the people claim to be Christian, therefore one would expect Christian rule (137). Theoretically, if all citizens were of one belief there would be no enigma or stigma. But that is clearly far from the case. The authors of the Constitution initially expected Christian rule, but along with current politicians and congressmen they later desired a firm policy of separation of church and state. The implication, not only in elements of society, but throughout government in America, is that religious persons are "different and more righteous" than others. Hopefully, that obsolete notion has now been sufficiently clarified by this book, one that constantly clashes with the process of fair and equitable treatment.

Deep religious belief involves a neural mechanism that just does not allow effective separation of church and state. Prior chapters explained why human minds do not shed their beliefs at will to a sufficient degree, and allow adequate judgment with full separation. This inability is often marginal, involuntary, and perhaps subconscious and therefore often undetectable. Belief in a god should provide no greater or lesser status than nonbelief, and application of the laws of state should be conducted accordingly. Righteousness and good works are not the exclusive qualities of people who are religious or claim to be religious. The laws were established to be universally applicable as though in a purely secular atmosphere. Certainly, times are changing and minds are changing with them. Since the laws are firm and often basic, religion must eventually change in such a manner that the laws of society can allow more uniform application and become easier to monitor and manage, not only in America but in all society.

An attempt to favor religious conscience is something like favoring one human emotion over another. The late John Rawls, a Princeton classmate and philosopher of note, believed that there is an overlapping consensus between believers and non-believers, but it is not obvious in the modern mind, especially with biased spiritual analysis and the advent of strong new atheism. The Judiciary will find it increasingly difficult to apply fair and equitable treatment under the varying and variable interpretation of the religious mind. The spirit of the American Constitution has been long questioned with respect to religion, and authors themselves may have been caught in the same web as all religious men who now seek fairness and independence in matters of separation, of legislation, of adjudication, and of governance. It is clear that the determination of truth can no longer and never has been obtained by laying a hand on the Bible.

An applicable article titled "Keeping the Faith, Ignoring the History," written by Susan Jacoby, author of *The Age of American Unreason*, appeared in the March 1, 2009 New York Times. It concerned the continuity of faith-based programs popularized in the previous administration. The important constitutional issue is whether or not the government can spend taxpayer money on religious initiatives in violation of the First Amendment, indeed, on perpetuating an imaginative practice. William Clinton had originally signed a bill allowing religious groups to compete for grants, and George W. Bush set up a White House office of Faith-Based and Community Initiatives through executive orders rather than a congressional bill. Large grants were said to have been provided for the Christian Right, including Fellowship Ministries and Teen Challenge, that openly encouraged religious conversion, including converts from Judaism. Apparently Dilulio, the first head of the office, resigned after eight months, saying that the program was political.

**The rule of law:** President Obama appointed DuBois in his new administration for this purpose, but required hiring to be vetted by the Justice Department. R A Mohler, Jr, president of the Southern Baptist Theological Seminary, criticized this procedure saying that the unlimited right to proselytize and to hire members of their own faith is essential if churches are not to "compromise their mission." Some argue that the First Amendment does not mandate separation of church and state but simply prohibits state preference for any church. Indeed, the First Amendment established freedom from government interference or involvement with religion in America. It also prohibited teaching religion in tax-supported public schools. It seems that religiously minded citizens are exploring every way that the principle of the law might be circumvented and government money made available to "keep the faith." To whatever extent the process is political, it cannot be freed from the religious mind, as explained in the earlier chapters. Although the pressure is strong on the part of the perceived gods of American believers, it is time to make an honest, clean-cut break between government and religion as was intended.

**Unbelievable confusion:** Several important works on the founding of America and the separation of church and state have been added to the spiritual fray (127). One, written by Stephen Mansfield, a former Evangelical pastor, spoke of irony in the state of present-day culture as one that "yearns to believe yet doubts those who do." The other, written by Forrest Church, a former minister of the Unitarian All Souls Church in New York, argued that our new country involved both secular and Christian foundations. The review of these works by Gordon Wood in the New York Review of May 1, 2008, presented an informative summary of how religion clashes with the constitution and how confused and inconsistent are the court decisions at times. It claimed infinite complication in the matter of legal and religious correctness and the "pervasion of stigma" among citizens of the United States. Wood wrote that, "The ten commandments can be displayed in some public places, but not in others. School prayers were said to be impermissible if sponsored by local or state administrations, but perhaps not if they are sponsored by the students." Legislatures can open their sessions with prayers, but public school football players may not open games with them. Observing religious holidays in the public schools is impermissible, but excusing students from attending schools on their religious holidays is permissible. In this perplexing atmosphere, teachers in public schools have become so uncertain about what is permissible and what is not that "they often avoid even talking about religion at all."

Wood also commented about the recent "new atheism" represented by the rash of anti-religious books referenced in earlier chapters. He said, "these secular outbursts seemed unduly angry and suggest an underlying apprehension of religion's continuing strength among large numbers of ordinary folk." Religious struggles in the United States used to be among the different denominations of Christianity. But in the past several decades the struggle has more and more become one between religious belief and "the growing secularism in American culture." His review claimed that many of the founders shared the view of a onetime speaker of the American Philosophical Society who abhorred "that gloomy superstition disseminated by ignorant illiberal preachers" and looked forward to the day when "the phantom of darkness will be dispelled by the rays of science, and the bright charms of rising civilization." He too confirmed the thesis of this book, the "growing secularism in American culture" and the bright charms of rising future civilization.

The core of this book is very much based on examination of "bright charms of rising civilization," and on dispelling "the phantom of darkness." But it applies not only to America where Christianity has still continued so strong, but to the entire world, where civilizations are not only being released from primitive ignorance and narrow tradition but are also being freed from the influences of leaders and purveyors of primitive beliefs. Religion is certainly continuing in strength among "ordinary folks" as was suggested, for "ordinary folks" have not yet had the opportunity to enjoy the luxury of a modern education and the privilege of understanding. They therefore must continue to represent a naive, vulnerable, and uninformed segment of world society.

**Will hope really influence the trend?** Growing secularism in America along with the rest of the world has clearly been demonstrated in these chapters, but it is challenged by gross lack of uniformity in the application of law. Meanwhile, in expressing hope that "faith will return to America," Pastor Mansfield acknowledged the "secularizing trend of recent decades," adding the hopeful thought that secularization is likely to "dwindle if not die." On the contrary, this book suggests that the secular trend cannot be reversed. Beyond all the wishes and hopes of the most devout scholars and theologians, beyond the daily prayers and sacrifices of the faithful, and beyond the problems of evil over the entire world, the still developing human mind will ever trend toward recognition of reality and therefore toward secularism. It is basic and fundamental with respect to human evolution, knowledge, and understanding in advancing humanity. Secularism is a magnet to which the "iron" of fundamental human character and advancement is naturally drawn, not by worldly or heavenly imaginative influences, but by the intrinsic underlying nature of man's evolving mind that must ultimately transcend emotion in his relentless search for factual guidance and reality.

**Ethnic and religious cleansing:** It was over a decade ago that a hundred thousand or more Muslims were killed in Bosnia, then a part of Yugoslavia. A recent update (138) indicated that an Islamic revival is underway. It has been traditional to refer to the three divided areas, Bosnia, Serbia, and Croatia, by their most prominent current religions, respectively, Muslim, Orthodox, and Catholic. Bosnia is undergoing Muslim expansion, apparently based on millions of dollars of financing by the Saudis. Although Bosnian Muslims are largely conservative, there have been violent episodes. There is great fear of conflict and even terrorism, and some seem intent on turning Bosnia into a Muslim state.

Islamic classes are conducted in state-financed kindergartens, a move that has been severely criticized, including criticism by the Organization for Security and Cooperation in Europe, an Austria-based group that monitors democracy. Even a liberal Muslim called the introduction of religious education in kindergarten a crime against the children. The area has practiced the Muslim religion stretching back to the Ottoman conquest in the fifteenth century. The most recent conflict was settled with the help of the Dayton agreement by dividing Yugoslavia into the separate national regions. The encroachment of festering religious conflict in southern Europe will indeed test the avowed advent of secularism. Current activity merely demonstrates the infectious power of religious contagion, as earlier suggested and causally explained in chapter two. However, it is only a temporary resurgence and in the course of time will find its place in the religious dilution of all European cultures.

**Unity must await rationality:** Because of their expressed rational views, many knowledgeable citizens, indeed, some of the most intellectual philosophers and scientists, are frequently called heretics, heathens, infidels, outcasts, atheists, agnostics, nonconformists, contrarians, lost souls, disbelievers, and the like. They are sometimes considered to be abnormal, liberal, free-thinkers, and often placed outside the sphere of accepted social behavior. They may be considered to lack morality, righteousness, and other favorable qualities, for many religious people still believe that one has to be "close to God" in order to share "his beneficial grace and goodness." These feelings of effective disrespect, and sometimes full-blown hate, are generated simply because minds have acquired different interpretations of belief or because one is a member of a group that collectively has a strong, differently expressed ideology than another. Only a century or two ago, even scientists were sometimes considered to function in a "different world." The terms "weird" and" crazy scientist" were quite common, for science typically involved "the unusual and the different" not understood or appreciated by the masses. Dislike or hate often exists without rational basis among widely different members of society, especially with respect to religious difference. The theme of this book is directed to accelerated unified belief and thereby to the reduction or elimination of religious stigma.

In recent years, religious conflict among nations, and among different denominations within them have, at the very least, been recognized with blurred visions of possible resolution. There is increased recognition of conflict among large factions, such as Christians and Muslims. A recent conference (139) between Catholic and Muslim representatives, held at the Vatican, pledged cooperation to "condemn terrorism, protect religious freedom, and fight poverty." They also agreed that symbols considered sacred should not be subject to mockery or ridicule. It seems that cooperation involving countries like Saudi Arabia, where non Muslim worship is not even tolerated, is currently a losing battle and, as with other competing beliefs, must patiently await the natural forces that are destined to bring unity within all camps. These forces are the non-human forces that continue to govern the reacting chemistry and therefore influence the circumstances of man. They are the unstoppable, invisible forces that pave the way of continuously evolving human life on its planet over cosmic time. They act on the inherent desire for human understanding, on the ever increasing desire for factual knowledge, on the gradual rejection of imaginative concepts, and on the eventual spread of secularism inherent in the latent vision of world unity. It is not the elimination of weapons of mass destruction that will most advance mankind, but success in the race between imaginative ideology and factual education. Despite continuing conflict, traces of such unity are more apparent that in any earlier age of man.

# Chapter Five

## Nature the Mechanistic God

When I became convinced that the universe is natural, that all the ghosts and gods are myths, there entered into my brain, into my soul, into every drop of my blood the sense, the feeling, the joy of freedom. The walls of my prison crumbled and fell. The dungeon was flooded with light and all the bolts and bars and manacles became dust. I was no longer a servant, a serf, a slave. There was for me no master in all the wide world, not even in infinite space. I was free—free to think, to express my thoughts—free to live my own ideal, free to live for myself and those I loved, free to use all my faculties, all my senses, free to spread imagination's wings, free to investigate, to guess and dream and hope—free to judge and determine for myself. I was free! I stood erect and fearlessly, joyously faced all worlds.

—Quotation by Robert E. Ingersoll relative to freedom from imposed imaginative traditional belief

One finds big nuggets of insight, useful to almost anybody with an inherent interest in the progress of human society. A vast ideological anatomy of possible ways of thinking about the gradual onset of secularism as experienced in fields ranging from art to poetry to psychoanalysis.

—THE ECONOMIST,
commentary on Charles Taylor's *Secular Age*

Where did the spiritual forces of natural magic and black magic begin? It was a question which perturbed Boyle. He was always fascinated by diabolic and magical phenomena, in 1658 sponsoring the English translation of a book by Francois Perreaud as *The Devil of Mascon*, and planning to publish a sequel to the collection of Strange Reports dealing with supernatural phenomena. But he pulled back. One should beware of magic, he reasoned, not because it was false, but because, being real, it was too hot to handle.

Natural philosophers were thus in a cleft stick. Convinced that true science would disclose God's agency in Nature (and thereby prove the soul), they sought to naturalize the spirit by bringing it under laboratory control. Via the air-pump and the test-tube, spirits could be made visible and so be rendered safe and effective: good spirits would toe the line.

—*Flesh in the Age of Reason: The Modern Foundations of Body and Soul,* by Roy Porter, Norton. Referring to the Hon. Robert Boyle, a founder of the Royal Society

A **perfectly natural universe:** Chapter three examined the two basically different forms of human belief in today's world. In the interest of simplicity, they were called imaginism and realism and their principles were clearly defined. This fifth chapter provides an expanded discussion of the basis of realism, the concept of a transcendent non-human force that we call "nature," and how it relates to the humanlike god concept that appeals so universally to the emotions of man. "Mother Nature" has often been a favorite expression of mothers and grandmothers in my time intended to explain much of the world around us. Therefore, Nature is used here as a simple descriptive word that represents a mechanistic power, rather than a humanlike god. It might be viewed by some as a form of "god" whose power was contained in the "big bang," a physical event in which the universe was said to have been "created" and set into motion as we now interpret it. Prior to the big bang, science becomes uncertain but retains its reliance on the properties of observed physical principles. There has been no valid indication of the existence of a humanlike god and in the larger picture there is really no rational reason to think there should be one. It should be remembered that life was not a part of the origin of the universe and appeared only through biochemical formation billions of years later. The mechanistic universe was never intended to have life and fragile life formed on one of its multibillion of orbs may in fact not survive. The evolved egocentric human mind finds great difficulty in appreciating this view.

**Time a measure of life and death:** No attempt has been made to speculate here about prior time, for the mechanistic character behind the origin of the universe is generally beyond present human understanding and therefore considered to have little meaning for present man. Since its origin fourteen billion years ago, the universe has appeared to follow the same mechanistic character as demonstrated at its beginning and there has been no evidence of any humanlike intercession. This book suggests that science has very carefully traced the nature and function of the universe and all its contents to what appears to be a non-human mechanistic character, suggesting that the concept of imaginative belief as described in this book is not only baseless but represents a certain futility for man. As fully explained elsewhere, no finding of man has at any time confirmed the existence or effect of a humanlike god, the framework that has been built around it by man, the response to any supplication, or the existence of any form of immortality. As much as man would like it to be different, all these notions must remain in the category of "human imagination."

**Knowledge is very recent:** The principal discoveries of science have been made in just the past half century whereas notions of a humanlike god have lived for millennia, through long periods of lack of factual knowledge and relative ignorance. During human existence throughout earlier millennia there has been adequate time to contemplate different highly desirable spiritual notions of man's humanlike god and they have gradually become more elaborate and refined, indeed almost real. Today, the average believer most commonly visualizes a god of human character, one that he can easily understand and relate to. He imagines that his god is benevolent and judgmental and exercises all the superlative traits that were described in previous chapters. He has also built an imagined framework around the god concept with respect to a sphere of heavenly location and judgmental function. This book has attributed the notion of a humanlike god to strong human desire for guidance and survival and to a well developed neural mechanism for infinite imagination. Although such imaginative belief clearly fulfills many of the emotional needs of mind processes, in the end it cannot provide any more to man than simple recognition of the reality of nature. Although the depth of imaginative belief is gradually fading in the modern world in favor of a mechanistic determination, the subconscious existence of a heavenly god is kept alive by strong emotion and by active tradition.

The first definition of "God" in Webster's dictionary is given as "the Being, perfect in power, wisdom and goodness, one who is worshipped as creator and ruler of the universe." This book does not recognize a perfect being with humanlike characteristics such as wisdom and goodness, nor does it recognize any effective means of its communication with humans. On the contrary, it suggests that modern science has traced the nature and function of the universe and its contents to the mechanistic character adopted throughout this text. The mechanistic function is expected to become clearer as more is known about man's universe. The new Hubble space telescope will soon furnish more spectacular pictures of the universe that will surprise the entire world, providing exciting new mechanistic characteristics of the universe we have yet to fully know.

**Man creates to please man:** This book has charged the early imaginative mind with creating anthropomorphic gods in its best interest, gods that are encouraged to live on as an object of worship and as a source of human inspiration and perceived support. Because the image has been raised to the height of idealism and perfection, god's "agents" have arisen everywhere, renewed generation after generation, and are committed to maintain faith in the god notion without a word of doubt or disclaimer. When my family routinely attended Sunday School, the Church presented a Bible that had a typed explanation pasted to its first page in large print. It told about its "translation in dungeons under the pain of death" and how it was a "message carried each day through steaming jungles and giant forests." It stated that the Bible is the place "where we find Christ or where Christ finds us," that God seeks us and loves us in Christ, and gives us his spirit that we may be growing members of His Church, that his biblical word becomes the agent of "his direct personal word to you." This type of introduction to a church seems to be quite common.

For many, this form of teaching creates early wiring of the mind which tends to remain intact. For others, the wiring may be diverted by reasoning based on the recognized reality of the world. Initial chapters explained how one wiring might influence another through cultural evolution, how it might go either way with varying intensity, but how learning results in a powerful engram, subject to the idealistic promises and to the advertised belief that the church is God's messenger and missionary. Earlier chapters also anticipated brain research that would allow a certain amount of "editing" of memory (140). Although this direction of investigation confirms the biological nature of learning and memory, it clearly has not yet developed to the point of reality recognition with respect to spiritual belief. It is interesting to note that, only in the past year, over five billion dollars were directed to neuroscience at the National Institutes of Health. The recognized importance of its objectives and the "wide open opportunities in brain science," will appreciably accelerate the pace of discovery and will no doubt begin to enhance understanding of spiritual issues discussed in this book. Learning and memory research are not intended to eliminate belief but to understand more about the nature of learning functions and the character of the memories produced. As stated frequently in this writing, man has a strong inherent desire to know all in structure and function. In time, he will know considerably more about the secrets of emotion and the neuroscience of retained memory and therefore more about the processes of belief and behavior.

**Mechanistic nature cannot throw dice:** Meanwhile, it is logical for man to interpret its god in terms of present understanding. A cold mechanistic basis of human existence would not be man's first choice of interpretation, certainly not for early uninformed civilizations. It is especially difficult for the average human mind to believe that man exists as a complex biological entity in a mechanically derived universe.

However, informed man has no alternative. "Mechanism" is defined as a doctrine that holds natural processes, such as life, to be mechanically determined and capable of explanation by the established laws of physics and chemistry. Over the long time of his development, man has eventually learned that nature, the most powerful time-force of the universe, follows clearly definable rules and laws which he can observe and monitor and from which he can explain much of the past and can reasonably predict some of the future. Man has long recognized and dealt with nature, often attributing it to an imagined humanlike god power. It is likely that many people think of nature and God in the very same breath, but only their God is given humanlike characteristics because man still largely thinks in those emotional terms.

Theological considerations are intrinsically tied to perceptions of the universe and to the existence of life. This chapter not only includes a brief review of some pertinent areas of factual knowledge about the universe in which man lives, but it also describes examples of human behavior and religious practices appropriate for rational reconsideration and critical examination in terms of this theme. This chapter additionally includes commentary about expanding and integrating certain ideas presented in earlier chapters, specifically, man's views of existence centered about a humanlike god, how those views become lodged in the vulnerable mind, and why they might now require updating and modification in terms of more advanced understanding of a mechanistic universe.

**The real universe of man:** The informed and knowledgeable rational mind currently sees the universe as an assembly of billions of separate heavenly bodies in virtually unlimited space, all in constant motion. It represents violent exchanges of mass and energy of infinite proportions. The fiery behavior of material of the universe may extend over billions of miles of space and over millenia of time. The universe is not an orderly arrangement of bodies, for its scattered pattern gives the appearance of remnants of a great explosion. The bodies are of many different sizes, and are generally spherical in shape since the force of gravitation pulls all nearby material together. Only the forces of motion keeps them apart. Gravitation also acts at a distance among the spherical bodies, some occasionally crashing together. But most bodies seem to exist separately in space or move, one around another, in an orbital pattern in which the forces of motion balance their gravitational forces. One might ask many applicable questions. Why gravitation, and why even a universe? Would the observed universe, as now reasonably known by man, be the first choice of any creating and designing omniscient god? It certainly doesn't seem so to many modern minds. Why do we have to think beyond human relationships, of globalization of the entire planet, indeed, beyond our simple flat world, of billions of unexplored worlds beyond our own?

**Man never asks to be born:** Although the emotional mind readily attributes the universe and its contents to the power of a humanlike God, more rational intellectuals increasingly perceive it to be a mechanistic phenomenon that no one yet completely understands. No human type mind needs to be involved and the human mind must eventually understand that although some highly informed theoretical minds, such as that of Stephen Hawking, (141), offer periodic speculations about pre-big bang time, it doesn't seem likely that man will ever understand its details, certainly not in terms of common human questions of meaning and purpose. In effect, man himself appears as a relatively insignificant parasite, clinging to his small planet Earth, wondering why he is there, and how his limited life should best be fulfilled. Following his involuntary birth, he may have a potential life of less than a century. But it can be a wonderful and rewarding life. Upon death he returns to the elements of the Earth as though he were never born. Man will increasingly become adjusted to this fact and will design his life more in line with such temporary coexistence and with contribution to ongoing fellow mankind rather than preparation for an ideal life in a heaven of imagination based on behavior and on the acceptance of his god.

The most modern concepts of origin and structure of our universe are largely contained in references in the appendix and will not be discussed in detail here. Among recommended reading is *Origins: Fourteen Billion Years of Cosmic Evolution* by Neil deGrasse Tyson, director of the Hayden Planetarium, and David Goldsmith. In contrast to the many religious publications based on imaginative beliefs and religious emotion, this scientific publication describes man's observed position and destiny. Michael Lemonick, author of *Echo of the Big Bang*, said the Tyson book makes the astounding astronomical discoveries of recent years "come alive in reader-friendly fashion."

**The sun as life's incubator:** All the material of the universe constitutes "chemistry" as commonly defined by man, much of it consisting of the simplest light elements hydrogen and helium, similar to the fuel makeup of the stars including that of our own sun. Our sun is clearly the most significant and essential central interest for man. Fortunately, it sustains a huge continuous thermonuclear explosion, an atomic fusion reaction, primarily between those same two light elements, providing the necessary warmth for life on Earth even though it is a hundred million miles distant. It was this radiated warmth that encouraged the chemistry of biologic life to form initially, and now contributes to its vital continuity. The sun is almost a million miles in effective diameter. In another five billion years or so it is expected to expand into a "red giant" about the size of the Mars orbit and then shrink to become a "white dwarf," according to the observed and calculated life of a typical star.

There are many eruptions on the surface of the sun and its hot winds constantly stream through the entire solar system. At its equator the mass of hot gases rotate with a period of about twenty five days and sun activity currently appears to have peaks with a "solar cycle" of about eleven years. The strong solar winds that impinge on the Earth could very well destroy all biological life. It is only happenstance that the magnetic field around the Earth, formed by its rotating molten iron core, protects man from its serious destructive effects. It is also only happenstance that life arose on a planet of reasonable size. Our planet could well have been a body of eight hundred, or even eighty thousand, miles in diameter rather than eight thousand miles. Such circumstances would certainly have changed the nature and prospects of human life in a manner far too complex even to contemplate. Speculation about other possibilities can be interesting, but it is best for man to recognize his lot as defined by observations of science and make the most of it.

**The critical fragility of man:** The fragility of human existence and the ongoing gradual development and change in man should be kept in mind as one reads these chapters. New knowledge is acquired day by day that contributes to man's improved understanding. It is not information obtained directly or indirectly from any god, but from the difficult processes of constant human sensing, learning, and from the intelligent application of his logic. The circumstances of origin and evolution of life on Earth, researched by man with great effort, are important prerequisites for appreciation of these chapters (142). A view of conditions of early Earth, suitable for life formation, was recently reported by Kenneth Chang (143). It is also described in several other listed references.

A group of nine spherical bodies, moving under the gravitational influence of the massive sun, is called the solar system, and our Earth is one of its relatively small planets. The concepts of its creation and the addition of human life by an act of a humanlike god are far from modern reality. The solar system was formed from gases and space material about four or more billion years ago, very much as we now see similar bodies being formed when we search the distant reaches of the universe. The entire solar system is situated inside one of four massive astral arms of our great disk-like Milky Way galaxy, one of billions of similar galaxies of the vast universe. Such galaxies individually contain billions of stars and orbiting bodies like our own sun and Earth and are said to have a powerful "black hole" at its centers. A black hole is considered a region of infinite density and gravitation from which little can escape, including light, and is detectable by measuring x-rays produced by entry of extremely hot gases. Man might imagine most of the universe without any life at all, without the organic chemistry that reacted to form life on our planet. The Earth certainly existed well before the chemistry of life appeared on it, and will probably end without any life based on man's present astronomical knowledge. Despite the vastness and age of our own galaxy, it is likely to contain relatively little other life if any. Did traditional religion have any meaning at all during the billions of years before intelligent life appeared on Earth?

**Man is responsible for man:** There is no immediate connection between the origin of the universe and the insignificant chemically based human life that came much later. Emerged man is the product of a well studied evolutionary process that acted on animal life and on the many simpler life forms that preceded it. If emerged man must have a transcendent force in the form of a "god", he must look to "nature." His notion of a humanlike god with all its assigned trappings may well satisfy his sensitive emotions, but it cannot satisfy the deeper requirements of the mind of logic and reality. Further, it does not add any known benefit except to the emotional mind itself. This book explains man's most likely position and his assumed responsibility with respect to ongoing existence and development. Having appeared on the planet, he is clearly expected to pursue continued factual knowledge and peaceful behavior. Society seems to be so busy with its promotion of reliance on an imagined god that it cannot effectively execute its responsibility to fellow man and to his planet. It is time to seriously consider responsible understanding of his position and to establish a system of organization that will contribute more fully to his best long-term interest. Man cannot rely on an imaginative god that has never, at any time, had any reality or positive physical influence in his past. It is clear that his concentration on imaginative beliefs has only been a positive effect on behavior through fear of a monitoring god and fear of certain death. It is a fear of something that does not exist, like childhood fears we have all experienced. Consideration of reality is now sorely required. The advancing rational mind should be capable of achieving full freedom from imaginative notions before the end of the next century.

**Rotating, revolving and speeding to nowhere:** Man seldom thinks deeply about his position of "virtual fantasy," residing on a huge sphere, afloat in the infinite universe, rotating and revolving in its long seasonal path around the sun. This representation of man's position is merely intended to provide a brief reminder of the reality of the human condition and its mechanistic character. If there were really a creative and controlling humanlike god, these conditions and the affairs of man would be quite different. Theologians, who seldom gain fully unbiased cosmological and physical understanding, find simplistic spiritual shortcuts to such explanations in the form of feelings and guesses, but informed scientists and intellectuals consider the origin and function of the universe to be a complex, mechanistic affair subject to well determined laws. There is no compelling reason to think that any humanlike mind was ever involved in designing or creating either the solar system or the universe or any part of either, since humans and "minds" did not even appear until billions of years later. "Design" cannot be a meaningful word in the "dictionary" of such a mechanistic universal force. It became a meaningful concept well after developed man. Despite incomplete information on origins, the most informed modern minds feel that attributing man and his universe to the work of a humanlike god is both primitive and irrational. It is what early uninformed minds might naturally imagine, and what modern, more informed minds must now reconsider based on factual observation and reason.

**Man a chemical automaton:** Years of research by the most brilliant of men have concluded that the chemical formation of life and the position of man are adequately explained in terms of modern scientific knowledge. The universe apparently appeared at the time of the big bang in a strictly mechanistic process. After Earth formed and cooled, cellular structures spontaneously appeared in its chemistry, some of which had the property of continuing chemical reaction called "growth." The form of growth is controlled by molecular arrangements in the cellular chemistry called the "genetic structure." As earlier explained, its genes function like the seeds of plants to determine its chemical form and character. In time, the Earth grew both fixed plants and dynamic creatures of various shapes and sizes, all part of the reacting biological chemistry now called "life." After billions of years of reaction, the chemical structures of living "creatures" developed means for mobility and some developed a neural structure called a brain with which movement and other electrochemical functions can be actuated. All chemical structures understandably react and change over time, the latest and most complex creature being modern man himself. It is the supreme force of nature that moves man forward and it is the chemistry of the brain that allows man to contemplate his structure, his history, position and a sense of his future. The process of adaptation in growth is called evolution and is characteristic of most forms of life. Who knows what the distant future will bring in terms of human form and capabilities? This study is largely limited to forces that directly or indirectly act on the present mind of man, forces that will change behavior, modify its beliefs, and proceed to a position of secularism based on eventual distribution and assimilation of factual human knowledge. The same forces are also likely to change the eventual shape and character of man and will probably eventually lead to other similar species. No speculation is made here regarding possible life on other orbs of the universe. It could be molecular, monstrous, or there may be none at all. In any case it would also have to be made up of the chemistry of the universe.

As with much of the planet's reacting chemical life, humans must constantly ingest new chemistry in the form of food with which to support the process of growth. Meat is one of the primary sources of protein, the per capita consumption of which more than doubled in only the past half century and is expected to double again in a similar period (144). The animals from which meat is obtained are generally fed with fish from the sea; however, the protein "pyramid" may very well be unsustainable as it is now viewed, threatening the very "foundation of ocean life." This observation again emphasizes the tentative nature and dependency of continuing human life. Humans have been around for about a million years but still face a changing world with which they must contend.

**Wired by teachers:** The unique ability of earthly man rests primarily on his relatively large and complex brain. It constitutes the basis of his very working life and his ability to reason and to contemplate his own function and eventual fate. The brain is capable of continuous electrochemical reactions, the dynamic function of which is called the "mind." The sensations felt and given various learned interpretations are merely the result of such reacting chemistry. It is the electrochemistry of the mind that most concerns us in the discussion of human belief in this book. As with many other creatures, humans have developed specific primary sensing mechanisms connected to the brain, as discussed in the second chapter. The electrochemical signals from these sensors are not only recognized by the brain in real time for determining status and for controlling the body, but are also recorded in memory for reference and future access. It produces a kind of "replay" of the original information or experience. Function of the brain produces a sense of "awareness" and in humans, a more complex process of "awareness of awareness." Since the brain functions by forming chains of electrochemical links in the learning process—the response to its inputs—the term "wiring" is sometimes used to describe the resulting functional neural circuitry. Earlier chapters described the wiring of the brain that determines all thoughts, feelings, and resultant behavior, including concepts and practices with respects to religious belief. They also suggested that such wiring is difficult to change or erase once firmly in place. "Rewiring" can not always be accomplished.

This simplified review of the human position and makeup introduces the body and mind as reacting chemical components of the universe, all made of combinations of the same everyday chemistry that man now well knows, chemistry subject to constant reactive change under the forces of nature. Earlier chapters explained the "learning process" of the human mind, how sensing processes result in memory records, and how that memory becomes your life and influences subsequent behavior, specifically how it contributes to the desire for a god of immortality. This additional description simply provides more background information to complement the explanations given in chapters two and three.

The total time period of human growth and function, called longevity, is limited, resulting in eventual death of the entire living organism. However, new microscopic cells are normally produced in the growth process, from which new human beings can develop and grow in the process of human reproduction. This chain-like sequential process of reproduction can form a series of generations similar in makeup and structure. Biologists tell us that, in the process of reproduction, special neural chemistry may even be formed in the reproducing brain that results in a natural attachment with offspring to assure proper nurture (34). The chemistry of the new offspring brain may also reflect attachment to the producing mother. New concepts of biological sharing between mother and offspring have recently been reported. This chemistry may influence the observed "god desire" and subsequent emotions of spirituality. As already stated, upon death, the chemistry and, indeed, the entire function of life, is effectively returned to the, Earth where it disperses as part of the planet and becomes available in some form for the growth of new life (87) or for some other purpose. Man will eventually become more aware that this process represents the well-known "human cycle" and would then become increasingly more concerned with behavior during lifetime than with imaginative expectations of an afterlife.

**Born on an insignificant orb:** In this chapter, the imaginative concept of a humanlike god, the assumed creator and guiding power for traditional believers, has simply been replaced by the more realistic non-humanlike transcendent mechanistic force called "nature" (145), a force more powerful than any humanlike force. This does not change the character or spirit of man, even if modern science does not yet have it all quite right. Man needs a sense of rational understanding without which he would intellectually "wither and die." It strengthens his mind and brings order and certainty to his life.

This modern, more realistic mechanistic concept obviates the need for unsupportable imaginative notions and represents a greater sense of rationality. It is the principal basis of the new atheism discussed earlier, atheism that generally includes agnosticism, humanism, secularism, and most all of the other non-imaginative concepts of prior chapters referred to under the term realism. It will contribute to an increasingly solid base of rational understanding, however pagan or heathen-like it might sound. One might say, "Thank God there is a more realistic alternative to the imagined humanlike God." One must now learn to understand it and live with it, eliminating all sense of enigma and stigma.

As earlier discussed, modern intellectuals, scientists, and many ordinary citizens have already rejected the humanlike god "notion" and believe that the universe is the result of the godless process described. Einstein, for example, suggested that it is not appreciation of the "imagined grace and benevolence of a humanlike god", but the observed "beauty and harmony of consistent and reliable nature" that rational man must feel and appreciate. With that view, Einstein frequently claimed to be more religious than the theologians with whom he frequently met. Today, there is no further need to imagine the existence of a benevolent humanlike god and the immortality of an imagined soul in order to make life happy, satisfying, and desirable. There is no implied spiritual danger of any sort in realism. Although there are means by which man can influence the effects of fundamental nature, his imagined conventional communication with God in traditional belief has always been without any verified effect. Man now realizes that he is subject to the fixed and invariant laws and has no control over major events of mechanistic function. This explains why man's attempt to reduce storm intensity or to achieve changes in environmental conditions by traditional prayer for example, has always been totally ineffective. The "rain dance" will undoubtedly be abandoned for this purpose everywhere, in the course of time.

Substantial knowledge and forward-looking vision are currently contained in only a limited number of talented and rational minds but are expected to gradually become the understanding of most minds. This has been the basis of change and progress recorded by the history of man in every age and at every level of knowledge, for in the final analysis, man will primarily seek factual understanding in preference to mere emotional satisfaction. Imaginism will slowly become realism and will radically change prevailing human thought and behavior. Behavior will no longer be subject to an imagined divine "judge and monitor" but will more widely fall into the province of interacting rational man. It will happen quite gradually, without fanfare or special recognition. There is no expectation of biblical "weeping and gnashing of teeth" along the way.

**The human phase may be limited:** This then reviews some of the basic functions of the life of man on his small planet within the vastness of the greater mechanistic universe of nature. It is essentially the "human phase" of the operation of "nature." Unlike the explanations of theology, the modern rational mind does not associate the function of nature with any form of humanlike god, but many relevant questions still remain unanswered. To date, biological life is known only on planet Earth, a bio-friendly planetary body of the solar system representing less than a mere speck in terms of the vastness of the total universe. The central neural mechanism of man not only serves the many functions already described, his functions of consciousness, learning, and memory constitute all of the life he knows. The mind utilizes learned and retained information in various ways to recognize and confirm reality and to generate both real and imaginary thoughts. The mind also responds to stimulus by various emotions such as desires and preferences as it strives for security, happiness and ultimate survival. The early chapters of this book described how emotions tend to realize those strong desires by generating imaginative forms of satisfaction and interpretations of existence called beliefs. They spoke of how the imagined existence of a powerful protective and caring humanlike god, based on faith and speculation of its omnipresent existence, has filled the principal emotional need, a preferential natural "subconscious" function of the constantly concerned "searching" mind. This book suggests that all these notions will completely vanish from the future human mind.

**Reality not as simple as theology:** The gods of man have been around about as long as emerged man himself. His assumed God constitutes a simplistic basis with which to satisfy questions of origin and purpose, prime questions that have haunted the mind of man, even before his development and well before he acquired sufficient factual knowledge to understand his origin and true position in the universe. None of his gods is soundly based. But the gods of man have frequently come and gone, especially in earlier eras of ignorance and uncertainty. Awareness of awareness tends to lead to a desire for divine leadership and guidance while the natural forces of change constantly press man to reconsider and update his interpretations. Man is certain to reduce his dependence on any god as he gains increased confidence in knowledge and understanding as projected in this book.

Mythology involved popular god forms about which much has been written. The Egyptians had Osiris, the Romans had Jupiter, the Indians had Brahma, the Greeks had Zeus, and the Teutonic people had Odin. These were imagined visions of gods, variously interconnected with humans in order to explain man and his universe in terms sufficient to satisfy restless emotions. Nonhuman idols such as golden calves, sacred cows and wild animals such as jaguars, were also given the position of gods in various societies (146). However ignorant and misguided, sincere belief was the principal sustaining factor. Although civilization has considerably advanced in knowledge, and has now migrated over much of the globe, imaginative belief systems still persist with explanations based on a variety of early concepts. The early mind could not be comfortable with the notion of a mechanistic transcendent power greater than man himself.

Whatever their source or validity, written records were eventually introduced in society as a basis for belief. The Bible and Koran are perhaps the best known and most widely studied records. Traditional religions, however in conflict, have developed around particular forms of a god, and some belief systems may even involve multiple gods. When man feels that he must have a "higher power" to look to, he should be aware that there have never been any demonstrated benefits to man relying on an imagined god. The natural transcendent power or force is now recognized by rational man but not a force with humanlike character. Man is clearly at the mercy, not of a god, but of nature. His life is subject to his ability to relate to the forces of nature, and at best, he is on his own to survive through the exercise of his developing understanding and ingenuity. It is the force of nature itself that has encouraged spirituality and has influenced the desire for a god outlet, but it is also the force of nature that is slowly changing the spiritual mind of man.

**Gods are changing too:** Understandably, all gods have now become more integrated and global. As knowledge grows and becomes more widespread, society will turn away from the idealized heaven in the sky and increasingly look to more cultural involvement and to greater social satisfaction. As recently as the nineteenth century, it has been said that the wife of Charles Darwin was very much troubled by the prospect of certain separation, fully convinced that she would end up in heaven and her husband in hell. Some people may still think in those terms but few seriously rely on it. Clearly, no past civilization could envision how its beliefs might be interpreted in distant future ages and man today has similar difficulty in projecting his spiritual future. But the recent atheistic "surge," based on the scientific and intellectual advancement of man, has already produced unprecedented change, a new direction toward more rational vision. Indeed, new human understanding, only within recent time, has given man his first true vision of the makeup and function of his universe as well as that of man himself.

**A tell-tale spaceship:** The human mind often requires an updated perspective with respect to his position in the universe. There is no better example than man's increasing direct knowledge of deep space. For many people, the recent invasion of space by NASA has advanced emotional feeling but has provided much practical understanding. As described by Timothy Ferris (147), it has now been about three or more decades since a space probe was launched to take a closer look at some of the outer planets of our solar system, transmitting reams of data and photographs back to civilization on planet Earth. That same vehicle has since been allowed to continue its journey into the vacuous space of our vast Milky Way galaxy and is now more than ten billion miles away, approaching the outer heliosphere, the very "edge" of our own solar system. Despite its phenomenal speed, it is estimated that it will be forty-thousand years before it passes anywhere near another star within our own galaxy, and it will be most of a half million years to near Sirius, the "dog star" that man has always considered the brightest star in the heavens. However, it will still be within our own galaxy. The galaxy nearest to our own is called Andromeda. But Andromeda is so far away from Earth that its image, as observed today in the form of incoming light, left it well before the first human mind even evolved and became functional on Earth, certainly prior to the existence of any known religious thought, before any known concepts of a god, and before any human knowledge existed at all.

**Galactic traffic:** The Space Agency recently reported a violent clash between two distant galaxies that were found to be moving toward each other in space, trillions of miles from Earth. But it is just by chance that our own planet is not involved in that same violent form of interaction which might erase all life in a heartbeat. We now know that the galaxy Andromeda is moving toward our Milky Way galaxy and will probably merge into it at some future time. These events represent typical space activity involving the deadly interplay of hot gases, radiation, and the forces of black holes as observed by scientists on Earth employing the latest telescopes, x-ray, radar and other observation technologies (148). It again demonstrates the tentative long-term position of man in his universe. Certainly a humanlike god would not amuse himself with such violent action and it has only been recently that man could even detect and understand it.

Our scientific watchdogs, not any benevolent gods, assure us that we are still on a relatively stable planet, free from immediate cosmic danger, and that we may have many thousands of such years ahead, years in which man has ample opportunity to seek a worthwhile purpose, a life of peaceful human coexistence and happiness. But we must remain alert, for the Earth has many old scars of impact, one of which is said to have completely wiped out the dinosaurs together with many other species. Further, early in the last century, a Siberian impact equivalent to multiple nuclear bombs destroyed a huge land area. Later in the same century, a major Jupiter impact was recorded and we have been told that our sister planet, Mars, may soon sustain a significant impact.

As earlier mentioned, about five billion years from now our own star, the Sun, about a hundred million miles away at the center of our solar system, will transform itself into a "red giant" that will no doubt engulf all of the Earth's inhabitants (149). That is only a little more than the age of our planet to date. Astronomers have recently discovered a similar depleting sun called Pepsi, that has already become a red giant with an orbiting "Earth" like ours, about forty-five hundred light years away, but few earthly inhabitants are well enough informed to understand universes, galaxies, light years, solar systems, spontaneous chemical life, evolution, and all their human implications. In fact, few men fully appreciate and accept the factual knowledge that has already been acquired, updated, and certified by man based on his million years of existence. Few have yet adequately learned about the structure and function of man himself, about his imaginative mind, about his material world, about his relationship with the universe, about the effects of transcendent nature, and about appropriate control of human behavior on his own planet. The mechanistic nature of our universe is still not a familiar concept. Man would be best served to replace faith in imagination with faith in reality as described by learned men of reason who continue to dedicate their lives to the establishment of factual human knowledge for all mankind (150).

**The last anthropomorphic god:** Many rational minds have already decided that the vision of the traditional humanlike god will also be universally rejected and replaced. This is probably man's "last god," for his mind is not expected to again revisit or support imaginative god notions. Opinions are perhaps more reliable today because of better understanding of matters that have spiritual implications. The new trend will most certainly not be the replacement of its god with a similar, more advanced god, as has been seen in history, but by increased recognition of the mechanistic "god of nature," a non-personal transcendent power that has existed at least fourteen billion years before man, one that can have no emotional concern for man, his welfare or his future. Man must create a suitable new "model," not for God but for man himself. He will undoubtedly retain a sense of spirituality long after the concept of a humanlike god has vanished. The change in belief system will require newly educated generations and increasingly realistic views. This book suggests that man is therefore facing the end of the "god period." He will instead continue his concentration on improved understanding of his universe and will pursue more productive behavior during his limited lifetime on Earth. Because of the slowly changing form and function of the mind, no one can predict the eventual nature of human spirituality with certainty.

The notion of communication with a spirit is, no doubt, as old as man himself. However, as man becomes more educated and knowledgeable, fewer persons in developed society will retain serious belief in such spirits and in the likelihood that any could be of help or harm. The limits of communication are now well known. Man has to resort to imaginary techniques or to sci-fi for its "extension" to reach a god. Only uninformed and still vulnerable persons will remain influenced by spirits and ghosts and continue to be emotionally sensitive to spiritual belief. The notion of spirits and ghosts (151) and the imagination exercised in belief systems apparently utilize similar portions of the mind. But spirits and ghosts will be increasingly limited to entertainment and games and given emphasis largely in motion pictures and television. Man is moving forward at a surprising pace, but he requires correction of the old before he can advance to the new.

**Political rhetoric sounds wonderful:** History records that in 1862, General John Pope "blundered his way" to defeat at the second battle of Bull Run. President Abraham Lincoln wrote, "The will of God prevails. In great contests each party claims to act in accordance with the will of God. Both may be, and one must be, wrong. God cannot be for and against the same thing, at the same time. In the present civil war it is quite possible that God's purpose is something different from the purpose of either party—and yet the human instrumentalities, working just as they do, are of the best adaptation to effect His purpose. I am almost ready to say this is probably true—that God wills this contest, and wills that it shall not end yet. By his mere quiet power on the minds of the now contestants, He could have either saved or destroyed the Union without a human contest. Yet the contest began. And having begun He could give the final victory to either side any day. Yet the contest proceeds."

**Revered but very wrong:** However much revered, Lincoln was clearly wrong, as he relied on his imaginative belief, on his early spiritual neural wiring. The battle was won by the better general with the better equipped and more able army. But it was stated in accord with what he believed and was intended to represent his position of leadership in terms of the belief of much of his country. Man, even presidents, have always explained events in terms of gods, but after the fact with a wide range of "adjustments." However omnipotent, no god would or could give victory to opposite sides of warring parties in response to their similar prayers. His flowery explanatory statements might have been in the best interest of leadership and symbolism as it might well be today in political terms, but no rational modern mind could accept his expressed position as fact with confidence and respect. If "belief" in Lincoln transcends rationality, it is merely a weakness or misunderstanding of the mind. He further said in his inaugural address that "Fondly do we hope—fervently do we pray—that this mighty scourge of war may speedily pass away. Yet, if God wills that it continues," it must be said "the judgments of the Lord, are true and righteous altogether." As might be expected, the Lincoln mind, however able and just, was not informed about the mechanistic universe and knew nothing about the function of the working emotional mind. He had learned the ways of a preacher. Like many current minds, it was strongly influenced by prevailing imaginative beliefs and by what seemed to be most righteous. It was in line with the teachings of most of the minds of his country a century and a half ago. Whatever Lincoln believed or said, it is clear that mechanistic nature is totally impartial and can have no special influence on the course of war. His views based on lack of understanding had little fundamental meaning.

**The Hindu sage tried his best:** This chapter promised a few examples of fading vestiges of past spiritual thought. One involves the reaction in India to its first unmanned spacecraft called Chandrayaan for "mooncraft" (152). This vehicle was launched among people who actually regard the moon as a god. It suggested how science struggles most everywhere to coexist with faith. The old Hindu astrological system is based on lunar movements which make the moon very important for them. The story spoke of the ancient Hindu sage, Narada, who tried to reach the moon with a ladder, but fell considerably short. For the most part, India has advanced measurably in science and technology in the past decade, and may be one of the first populous nations to show major progress under the changing influence of modern education, perhaps in only two or three generations. But it is not there yet.

**Reality might pale against creation:** It is clear that the weaning of society from its strong imaginative visions will be a function of world educational processes and may take years and generations. However, this book acknowledges imaginative processes as being quite natural and regards them as common influences that are reasonably understood and subject to modern scientific explanation. This writing is not intended to convert all errant minds to realism, but only to provide better understanding of belief mechanisms and their effects on society. It repeatedly suggests that conversion to realism will take place quite automatically in the course of time, simply through the ongoing recognition and dissemination of factual knowledge. Although the human mind has seemingly unlimited power of imagination its judgments must all successfully transit the neural minefield that protects factual information. The natural forces of the universe that impinge on man are believed to encompass the desire to learn, the desire for realism, and therefore, must result in eventual secularism. Every effort should be made to encourage the teaching and promotion of modern knowledge and to require disclaimers for teaching any unproven and unverified imaginative notions. Use of the Internet and modern computer games like "Spore" are expected to be helpful in expanding world education and in the encouragement of reality. Electronic communication is a welcomed new methodology, one that has never before been available to man and constitutes an explosive powerful new tool, the effects of which will be increasingly far reaching.

With all their conflicts and differences, imaginative religions have carried a moral imperative to improve and assist mankind. They have been shown to satisfy emotions and to influence good behavior and prosociality (153). How else could man have successfully proceeded in his past development? However, any new belief system that replaces traditional systems should similarly encourage improvement in human behavior and more orderly world organization and cooperation. It should encourage direct support for the disadvantaged, sick, and needy, as have many traditional religious institutions in the past.

**Cultural evolution a most effective force:** Beliefs have developed in the long generational chain of most human cultures in a process earlier called "cultural evolution." Very few believers have either become religious on their own, have seriously questioned the logic of their sources, or have analyzed their own feelings and beliefs in terms of modern reality. It is quite natural to contemplate life, origin and eventuality but few have acquired knowledge of the function of their own world or their own minds. The average believer simply goes along with the belief of others, accepting the established notion of a humanlike god, the framework built around it, the associated principles of good behavior, literal belief in the Bible or Koran or some equivalent writing, and more or less regularly attends or participates in some kind of services. The average believer may participate in public prayer, but seldom penetrates the casual surface position, and is not likely to offer sincere monastic prayer to his god in the quiet of his privacy. There is a strong tendency of society everywhere to just go along with some "established practice."

The typical modern believer therefore lives his life in some form of "imagined divine involvement," maintaining a fear of recognition of reality lest he lose his "connection with God." He expects forgiveness for his transgressions and hopes for continued life after death, but he nevertheless exercises many short-cuts with respect to the formal tenets of his belief. Practical social issues, such as the use of contraceptives, the practice of abortion, the manipulation of human stem cells and cloning, artificial insemination, homosexuality, special dress or dietary observations, where they might be mandated by religious laws, or required church attendance are not always taken seriously today. The degree with which religious life is observed by average citizens covers a wide spectrum, a range that has expanded considerably, becoming much "looser" over just the past half century. This book has already shown that the depth of religious practice is gradually diminishing, asymptotically approaching an effective level of "no religion" if projected sufficiently forward in time. Only at times of births, marriages, funerals, sickness, or in times of other serious trouble, is religious thought more strongly exercised now and often temporarily renewed.

**Believers are now more considerate:** A recent article (154) discussed the new, "more practical" view of Americans on religions other than their own, as clarified by the Pew Forum on Religion and

Public Life. Titled *Religion for the Godless,* it said that its findings "threw the Evangelicals into a tizzy." The deep believers, who take the Bible literally, believe that "no man can cometh unto the Father . . . but by Jesus", that heaven is reserved for special believers. But we have learned that the world is no longer flat and that the Earth is no longer the center of the universe and that times and viewpoints have radically changed. Half the respondents now think that people with no religious faith can also go to heaven and achieve eternal life. Far less than half of the Christians believed that the Bible is the "word of God" and most said that one could achieve eternal life by just being a "good person." This brings belief closer than ever to a secularist position, but eternal life remains as an impossibility. These sentiments demonstrate a major new relaxation of prejudice and strength of belief. Society is indeed slowly moving toward more rational spiritual understanding.

Having set the stage in this chapter for the mechanistic basis of natural belief and the perennial search for earthly pleasure, satisfaction and advancement by mankind, it would be useful to revert once more to the published experiences surrounding traditional beliefs. This book began with a reference to the famous work of William James and its stated intention to fill the missing role of causality as a "sequel." The religious experiences described by William. James in *The Varieties of Religious Experiences* were so vivid and so personally consuming that the reader might easily become envious of the state of religious bliss, a deep conviction without a touch of doubt. His discussions of religious experiences frequently suggested that the minds of his subjects were in direct communication with their god. But we now know rationally that such communication is not possible. However, the mind has far reaching capabilities and, in extremes, can even produce a trance-like effect such as sometimes seen in religions like Buddhism, Taoism, and even in Pentecostal services (155), as well as in various forms of religious meditation (156). Mental exercises of this sort can also have physical influences on the body, they can affect the hormonal and immune systems and therefore introduce a functional feeling of realism, but these mental exercises have the same beneficial physical effect with or without an imagined god.

**James too early to know:** At the end of his study on religious experiences, James also said, "the only thing that it unequivocally testifies to, is that we can experience union with something larger than ourselves and, in that union, find our greatest peace . . . . Philosophy, with its passion for unity, and mysticism with its mono-idealistic bent, both pass to the limit and identify with something, with a unique god who is the all-inclusive soul of the world." But William James never seemed quite finished or satisfied with his descriptions and he ended his book with the expression, "But all these statements are unsatisfactory from their brevity, and I can only say that I hope to return to the same questions in another book." As suggested in the introduction, this study is intended to be that "other book" based on new causal knowledge and on human development of another century. Without advancing to dreams of Utopia or to greater idealism, the concept of modern realism is intended to fully reconcile science and religion and in so doing, fully resolve the underlying enigma. Despite his close observation of the deep spiritual effects on man, James could never bring himself to feel the need for theism. This book has explained that all emotional feeling is a matter of neural chemistry and subject to change. It is the same type of mind chemistry with which offspring may reach a relationship of bliss with its mother, or mother with father. Emotional response is certain to change over time as the prevailing chemistry changes. The controlling chemistry will change as a result of learning, adaptation, and natural genetic development.

**Mysticism is now largely poetic:** In the course of the next several decades the science of the mind will surely advance to achieve far greater understanding of beliefs, concepts that have been interpreted and misinterpreted for many centuries without sound basis. The human mind, through contemplative techniques and meditation, can now enter various stages and forms of "mysticism", a condition sometimes encouraged with various mind altering drugs, reasonably well understood. The influence of intense belief and the contribution of natural drugs on religion was described in McGraw's study called *Brain and Belief* (44). It is an effect found in many past religions. In Judaism it is called Kabbalism, in Islam it is called Sufism and in Hinduism it is called Bhakti. Some churches even believe it represents a form of special "grace" bestowed on man by his god. Following the first century, Neo-Platonism gave rise to Christian mysticism. It continued to be recognized in various forms throughout subsequent centuries and was said to be later expressed through authors like Keats, Huxley and the Catholic monk, Merton. It strongly influenced Oriental thought and its remnants are still found in various Asian religions such as Buddhism and Taoism (157). This study has concluded that now it all can be translated with understanding.

Mysticism has now begun to suffer from the explanatory influences of modern education and new understanding of mind function. New brain imaging and other techniques, not yet fully perfected, are beginning to contribute much to improved understanding of consciousness and "inner" thought processes (158). Much of general brain function is already understood, but its activity is highly complex and yet escapes full detailed explanation. This is primarily because the brain consists of billions of elements, many of which are interconnected and also connected to other parts of the body outside the immediate brain area. Panksepp (34) and many other capable neuroscientists have clearly described basic brain function and there are realms of more recent data still to be analyzed and considered in looking to the next major phase of brain understanding.

**The next form of brain:** An interesting aspect of human brain structure is the expectation of continuing brain evolution mentioned earlier. The division of the existing human brain into separate temporal parts, the "lower primitive", the "mid emotional" and the "prefrontal judgmental" areas, constitute clear evidence of time stages of evolutionary development. Its ongoing integration to perform various tasks is of special consequence in examining the telltale forces of human development. Again, change occurs, not only in slow biological evolution but also in radical, more rapid change, one that we earlier called cultural evolution. One need only look back to the nineteenth century, when political candidate Weaver proposed a national policy of an "eight hour day," "women suffrage," and "racial equality," and was summarily ignored and repudiated for such "outlandish" ideas. Man was just not ready to face those new issues at that time and had to await further development, experience and change. And it soon came. World secularism is also an issue that must await its time. This book is not intended to analyze details of modern brain research, but it is well worth the reader's effort to become informed about its principles in order to envision how it all manages to relate to mind set and to the protection of its emotional interests through the exercise of imaginative powers as described in these chapters.

**Religion had an imperfect history:** Based on historical human advancement, this book has reminded the reader that each generation assumes a somewhat new and different interpretation of the meaning of religion and its many spiritual ramifications relative to that established by the less informed minds of prior centuries. Some pursuits are accelerated while others are turned off, depending on the influences of cultural evolution. Religion, by its conflicting nature, may have actually served as a drag on human progress. There are some religions in which freedom of education is still severely restricted by the provisions and practices of their own tenets. Members of the faithful known as fundamentalists seem most vulnerable with respect to emotional extremes. Other fundamentalists can be equally narrow. The imaginative mind may readily become unreasonable or even psychotic in function and can seriously interfere with the conduct of peaceful society (159). Although generally of aggressive character, religion may well have only an outward appearance of strong and sincere belief. It may be combined with visions of earthly power, control and domination far from the expectations of any benevolent god. When this occurs in large organized groups under the influence of mass behavior, it can become violent and destructive. Strong imagination of the proper role of God's warriors, defenders, or missionaries can become especially dangerous under some conditions.

"Fundamentalism" was a name originally applied to a twentieth century movement in Protestantism that emphasized literal interpretation of the Bible as the basis of Christian teaching and behavior. The term is now applied to other religions that require strict literal adherence to its expressed principles. Fundamentalism often creates major continuing problems in a society that does not agree with the principles of its movement. The theme of this book rejects all belief systems and principles that are inconsistent with factual knowledge and anticipates a steady movement toward conservative principles of secularism, a position which is considered free from irrational mysticism and the uncontrolled extremes of religious emotion. It is the fundamental and factual findings of man that must serve as the invariant base for his life, a uniform and verifiable base on which all man can rely without unnecessary fear and without the uncertainty of imaginative interpretation. It should be a sound basis on which he can comfortably exercise his clear responsibilities while seeking maximum enjoyment in living his limited life with "nature."

**Imaginative religion is like a dream:** An article of interest, entitled "*Ways of Knowing*" (160), was referenced in a recent issue of the Global Spiral of the Metanexus Institute with the following descriptive commentary. "Religious people have often reported meaningful encounters with all kinds of non-human beings including, for example, God, angels, demons, ancestors, sentient trees, or mind-reading animals. It has been difficult for scientists to adequately interpret these experiences within the scientific paradigm of a mechanistic and entirely materialistic cosmos. Consequently, religious experience has often been ignored by scientists or dismissed as emotional, irrational, or pathological." Reported encounters of the examples given could hardly be ignored by modern scientists or even by rational non-scientific minds without a great deal of public discussion and explanation. Such encounters, verified by accepted scientific methods, would be given immediate recognition and wide airing, for they would certainly be phenomenological "firsts." Placing God in the same category as a sentient tree or a mind-reading animal merely suggests that all must be of an imaginative nature. Views of this type provide another transparent window into the deeply imaginative religious mind and add increased significance and support to the analysis included in the early chapters.

**Man will be unified when gods are gone:** Another commentary appeared in the same issue of the Global Spiral with respect to a review by Varadaraja Raman of *Darwins Cathedral*, a book by David Sloan Wilson. It said, "Just as individual organisms (creatures) adapt themselves to the environment for their own good and propagation, so does a group of people, such as the players on a football team, or the members of a religious community, for the constituting members are organically interconnected." The interconnection of most religious groups has a special character, indeed, but not often recognized from within. Such a group is usually formed and bound by the same imagined visions, one lending support to another with little interest in self analysis or rational appraisal of origin. A problem arises when there are many such groups differing and competing. As explained earlier, the spiritual mind tends to look for other minds of similar thought or similar objectives with which to gain reliance and satisfaction. Mass psychology is often not only misleading but dangerous. The entire world society must begin to act in greater unity in the common interest. Man has indeed reached a point where differences and separatism are no longer practical. Planetary society must now view itself as one and begin to think accordingly despite continued separation into nations, cultures, religions, colors, lands, and languages.

**Inescapable belief:** In his new book *A Secular Age* (161) Charles Taylor, a Roman Catholic emeritus professor of philosophy at McGill, argued for the "deconstruction of the death of God." For this he was awarded the prestigious Templeton Prize. He demonstrated a current quarrel with secularism, the idea that, as modernity, science and democracy have advanced, concern with God and spirituality has "retreated to the margins of life." He suggested that we look at the "right places" and allow the mind to open itself to moral inquiry and aesthetic sensibility rather than traditional theology as the gateway to religion. He further argues that "modern secularity" is the fruit of new inventions, newly constructed self-understandings and related practices and cannot be explained in terms of perennial features of human life. He claims that God is "sanctifying" us everywhere in ordinary life, in our work, in our marriages and elsewhere. The prestige of the Templeton Prize must be well respected, but such sanctification must be subjected to more rational explanation. This book fully explains secularism in terms of "perennial features of human life." It also looks to "modern moral inquiry and aesthetic sensibility rather than to traditional theology."

There does not appear to be any credible evidence of social damage by realism. These comments are assumed to be the work of the imaginative and emotional spiritual mind as described in earlier chapters. Taylor's book is reminiscent of a hope or a dream, that human history is not the preservation of life, but the salvation of the soul, not the right to labor and own property, but the duty to abide by moral order, not the greed of the market, but by the "grace of the cathedral." He suggested that spirit lives on in the imaginations of mind, whatever the material forces of secularism, and that a new poetic language can serve to find a way back to the "god of Abraham." He acknowledges the creeping clouds of secularism and a major change in old theology, but places his faith in the continued human "expression of spirituality." This study finds those admirable views based on an imaginative ideology not only to be contrary to reality but also contrary to the long-term direction of human change. It is, however, correct in its recognition of secularism to the extent treated in this book. There is a growing distinction between expression of spirituality and subscription to religious belief. This distinction is now frequently acknowledged in the new atheist and secular movements.

**Award for religious encouragement:** As John Patrick Diggins wrote in his December 2007 review of that same book, "There are many reasons to read it but, waiting for God to show up is not one of them." It may be in order to suggest to the respected Templeton prize committee that no amount of promotion or application of funds will "stem the tide of secularism" based on the natural direction of man. This contrary vision should, in fact, be recognized with future awards in the interest of educating man. Although such publications may delay the process, it is clear that the era of imaginism is gradually fading and the search for real truth rests in the factual determinations by man as part of the advancing bank of human knowledge. The Templeton prize should be shared among those that effectively interpret the relationship and compatibility between science and religion, not promote religion, including the tell-tale science of the human mind that must bear the entire responsibility for its steady march toward reality. It certainly should not encourage proselytization by minds of imagination. It should reward those who muster the necessary understanding and courage to resolve the endless enigma, those who are determined to eliminate, not perpetuate, the apparent mysteries, seeking the realities of the universe, and human spirituality. Man, having lost faith in the irrationality and uncertainty of his conflicting imagined religions, must now rely more heavily on realistic determinations if the modern, still evolving, human mind is to continue to advance.

**Progeny will eventually adjust:** A recent article by Steven Weinberg, a Nobelist in Physics at the University of Texas, appeared in the New York Review of Books entitled *Without God* (81). It was an excellent commentary of "living without belief" in a dependent traditional humanlike god. He first observed that Ralph Waldo Emerson, a recognized scholar of the nineteenth century, grieved over "what he saw in his own time as a weakening of belief, as opposed to mere piety and churchgoing in America, and even more so in England." Among many other scientists, he recognized the incompatibility between the teaching of science and claims of traditional religions. He offered a number of suggestions about questions that had been raised throughout society by the consequent decline of belief, the principal question being, "How will it be possible to live without God?" Referenced in earlier chapters, Zuckerman (4) tells us how routinely and successfully some societies already live well under secularism. They must now learn how to live differently, just as early man created a god and learned how to live with it.

Weinberg acknowledges that not everything has been explained by science, nor will it ever be, but that nothing has been observed that seems to require supernatural intervention for its explanation. He agreed with the conclusions of this writing, that we have to accept the fact that our home, the Earth, is just another planet circling the sun, that our sun is just one of a hundred billion other stars in our galaxy which, in turn, is just one of billions of visible galaxies. Most important of all, he emphasized the ordinary pleasures of life which have been "despised by zealots, from Christian anchorites in the Egyptian deserts to today's Taliban and Mandi Army." He reminded us that, visiting New England in early June "when the rhododendrons and azaleas are blazing away," shows one how beautiful spring can be, and he adds, "do not dismiss the pleasures of the flesh."

Weinberg also recognized that some of the art and literature that came from sincere religious inspiration will gradually disappear and that more recent great works were the result of inspirations rejecting religion. Needless to say, he regrets, with many of us, that the promise of life after death cannot be fulfilled and that "it makes cowards out of us all." He ends his article with the note, "Living without God isn't easy. But its very difficulty offers one other consolation, that there is a certain honor, or perhaps just a grim satisfaction, in facing up to our condition without despair and without wishful thinking, with good humor, but without God." This book suggests that the long-term forces of human adjustment and survival will absorb these concerns, however basic, as they always have over the history of man.

There have been many critiques of the Bible yet it stands prominently as the traditional reference for most Christians. The Moslem religion is similarly based on the Koran. They are documents that are thought to contain ancient and therefore, for many, more "believable" information. Although this study claims that the minds of the world will gradually move toward reality, the process must be given a great deal of leeway. One must consider a maximum acceptable rate of change among deep believers, and particularly how much existing tradition might safely change over a single generation. When one reads the faces of great theologians, of preachers and pastors and deeply religious persons, of sincere leaders who have gone into the ministry "for a noble purpose," when one hears their programs of service and of overt acts of righteousness, when one sees their efforts and sacrifices, when one hears the wonderful religious music and witnesses the most impressive pageantry, it is almost impossible to argue in favor of reality. There then seems to be little purpose in being right, in pursuing reason, in exercising greater knowledge and insight. How indeed can the long established traditional theistic belief's be erased or replaced in established theology without some "emotional damage"?

**Struggles with the imaginative:** A feature article of related interest was published under the title "God has always been a puzzle . . . . for Scot Atran." Atran (187) said, "Call it God, call it superstition, call it belief in hope beyond reason, whatever you call it, there seems to be an inherent drive to believe in something transcendent and unfathomable." He was said to have been struggling with the god problem since the early age of ten. At age fifty-five, he no longer believed in God but wondered why so many others do, most everywhere in the world. He received a doctorate in anthropology at Columbia University, and at Cambridge University, he studied the nature of religious belief. He was interested in why humans evolved to be so religious. He felt his interest to be, not in whether God exists, but why "belief" does. He also felt his interest to be quite different from those of Dawkins, Harris and Dennett and other "new atheists." The author of the article questioned, "Which is the better explanation for belief in God, evolutionary adaptation, or a neurological accident"? This book has already answered such questions in the first two chapters and has predicted that the constant changes in man and his developing knowledge will gradually result in a more widespread sense of secularism, irrespective of the past. The "sense of guilt" and the influences of past tradition will be completely lost in the process of human advancement and in the newness of man's realization and appreciation of his true position in the universe. But, when we speak of future man we always mean the progeny of existing man.

Again, the somewhat careless irrational beliefs in spiritual events will continue, but will automatically move toward greater reality of interpretation. Witches, angels, demons and the like have often plagued the uninformed and fearful mind of mankind in the past, but the modern mind is far less apt to retain such serious belief. It is true that belief in the afterlife and the positive effects of prayer are still widespread, for they reflect the strongest fundamental emotional needs. A Harris poll taken a few years ago in America, suggested that belief in heaven and hell was still widespread. Belief in a personal god was even more prominent. It is certainly not caused by a "neural accident" but by a combination of all the many reasons enumerated in this study. Scot Atran, or any other investigator of religion, like Turgenev in an earlier century, should feel, however, that the "god problem" is no longer an unusual, unresolved puzzle.

**Brownie points through personal art:** It is said that art is often used as a form of prayer. This book indicated earlier that the emotional religious art of the Italian Masters is no longer being produced as in the period of the Renaissance. However, in the depth of theological study we find occasional expressive desire to reach closeness to the god in which one has firm belief. One author of an article on Iconography (162) recently said, "I had no idea that it would answer my hunger for God . . . . My rational and intellectual self was saturated, yet I still felt distant from God. I was seeking a 'knowing' that is beyond the intellectual and rational mind." "Iconographers believe that because of the Incarnation, matter has been imbued with divinity. The icon is an example of the process of the Creator interacting with the creation. The act of 'writing' an icon is said to be an act of revealing an already present image; the goal of iconographic writing is to reveal the incarnate God." "Each time I pick up a paintbrush to write an icon, I take one step further on the ascent to meet God. As I ascend, God descends to meet me . . . . The image of God is with me throughout each day and night."

Icons, traditionally products of the Eastern Orthodox Church, are usually images of Christ, the Mother of God, prophets or evangelists or events from the time of Jesus. Painting an icon is called "writing" an icon because it follows established rules as for a poem. It is said that the act of creating makes icons different from western art since it engages the artist in the act of "creating" which gives icons meaning. This description takes man back to an earlier but now rapidly waning era of deep imaginative belief.

**The meaning of a mandatum:** In his most recent visit to America, Pope Benedict XVI was expected to address the presidents of Catholic colleges and universities. He is considered to be a "professor" looking for fidelity to church teachings. His predecessor tried to assure the identity of all Catholic institutions by insisting that Catholic theology professors sign a document called a mandatum, affirming their fidelity to papal teaching (163). It is not clear whether this was intended to refocus wandering ideology but Catholics are counting on the new Pope to enforce this. There are many new arguments about the identity crisis on Catholic campuses that has apparently existed for half a century. It seems to have as much to do with changes in Catholic students and their parents as it does with faculty members and administration. This is in line with information on the changing spiritual mind discussed in earlier chapters of this book.

**Decline in Catholic Colleges:** Only a few decades ago half of all Catholic children attended Catholic grade schools and high schools and the ten percent that went on had about three hundred Catholic colleges and universities to choose from and they were expected to enroll in one of them. Those who wanted to attend an Ivy college or another non Catholic school were expected to obtain permission from their pastor. Today, a Catholic institution is less prized by parents in favor of the higher rated schools if the student has a good enough record to qualify for admission. The American Council on Education issued a study that failed to uncover a single Catholic university with a distinguished or even strong graduate department. John Ellis, a leading Catholic historian, suggested that American Catholics should support no more than three Catholic universities, one on each coast and one in between.

The number of Catholic colleges and universities has now declined by a third. Many have become secularized in accord with the trend suggested in this book, cutting all ties to the church in order to survive. Others, especially those for women, finally closed their doors for lack of applicants. There are apparently more Catholic students at the big public universities in the Midwest than at any Catholic college, and they are there by choice. Catholic education seems to be in lock step with the changing minds of man. Students approaching the college or graduate level are increasingly questioning their belief systems, and all institutions are pressed to take this into account. The upcoming students, the new generation, can no longer be coerced in their teachings. They must follow the natural forces of development and change. There is now a special problem at the university level relating to their simultaneous teachings in science and religion that contain conflicting principles.

Universities have difficulty in maintaining the earlier underlying basis of education, that is, the building of factual knowledge together with religion. This problem may gradually be resolved with the movement toward secularism but may leave some temporary pedagogical damage in its wake.

However, the power of acquired imaginative belief instilled so deeply in hungry minds for so long is still strongly with us. A story (164) appeared not long ago that provides clear evidence of this. It tells of a family of four children and their parents who were determined to see Pope Benedict XVI on his recent visit. They made a forty-two hour trip from Spring, Texas, leading a dozen Texas neighbors in an accompanying van, and eventually got lost in reaching the Washington, DC area. They were members of an organization called The Neocatechumenal Way and, en route, they sang hymns, said the rosary, and discussed the mysteries of Jesus Christ. One member wrote in her diary and counted the giant crosses on Baptist churches. They were finally set straight by local members of Neocatechumenal and gathered outside the Apostolic Nunciature, the Vatican Embassy in Washington, prepared for a musical welcome to Pope Benedict. Reminding themselves that the Lord provides, they said it was not a vacation or a pleasure trip but a pilgrimage, a trip of faith. Whether in agreement or not, one has to recognize and admire the power and influence of such strong belief. It is a remarkable property of the complex human mind described in earlier chapters.

**Man created Saints:** One of the end products of imaginative religion is the creation of "Saints." A large number of persons in biblical and religious history have acquired the title of saint, but little is known about the circumstances of creating sainthood. The Vatican issued a new directive calling for greater "rigor" in its own saint-making process as reported by James Martin, a Jesuit priest (165). It was contained in a 45 page document issued by the Vatican's Congregation for the Causes of Saints in response to concerns that canonization procedures have been "watered down." During his pontificate, Pope John Paul II beatified over thirteen hundred persons and canonized about five hundred, exceeding similar acts of all his predecessors combined since the current procedures were introduced in 1588.

Canonization has long been an arduous procedure which involves the gathering of evidence for a life of "heroic sanctity." One medically certified miracle is required for beatification when a person is declared "blessed," and one more for canonization, when a person is declared to be a "saint" worthy of public veneration. The standards for verifying miracles is said to cause "eye rolling" among agnostics and atheists. The Congregation draws on teams of doctors who assiduously rule out any other cause for healing. Typically, the person cured will have prayed for the saint's intercession. Any miracle must be instantaneous, permanently and medically verifiable. Those cured cannot simply have improved, cannot relapse and cannot have sought medical care. The process can therefore take decades. The church claims that it does not "create" saints but simply recognizes them, that the process is a serious scientific business difficult for agnostics and atheists to disbelieve and easier for believers to believe. The true scientist in medicine (166) has great trouble with this entire process and feels that it will soon be abandoned as the world continues to progress along the secular path. The reader is referred to *Blind Faith* by Richard Sloan (38) for a scientific discussion of faith and medicine.

**The role of education:** Beyond natural changes in Catholic education discussed above, this book has been deeply concerned with the lack of adequate progress in factual world education. Proper education can mean the difference between accepted imaginative promises and established rational thought. The constantly developing human machine has failed to recognize the true position of man and is slow to accomplish its implied obligations. Society has failed to meet its educational responsibility very much as it has failed with respect to world hunger and medical care. The mechanism of encouragement of human advancement over other species lies with factual learning and teaching, a process whose technology was already described in initial chapters. It is the major gift to man by nature and must constantly be given maximum effort and support. One merely has to look at the condition of America to understand the condition of human education. America, considered one of the most advanced countries in the entire world, is barely a step ahead. A recent New York Times editorial (167) admonished that the Nation's future may depend on how well we educate the current and following generations. This viewpoint is also reflected in an article reviewing the twenty-fifth anniversary of a major educational report by the National Commission on Excellence in Education (168). No one seems to have the necessary understanding and the will to effectively engage the educational challenges.

**Vanishing students:** More than a million students, one every twenty-six seconds, drop out of high school every year, a sign of big trouble in an era in which a college education is crucial to maintaining even a middle class quality of life, not to mention for the good of the country as a whole, in a world that is becoming more competitive every day. Its shortcomings may well be responsible for losing the top position in world leadership. It claims that ignorance in the United States is not bliss, it is a widespread disaster. A survey by Common Core determined that a fourth of students could not identify Hitler, a third did not know that the Bill of Rights guaranteed freedom of speech and religion, and less than half knew when the Civil War took place. High school is no longer enough. A third of high school students graduate but are not prepared for the next stage of life, either productive work or some form of secondary education. It is scary and alarming by any standard. Although educational materials are available, the results of their use are disastrous. How long will it take for these students to fully share in the framework of human knowledge? Is it no wonder that the American mind adheres to primitive emotions and to imaginative belief? How can maturing students without a full education absorb anything other than the most simplistic imaginative thoughts in response to spiritual emotions? It is far easier and feels good.

Bill Gates, founder of Microsoft Corporation, offered a brutal critique of the nation's high schools a few years ago, saying, "When I compare our high schools with what I see when I'm traveling abroad, I am terrified for our work force of tomorrow." The Educational Testing Service cited several powerful forces that are affecting the quality of life for millions of Americans and the nation's future. These included the disparity in literacy and math skills that vary widely across racial, ethnic and socioeconomic groups. It suggested that we are not even coming close to equipping an educated populace with the intellectual tools that are needed. This book suggests further that the nation is not even close to recognition of reality in spiritual beliefs as well. However, it describes the underlying natural forces that slowly moves man forward. Hopefully there will be a natural acceleration of education encouraged by the recognition of the need for human competition over the next few decades, leading to a period of greater enlightenment.

**Belief in the Devil makes it real:** There is no greater example of imaginative thought than the forces of evil recorded throughout the history of man. Belief in the existence of the Devil has already been mentioned as commonplace for many minds. Yet there has never been any evidence of such existence and accordingly, that form of belief is said to be slowly vanishing in developed areas along with imaginative belief. Books such as *Evil Incarnate, Rumors of Demonic Conspiracy and*

*Satanic Abuse in History* (169) take one to the far reaches of demonology and remind us of how imaginative spiritual thought can easily get out of hand. Its description aptly states that Satanic atrocities are both inventions of the mind and perennial phenomena, not authentic criminal events. Evil has nothing to do with a humanlike god or religion. It is an inherent characteristic of evolving man that must be recognized and tamed with understanding.

Imagination in connection with religious thought is limitless. Religious thinkers continue to speculate on possible occurrences. A news article some time ago discussed the number of Americans who believe in life after death. Ten percent indicated they return in different form, about the same number believe there is no continuity of life in any form. About a quarter of people believe that they have a soul that lives in a different place depending on past actions. That notion is similar to the belief of about half who think that humans go to heaven or hell depending on confession of sins. These numbers are truly surprising in view of the modern explanation presented in this book. There are always wide continuing differences in imaginative speculation. Why is dissemination of nonfactual information so freely allowed without a disclaimer? How long must man suffer imagination and uncertainty? Plato and others argued that the body may decompose but that the soul is imperishable, a belief long since past its purpose. The news article reminded us that there is actually little mention of an afterlife in the Old Testament and that the rewards and punishments invoked by Moses were to take place in this world and not the next one.

**An eventual mind machine:** Knowledge and experience recorded in human memory, as described in early chapters of this book, exist in biochemical form, not yet directly capable of isolation, recording, and observation. In my earlier book, titled *The Man Who Created God*, I spoke of a hypothetical invention for unfolding the brain electronically, scanning, recording, or modifying its contents. This imaginative notion represented the mechanistic and chemical character of man. The existence of information retained in the mind constitutes the very life of man, his guidance, judgment, and his spiritual belief and concept of existence. How advantageous it would be to add the contents of one mind to another to provide a chain of experiences that do not have to be restudied, relived, or added by subsequent generations. How advantageous it would be to inject factual information in all minds to serve as the basis for good judgment. Since information in the mind is the essence of a person, my machine would, in a sense, result in possible recreation, a cloning of minds. Although it all seems possible looking forward, it would be inadvisable to hold one's breath while awaiting the perfection of such a device.

Previous chapters reviewed human mind function as it concerned religious belief and introduced the idea that the simpler mind of primitive man could only manage the simplistic god concept, and because of its appeal that simple notion has always remained without a great deal of competition from factual knowledge. But modern man now has the capability for more complex research and understanding and it is time to update concepts of the mind based on careful study and to shake off other useless, unsound, and less logical notions. This chapter is intended to discuss the existence of a transcendent force called "nature" as the prime driving force of the universe and how its presence continues to impinge, not only on all the material of the universe but also on the chemistry of life. This chapter further explains that the concept of a mechanistic universe with the spontaneous appearance of basic life from the chemistry of Earth may be a very difficult reality for human minds to understand and fully digest. Man is familiar with the notion of origin but has had a problem with relating his existence to any nonhuman creation. He can better understand changing events that, in some manner, relate to his own capability and to his own understanding, processes of origin and design with which he is familiar or which he believes have been processed by a benevolent humanlike god. He could understand existence better if there were a god and if it were responsible for simply creating life by waving a wand. That is the most simplistic religious interpretation. He can understand the processes of chemistry and even the more complex processes of biological chemistry. He might even understand the variations in chemistry and its role in growth over time and the related concept of evolution that involved changes in biological chemistry that led to different shapes and sizes of life. These considerations, however, involve more factors and make the entire subject more complex. It is far easier to revert to traditional religious belief. He might even understand the eventual biological growth of the brain and its essential control capability for creatures that have the good fortune to be equipped with such neural structure. With a brain and its connected sensory chemistry, one can acquire and remember experiences and learn and remember information about the world and how processes function. Today, even if structure and function is not entirely understood, the existence of more complex technology for everyday life serves to increase acceptance.

**Fallible imagination:** The expression of "religion as reality" is represented by a book titled *Religions*. It is one of many that claim "infallible proof" for authenticating religions, revealing the true nature and attributes of God. It is intended to expose the "hidden realities" of principal world religions. This viewpoint was discussed in early chapters as the result of "different" neural wiring compared with wiring that recognizes only established human knowledge and scientific fact. The reference to neural wiring was based on processes of learning and memory formation, the result of early exposure to the teaching of religion and to related religious experiences. Firm wiring of the mind as the result of highly desired emotional response does not make a belief factual.

The concept of "proof" does not comply with the methods of proof employed in the sciences and is therefore rejected by minds other than of similar emotional bent. If I claim there is a valuable diamond in my tightly closed fist how would one know for sure. Would the word of a stranger be satisfactory? Would the gem have any value while it remains in the fist? These are all questions equivalent to religious claims. This book has recognized the aggressive manner of leaders of belief and their constant effort to "prove" its rightful place and meaning to man. It has explained the intractable nature of the positive neural feedback effect and suggested that several generations of exposure to the more advanced information of modern science and human knowledge may be required to recognize and understand the difference between hard reality and imaginative short cuts. Indeed, the existence of belief can be proven but not the existence of a humanlike god. On the other hand, one can prove the existence and power of the superior forces of nature as well as measure its characteristics.

**Man is only beginning to understand:** We see that man is only now beginning to come to terms with his real world. No sign of life of any form has yet been found anywhere else in the universe. Living biological chemistry, successful on Earth, has not been able to form in any other of the explored atmospheres. Man on his planet should therefore be considered unique and regard peaceful coexistence with fellow man his first responsibility. The time has arrived when his true position is recognized and appropriate homage must be shifted from an imagined heaven to the reality of Earth and to underlying nature. This book therefore again suggests that the achievement of unity of spiritual understanding would strongly influence long desired human appreciation and satisfaction in the certain processes of further globalization. It has never been more important to the success of future man than at this difficult juncture. Unity can never be a successful product of the mind of whimsical and highly varied imagination as in the past, but must be derived from an extension of factual human knowledge and application of reason. It must be based on a fixed and uniform reference of new understanding. This important study describes how ongoing human evolution and appropriate succession of increasingly able minds must and will move society to converge naturally toward that objective. Man must escape the fear of nihilism and lack of purpose associated with the reality of his origin by simply breaking the bubble of an imagined humanlike god and proceeding on the road to suitable appreciation of reality. His true position is now well defined within the accepted framework of knowledge and he must therefore utilize his increasingly capable talents to convert his world into the realities of pleasure, satisfaction and happiness with appropriate recognition of all fellow earthlings with a sense of mutual contribution and purpose.

**The dramatic position of humanity:** We do not entirely understand why the position should be as dramatic for earthly man as we have now determined it to be, but we are an involuntary part of human life and part of its ever advancing and changing intelligence. The full story of origin of the universe is today clearly beyond a detailed understanding of man, for he is but an insignificant recent observer totally unrelated to its mechanistic origin, structure, and function. The notions of gods have always been imaginative products of the emotional minds. The entire knowledge of man is represented by the descriptive language of his factual science but he may always remain oblivious to the fundamental character, origin and purpose of transcendent nature. To the extent that more will be learned about the planet and the universe, advances are expected to be entirely in the province of science and not in that of theology. But none of the profound theories of the universe will be of practical contribution to the welfare of advancing, yet struggling society, only now beginning its true globalization phase. The successful interplay and welfare of man must be determined by man in concert with nature and not by any god. It is therefore urgent for man to pursue an appropriate course of realism with "religious" fervor.

**The alternatives are imaginary:** Not only does man live in a very dynamic universe in which heavenly bodies, space material and gases are constantly in motion, constantly reacting and changing in makeup, everything on Earth is also changing and reacting, growing and dying in a sense. Man has learned a great deal about his environment and about the laws of nature and has reasonably adjusted to them. However, he has learned less about his spirituality and about his emotions indeed, about the god he imagines. This book anticipates that he will know far more in time, and in that way will eventually find reasonable surcease and peaceful coexistence. It is for that reason that the solution of the god problem of Turgenyev is so important. It is for that reason that the theme of this book, and those of similar publications, must be seriously considered and distributed for all to read, for it is indeed the road ahead.

**Oh but to have a benevolent god!** There is surely no man alive who would not prefer a benevolent humanlike god to the reality of his seemingly cold and lonely universe (85). He has diligently sought such a god for millennia with every means at his disposal. He has meanwhile imagined and taught about many different gods. He has defined, modified and defended many imagined gods. He has honored and condemned imagined gods. He has bought and sold imagined gods. He has reproduced and murdered for imagined gods. He has sacrificed himself and others for his imagined gods. He has engaged in vicious battle for imagined gods and has even fought competitive imagined gods. He has become a lone hermit and has gone mad for imagined gods. He has built high levels of theology around his imagined gods. He has produced great art and sculpture for his imagined gods. He has built the most magnificent of great edifices for imagined gods. He has become wrenched in high emotion and bowed in desperate supplication to imagined gods. Unsuccessful and unsatisfied in his real world, he had to resort to his only means of appreciation, his keen imaginative powers. Indeed, the traditional religious image is one that best fits the daily life of man and has therefore been fashioned and retained in a form most desired.

**The increasing pursuit of reality:** As recently as November 2006, a group of scientists and intellectuals convened a significant forum called "Beyond Belief," at the Salk Institute in California (170). Its purpose was to discuss the position of science and religion, somewhat reminiscent of the classic arguments of nineteenth-century England following the publications of Darwin and other then emerging scientists. Discussions were based on seriously considered observations, a mass of appropriate research and logical conclusions. This was indeed only the beginning of a new, more organized direction of inquiry and expression, and there will soon appear many more books about non-belief and many more such forums. They are no longer in the category of religion bashing but are serious attempts to restore basic reason and forestall deception, mind control, and the dangerous consequences of misdirected beliefs by extremists and militants through exploitations of human emotion with utter disregard for factual reality.

There has never been a more appropriate period in history to unlatch closed minds and look, without bias, at the arguments made by men of great understanding and intelligence. They are the recognized leaders and teachers of men, they are men of initiative, of discovery and of invention without whom the world would wallow in ignorance.

We see in them, not the end of imaginative belief, but the first effective phase of a more rational movement toward the meaning of belief, albeit slow and gradual. The movement may accelerate just in time to encourage greater understanding of the modern mind of behavior. This book expects the mind itself will gradually change in function. It may well avert new and more disastrous conflicts among men of deeply opposed minds in an era of weapons of mass destruction. As stated earlier, it is increasingly apparent that some religious beliefs are experiencing weakness and indifference while others are tending toward increased distortion, a potent mixture of narrow fundamental belief and political unrest, some suggesting violent militaristic ambitions for unclear objectives. Still others, infiltrating settled and conservative religious areas, border on unbridled extremism and fanaticism without logic or reason even in religious terms.

The feeling and expression of awe and wonder about human existence and meaning that we call spirituality is a natural part of most every thoughtful citizen. Dean Hamer, a prominent geneticist, believes that a "god" gene actually exists in the human genetic structure known as the VMAT2 gene (171) that might explain such emotion. It is thought that this gene is associated with spirituality. Although it would not be expected to induce a particular belief, it might well structure the mind for spiritual proclivity. In today's technology of biological gene selection, one might then produce children with assured spirituality or with an opposite inclination. But would it also encourage goodness and righteousness? That is doubtful, for those characteristics are not exclusive to the religious. On the other hand, applying gene "knock out" techniques using genetic engineering, one might expect little or no interest in religion? This discovery is yet to be further studied and confirmed. It would be of considerable interest to know if our animal ancestors had the "god gene" but were unable to express it. In any case, this book believes that genetic expression may well produce emotional sensitivity to hope and desire, sensitivity that may lead to imaginative notions but it foremost sees long-term change that favors unity of religious thought and improved human behavior.

**Is the religious mind more limited?** The pervasive desire of the present human mind to embrace religion is most certainly a tendency common to much of society. Those who are not inclined to be affected by imaginative notions and tend to embrace secularization, must be structured somewhat differently as to neural detail. The greatest concern, already addressed in earlier chapters, is whether this difference affects the ability and capacity to perform frontal lobe tasks. Genetic coding variants do affect amino acid sequences. Some change the activity of important enzymes with their most pronounced effects applicable to the frontal lobe, the most critical region for negotiating complex environments. It is the seat of reasoning, a function that has been emphasized in initial chapters with respect to rational choice of belief.

Over time, religious frameworks have settled down to somewhat similar forms. But none of the traditional visions have yet been able to withstand the tests of factual knowledge. Religious teachings simply represent desires and needs of man which are readily absorbed without any special education. The viewpoint of common man is very much limited to his particular experience and education and therefore remains quite narrow. In the final analysis, most human belief is influenced by others and formulated in accord with convenience, usually ending with some acceptable style of organized religion built around whatever rites and practices best suit particular adherents. Some believers convert from one form to another looking for just the right one, but once questioned, there is seldom a good "Goldilocks" form of lasting consequence. Others are subject to second thoughts after considerably broader education and maturity, dropping all imaginative aspects and joining the ranks of natural religion. One can hardly imagine that some of the giants of disbelief were once religious. It is my understanding that even Richard Dawkins was once an Anglican.

**Natural sensitivity must be forgiven:** No man can currently be criticized for his sincere religious views, for sensitivity is a consequence of his current natural makeup. Certainly, early ancestors and ancients could not be criticized for ignorance of their complex and confusing world. However, in the modern world one may well be admonished for lack of tolerance, for lack of consideration of alternative views, and for refusal to maintain pace with expanding human knowledge. Today, most humans freely express their belief in a superior being. It is implied that, if humans are not in tune with the assumed goodness and righteousness of God, they must be evil or without goodness. This is an unfortunate expression of bigotry prompted by the very manner in which religion has been proliferated. In its representation of direction of modern belief this book visualizes a gradual change in that viewpoint based simply on the gradual application of improved underlying human logic. This book speaks of a time of unity in spiritual interpretation, a time when stigma may actually be erased, a time when social stigma directed toward disbelievers may well be fully reversed, redirected toward imaginative believers.

**Old beliefs shattered:** The time may come when the concept of a humanlike god in the superior capacity described may actually seem foolish. However, man may not be far enough along the learning curve to appreciate this potential position. It sometimes takes ages for full recognition. A six meter statue of Giordano Bruno was recently unveiled in Berlin's Potsdamer Platz station. Bruno was a Dominican monk who was burned at the stake in 1600 for denying the divinity of Christ and for supporting the now proven Copernican model of the solar system. It was suggested that passersby would reflect on the role of greater human reason in encouraging world improvement. Many Europeans particularly understand the injustices, not only with respect to the religious dogma assuming knowledge of the function of the physical world but also denying the expressed conclusion of a learned man after years of theological contemplation.

Meanwhile, in their nearby garden the Vatican plans to erect a statue of Galileo, the pioneer astronomer who was also prosecuted in the same era for violating the then prevailing view of the universe. He stood trial in about 1633 and was forced to recant his discoveries and remained under arrest until his death. Not long ago Pope John Paul II encouraged the clergy to reconsider the episode in order to make public apology, reaching new factual understanding, closing the Galileo affair after four or more centuries. In this matter the Church had no option, for its misjudgment was clearly a violation of factual knowledge. It is said that this retrospective correction was made to encourage the relationship between science and religion so that the Church might concentrate more favorably on current issues such as stem cells, contraceptives, euthanasia, abortion, and similar unresolved and disputed considerations. It is unclear that scientists will be deflected in any manner, for science, unlike religion, has no place to hide.

These are but two examples of acknowledgements of dogmatic beliefs gone awry in the same time frame. This book is directed to the clarification of a more fundamental issue, the eventual correction of purely imaginative belief that will, no doubt, leave many souvenirs and scars after an equivalent time period. Future gardens and museums are expected to be filled with similar apologies and recantations. The imagined divinity established by limited ancient minds appears now to be terribly wrong. An imaginative transcendent world of no reality, a life of uncertainty and conflict, concern and emotion of wasted effort in which misguided faith and belief has caused "disarray" in the natural struggling mind, a mind which has created something that is just not there. The long distorted mind may have to be reviewed, repaired, and restored. It has much to do to understand and improve its earthly world, to offer contribution, attain satisfaction, and enjoy a limited practical life. Its new factual belief cannot any longer be misjudged or shattered.

**Belief in early America:** With respect to expectation of greater reality, something should again be said about the changes in religious thought from the time of the Revolutionary War, the age of America, now about 230 years. I again refer to reviews by Gordon S. Wood of important books on religion and the founding of America that appeared in 2006 and 2008 issues of the New York Review of Books. He commented that "Even in the face of the relentless secularization of the twenty and twenty first centuries, religion in America still flourishes. Except for being freed of tax, it flourishes without substantial support of the government or state, without any of the traditional establishments that have maintained religion elsewhere in other nations." Could the more recent changes signal the pending last gasp of imaginism?

Despite the progress of early science in pre-America, religion was still the way people explained the world. The original settlers could not conceive of separation of church and state for they assumed that the world was fully explained by a singular form of belief. Now, with the existence of over fifteen hundred different religious groups, the current generation has exercised an obsessive effort to keep religion and affairs of the state separate. However, in almost every colony of the times all except Protestants apparently suffered political discrimination. With the Revolution came the Enlightenment and its liberty of conscience, and the close connection between church and state eroded. Colonist Paine was quoted as saying, "of all the systems of religion that ever were invented there is none more derogatory to the Almighty, more unedifying to man, more repugnant to reason, and more contradictory in itself, than this thing called Christianity." Many of the founders were Christian deists as defined elsewhere in this book. Although the Constitution of 1776 affirmed the belief in religious freedom of the Enlightenment, most of them retained their religious roles. Some states required officeholders to be Protestant. Some required them to believe in one god, heaven, and hell and still others in the Trinity.

In 1811, jurist James Kent, the chancellor of New York, who was involved in a famous blasphemy case, suggested that to defile with contempt the Christian religion was to strike at the roots of moral obligation but, in private, he was said to call Christianity a "barbaric superstition." The Wood review, perhaps the best about the time of America's founding, demonstrates the dependence on human thought and the expectation of religious change such as emphasized throughout this book.

**Modern Egypt:** A description of writer Alaa Al Aswany spoke of the failure of Nasser's government to build a modern secular Egypt. The illegal "Brotherhood" party won twenty percent of parliamentary seats in 2005 and became the largest opposition bloc to the ruling National Democratic Party. Aswany represents a young and interested group of new Arabs opposing the repressive policies of Islam and dictatorships worldwide. He is pursuing secularism and democratic rule, a dangerous and courageous move in current Egypt. His objective is to understand and explain the physical and moral lot of contemporary Egypt.

Aswany claimed that he was writing for ordinary people. His confidence and pungency were based on his deep belief in Egypt, which had such rich cosmopolitan layers of civilizations, Pharaonic, Roman, Islamic, Mamluk, and Ottoman, that had been severely held back by successive dictators from reaching its destiny as one of the world's great nations and allowing Cairo to return to the position as the capital of the Arab world. With his increasing public recognition and his influence with his young followers, he may well become another contributor in the Islamic world to the secular movement described in this book. He claims that, "We want democracy in the Middle East, that the West is obsessed with terrorism, but if it supported democracy here, there would be no terrorism." His emphatic opinion is expressed succinctly, "and things are so bad now that they cannot go on like this. They have to change. I think we are in for a big surprise." Will these efforts lead to the return to the once renowned center of factual science in Alexandria and ultimately to modern movement acknowledging a nonhuman god?

The infinite power held to be in the sky must be brought to Earth. There must be a powerful central governmental force that transcends borders with respect to human rights and the plight of citizens. Differences must be fairly negotiated and mass ravaging and killing brought under reasonable control. The centers of medicine must be expanded at once to protect civilization with the most modern means. Corruption, greed, and misuse of power so prevalent must be considered in choosing leadership. Leadership that violates the interest of its people must be limited. Leadership that depends on the guidance of an imaginative god must be replaced as unworkable. The best and most caring minds of man must serve the world interests, not in some ideal manner that falls into the known traps of past world organization, but in a well-organized practical approach to resolve the major problems of society without the baggage of imagined gods and unreasonable religious laws. Man can theorize and speculate and imagine, but only his reasoned actions must have meaning and represent the maturity and knowledge of advancing society. He must form a society of monitored laws and self-controlled behavior dedicated to providing happiness and success to the many lives of limited duration while sharing the products of nature. However ideal the organization of society may sound, it is within the capability of man to create both leadership and circumstances that come reasonably close. Man has now reached reasonable maturity and if reason cannot be exercised to achieve a much improved measure of social success, he will have succumbed to failure.

**Man developed from scratch:** Through the exercise of logic, men of vision are already aware that it is the transcendence of a natural force that is responsible for all of what we see and know. We know it is a force of origin and consequence and that man is insignificant and powerless in the larger view of the universe. He has evolved from its chemistry and, by similar processes, has acquired a means of thought and action with the ability to communicate with each other. He exists in a purely material world that appears to have had a material beginning. Man cannot communicate with material things, nor can he communicate with nature. However, mechanistic systems can function "forever" without man. The mechanistic system of nature has evidently functioned for more than fourteen billion years, over thirteen billion years before man himself appeared from the biological formation of life on planet Earth. Man should be extremely appreciative that he can finally begin to understand the complexity of his existence, that he can observe and measure nature even though he will never understand its full meaning or purpose. As with many other complex matters his present mind finds his existence difficult to visualize. This study promises that advancing capability will make it all easier. He should meanwhile appreciate and admire the beauty and harmony of nature, as Einstein often told his peers. But even Einstein was fundamentally limited as to full and complete understanding for science of the mind has since advanced. I have long written about the contrast between religion of a humanlike god and the god of nature (85) for it has indeed again become the principal challenge in human thought in the advancing modern world. It is not a story of mystery and magic but one that requires the study of facts and careful exercise of reason.

**The facts of life:** The rational and intellectual minds of modern understanding find no need for evoking the blessings of an imagined god. They have considered the terms of existence and have found satisfaction and contentment. They have found no doubt in their position of realism. They have accepted reality without compromising the ability to experience all the beneficial emotions of life. In ending this chapter on the "god of nature," it should again be emphasized that there is no factual evidence or rational basis for belief in the existence of a humanlike god as assumed in traditional religious frameworks. On the other hand, the concept of a mechanistic universe is based on an ideology that fits observations and measurements as far as they can be determined by existing factual science. These few salient points are perhaps most central to this chapter on the transcendent power of mechanistic nature with which communication by man is not possible:

1. The bodies of the vast universe are the product of an event described as the big bang with its parts moving away from each other in what is called inflation. Scientific measurements now confirm the effects of such occurrence.
2. The universe consists of an almost limitless number of orbs of different forms propelled by initial forces which follow laws of mechanical motion. It includes billions of galaxies like our own Milky Way galaxy increasingly apparent with the new space telescope. The universe is about fourteen billion years old.
3. The Milky Way galaxy is a spiral galaxy of billions of stars about a hundred thousand light years in diameter in rotation

about a black hole of almost infinite density. Our solar system is located about thirty thousand light years from its center near the end of one of its four great arms. The solar system is between three and four billion years old and contains nine planets the motion of which is well known to man. The sun's fuel is expected to fail in from five to six billion years when life will become extinct.

4. Biological life formed in the chemistry of the Earth, one of the planets of the solar system whose atmosphere is capable of supporting it. Cellular life appeared within a billion years after cooling of the planet and its chemistry grew and changed through many forms in a process called evolution leading to the present human form.
5. The biological form of man included a nervous system associated with a nerve center called the brain. The complex brain evolved with the evolution of man and is still in the process of change and development.
6. The function of billions of connected elements of the brain includes sensory and motive areas that control muscles. The brain has also developed the ability of self awareness in a continuously interconnected neural system the function of which is called consciousness. Conscious brain function is called the mind.
7. The neural system developed five principal sensory circuits with which man can determine his place and condition. The system is sensitive to thoughts and stimuli eliciting feelings of emotion. But the life of man, continued by reproduction from previous life, is limited to less than about a hundred years before cell failure.
8. The neural system developed the capability of acquiring information called learning and of recording experience from

the sensory circuits in what is called memory. Memory can be accessed at times. Feelings and emotions can be activated by either real time sensory stimulus or stimulus from memory. A combination of stimuli and memory is responsible for judgment and behavior.

9. The early undeveloped mind of ignorance about himself or his world felt that there must be some extraordinary ruling and protective god, omnipotent, omniscient, and omnipresent similar to leaders among fellow man. It attributed human character to his god, judging and punishing misbehavior and providing a sense of immortality. Such belief required no proof, resting on faith alone but could not be easily ignored because of instincts of desire and survival.
10. The ideology of current man is a mixture of that faith and the factual understanding outlined above. Ideology based on faith is called imaginism whereas that based on measurement and observation is called realism. Imaginism has developed as human tradition because it best fits human emotions and desires. Realism is gradually replacing imaginism because the human mind has advanced to the point where it desires factual belief and is able to face reality without requiring an imaginative idealized alternative.
11. No sign of any other chemically formed or evolved life has yet been detected in the universe beyond Earth, but search capability has thus far been limited. The process of natural formation of cellular chemistry of growth and reproduction may be quite unique. Man must now readjust its ancestral animal instincts and seek peaceful coexistence. Man must live primarily for fellow man rather than for building formal systems and organizations around imaginative beliefs.

12. In their best interests, humans must abandon imaginative ideas developed and marketed by emotional leaders, claiming it to be the desire of their god to do so. They must concentrate on understanding and beneficially utilizing their relationship with nature and its function. Man must also consider that persistent imaginative thought, conflicting with reality, may have detrimental effects on further evolving reasoning capability and may lead to missteps in the development of his optimum life.

# Chapter Six

## Contrasting Powerful Minds

I maintain that the cosmic religious feeling is the strongest and noblest motive for scientific research. I am of the opinion that all the finer speculations in the realm of science spring from a deep religious feeling . . . . I also believe that this kind of religiousness . . . . is the only creative religious activity of our time.

I cannot conceive of a god who rewards and punishes his creatures, or has a will of the kind that we experience in ourselves. Neither can I, nor would I want to, conceive of an individual who survives his physical death; let feeble souls, from fear or absurd egotism, cherish such thoughts.

—Albert Einstein, as quoted in *The Quotable Einstein*,
Alice Calaprice, Princeton University Press

I began to pray daily, hourly . . . I took long rides out into the desert where I could be alone at prayer. I prayed with my wife in the evening. As I tried to understand my problems I tried to find God's will in acting on them. The rocket genius is a brilliant conversationalist, extremely handsome and socially charming . . . his lucid conversation covers everything from the atom to God, who he believes in deeply.

—Notes about Wernher von Braun from *Von Braun, Dreamer of Space, Engineer of War,* Michael J. Neufeld, Alfred A. Knopf

**The most reverent nonbeliever:** In studying the reverence of Albert Einstein, I found that he followed a religious path similar to my own in his early life. He abandoned Christian views that he acquired in his early mandated Catholic education in Munich as he became more knowledgeable about the world of reality. This transitional process seems to be quite common in the lives of many intellectuals and scientists since parents invariably introduce religious ideas to their children very early, well before they reach the age of independent reasoning. Scientists are not automatically atheistic or anti-religious but tend to formulate independent judgments based on accumulated hard factual evidence and reason. Their professional accomplishments in science actually depend on their ability to be imaginative but make a distinction between factual and imagined conclusions and on rejecting concepts based largely on emotion. Once established, the transformation of a deep-seated ideology is not easy for the reasons explained in early chapters. One can well appreciate the mental agony of Mother Teresa, sometimes wavering precariously on the spiritual fence. A rational mind must coldly and courageously abandon an imagined ideal framework of heavenly attachment and immortality for that which it has determined to be real and factual. However, it is important that all minds gain an understanding of religious processes as well as a sense of concern about biological and neural frailty with which it is related. My early religious experience was essential to my present understanding of the complexity of religion.

Contrary views: Earlier chapters discussed the nature of religious belief, how the emotional mind creates an aura of desired protection by a benevolent god and how human imagination often contributes to a completely irrational conclusion. Mind structure and function have already been implicated in all religious processes. However, a discussion of religion as the "world's greatest enigma" would not be complete without reference to a well-known technical mind that did not favor realism as defined in chapter three. I chose to compare two minds that appeared to be similar on the surface but completely opposite in terms of function relative to religious belief. It demonstrates that, despite diametrically opposed beliefs, two minds can appear similar and can equally contribute to important technical and scientific progress, making unique and lasting contributions to society. It seems as though spiritual and technical functions may be independently assigned to separate neural channels.

I have observed hundreds of minds in business, legal, medical, scientific, and in other fields. It has been particularly interesting to examine the relationship between religious belief and professional practice. I have often found some suggestion of a logical relationship but no regular pattern. For purposes of discussion here I have selected the spiritual minds of two famous men whom I have known. Both have touched my life in some way, Albert Einstein, a former neighbor in Princeton, New Jersey, and Wernher von Braun, in a research relationship at White Sands Proving Ground, New Mexico. In terms of my definitions of belief of chapter three, I considered Einstein to be a staunch believer in realism and von Braun a staunch believer in imaginism. Apart from religiosity, both minds were highly motivated with similar creative objectives that would contribute uniquely and irreversibly to advancing humanity. Both minds were technically trained and highly qualified, continually pursuing extraordinarily ambitious objectives, primarily under pressure from within.

**The most famous mind:** As a well-known world figure, Einstein had the reputation of being perhaps the best known creative mind in modern history. He had given the world a totally new way of interpreting motion in space and time, largely from purely theoretical considerations and therefore was widely respected as a mind of special "logic and reason." His scientific contribution was indeed unique and without precedence. At the end of the Einstein century, there were few educated persons the world over who were not acquainted with his name and his contributions. Because of his keen analyses and revelations about the workings of the universe, combined with his expressed interest and concern about world peace and his expressed concern about the welfare of the ordinary man, he was placed on the highest of pedestals never to be displaced. His field of theoretical physics dealing with fundamentals of the universe was frequently related to theology. Theologians everywhere sought his opinion about religion and letters of inquiry continuously flowed into his Mercer Street home, many from ordinary citizens and some from prominent theologians. He responded by scribbling appropriate replies in German on the envelopes for his secretary to transcribe and mail on his behalf. He labored diligently to bare his opinions and inner feelings to all inquirers with humility, clarity and firmness. The voluminous archives and biographical publications tell his amazing story very well.

My special interest in Einstein's life had been directed to spirituality and religious belief for some time. On the occasion of the recent celebration of the centennial of his first great scientific contributions called his "annus mirabilis," much was written about his life. I took that occasion to write about what I had carefully learned of his belief. It was presented in the form of an op/ed published in the *Philadelphia Inquirer* in January 2006 (18). The editorial consisted of a single page review of the merits of his so-called cosmic belief, a subject frequently discussed in books and articles referring to his unique concepts of God and religion. My editorial was partly intended to correct public misconceptions of his alleged belief in a humanlike god as well as his frequently assumed position as a "cold" atheist.

I cannot think of any famous person in history whose thoughts on religion were more widely sought and analyzed. People everywhere were convinced that, if any mind were to know the truth about God, it would certainly be that of Professor Einstein. His spiritual views however were always in great contrast to prevailing traditional religions. But, despite the contrast, widespread respect and popularity with both heads of state and ordinary citizens never faltered, most of whom never fully understood either his true religious views or his complex theories of relativity. Theologians everywhere strove, without success, to bring his beliefs into some focus with their own. Writings referring to his frequent references to God suggested to many that a humanlike god might possibly be part of his belief. I am quite certain that his use of the word "god" or "soul" must be dismissed as a "facon de parler" for, in contrast to traditional religious usage, he frequently denied belief in an anthropomorphic deity of any sort as well as the concept of a human soul. The following paragraphs contain the essentials of my editorial.

**Deeply considered reverence:** Albert Einstein gave the world an amazing new account of the laws of the universe together with a clear understanding of the place of God in human life. Most physicists understand his concepts of relativity but few people understand his concept of God. Although he frequently assured the world that he was more religious than most others, what he meant by that was not what many people might think. World events and history suggested to him that man may have taken a wrong turn somewhere and that society must again take stock of its assumed purpose and direction. He felt that spiritual realignment may be required together with efforts to improve society and promote understanding and peace among people.

Einstein's first step may distress some people, perhaps many. Much of society still looks to a personal god for comfort and teaches its children to do so as well. Einstein firmly announced, however, that there can be no anthropomorphic god. He found no meaning in notions of efficacious prayer to a supreme being or of immortality of the soul. Nevertheless, he deeply appreciated "spirituality" and repeatedly offered to explain his "cosmic" or natural religion. He found his unknowable god in the beauty and harmony of the transcendent forces of nature. These he beheld with awe and reverence. And many of the values he found in revering his god were all but indistinguishable from traditional religious values. Einstein felt that religious awe toward the forces of the universe had several advantages over traditional religiosity. He felt that his belief was truly religious and left him with a strong sense of duty and justice, making other people's lives better. Yet, unlike much imaginative religious belief, it was also in accord with advancing scientific knowledge. It made no impossible promises. It created none of the fear and uncertainty he found in so much traditional belief, fear and uncertainty he thought had harmed and would continue to harm ongoing human development.

For Einstein, fear and uncertainty were the sources of superstitious beliefs that had effectively paralyzed humanity. He saw a great opportunity to address these toxic wellsprings, feeling that today we have less to be afraid of and less to be unsure of, and that our ancient ghosts are less significant in the modern world. Another advantage he saw in his beliefs was that they were fully in line with fact and reason. And because he thought logic was common to all human minds, he felt his authentic cosmic reverence could provide a unity never before achieved among the world's people. Common spiritual understanding had always eluded us as the key to a more ideal peaceful coexistence, yet we never quite achieve it. Again, Einstein saw an opportunity here. He felt there was a correct interpretation of the wonder and holiness of reality and would not the correct interpretation be the same for all? Belief, ecstatic belief in the truth, in accord with the wonderful revelations of nature and our own minds could drive away forces that separate us.

Einstein rightly saw a double imperative for society; to uphold and preserve reason and the reverence for fact while celebrating and defending personal freedom and human rights. In Einstein's view, all these were precious objectives and all were religious values. But there remained deep resistance to such ideas. And that is odd, as if a man who was so right in one arena must be wrong in another. But few periods of human history have ever had more reason to reinterpret existence than the present. World conflicts are many. And a great conflict has been under way in our culture, not so much between religion and science as between human emotion and reason.

All throughout history, narrow religious beliefs have resulted in disastrous wars and inhumane persecution. Today these beliefs have a new significance as adherents try to inject them into the conduct of medicine, politics and economics. Of even greater concern, our national leaders, legislators, and jurists may fall into the same web. Such narrowness of belief is on a collision course with the rapidly globalizing new world. That world requires a mind, as Einstein foresaw, that reveres fact and reason but combines reverence with a vigorous expansiveness, openness to experiment, tolerance of dissent, celebration of human rights, and an appropriate system of justice that accords fair and beneficial treatment to all.

A narrow-gauge mind addicted to distorted illusions, will be ill-equipped to make decisions on topics such as stem-cell research, abortion, artificial insemination, social and economic equality, contraception, same-sex marriage, evolution, euthanasia, assisted suicide, animal rights, jihad, suicide bombing, homosexuality; and the place of spiritual education. Albert Einstein gave us more than a revolutionary view of the universe. His great mind also gave us much that has improved our lot. But if we refuse even to consider his beliefs, or if we reject them, content to embrace illusion, or if we insist on seeing more than he did, we may lose his most important contribution of all.

In my personal observations I found that Einstein was never one to be ambiguous. His position was especially firm and intractable. Although he frequently admired the marvelously arranged universe, its "beauty and harmony" and its manner of function under observable laws, he attributed it to nature, a transcendent mechanistic power not fully understood by man and never to a humanlike god. He did not consider himself to be a pantheist who regarded all of reality to be divine with god present in all of nature and the universe. His god was simply the transcendent power and force of the mechanistic universe functionally independent of man as described in earlier chapters. With his deep admiration of the beauty and harmony of the universe and its manner of operation according to mathematical laws, he frequently claimed to be "more religious" than most others around him. I always had the sense that he wanted to stop there and not be pushed for greater definition. He did not want to mince words or become involved with philosophical descriptions that could very easily be misconstrued. He was sensitive to misinterpretation by those who would have liked him to believe in a traditional god.

In his later years, Einstein fully realized that he had the ears of the entire world, not only those interested in theoretical physics, but also those with an extreme desire to know the true spiritual visions of his exceptional mind. They encouraged traditional religious views and relentlessly attempted to extract an admission that he too was a believer, but without success. Every time he used the word "god" in his writing or in his conversation, they pounced on it heavily trying to fashion it into some sort of spiritual confession. But Einstein, keenly aware of this religious trap, was always firm and direct in his expressed interpretations. Though his beliefs were fixed and invariant, he never wanted to influence others as to what they should believe. He not only met individually with theologians, he became the central figure in several formal conferences devoted to questions of religion. He never dodged the issue. His views were expressed with great conviction and also with great care for fear of unnecessarily antagonizing theologians who increasingly became his friends. He openly continued to publicly express his belief that there could be no humanlike god or any related spiritual framework as expressed in traditional religious teaching and in widely established practices.

**An updated Spinoza:** The Einstein belief was clearly realism in contrast to imaginism, as defined in this book and, at times, he referred questioners to the published philosophy of Baruch Spinoza. Spinoza was a seventeenth century philosopher who, in turn, followed much of the thinking of Descartes. Spinoza, however, first believed in a "rationalistic pantheism," a belief that regards all of reality as divine, that a god is present in all of nature and in the universe. This idea is present in various religions but, unlike the Einstein belief, most gods are given a humanlike form. Like Einstein, Descartes admired the harmony and beauty of nature and felt that its basis was a transcendent force that had no relationship with, or interest in, human life. Although he respected all forms of human spirituality, he was not sympathetic to any that included a benevolent god who monitored and punished the behavior of man.

**A special kind of brain?** Einstein's education was largely devoted to inquiry and explanation of the natural laws of the universe. Many biographies have been written about his life, the most recent and most comprehensive being that written by Walter Isaacson (174). Indeed, the work of Albert Einstein is not only regarded as a "major contribution to human advancement," the unprecedented interpretation of the relationship between man and his universe, but today, more than a century later, he is still regarded universally as having the "greatest human mind." His preserved brain continues to be the object of periodic analysis and discussion. New studies, such as those by Sandra Witelson of McMaster University and Dean Falk of Florida State, claim to have identified unusual features of his brain. His parietal lobes were said to be substantially wider than most and a rare pattern of grooves and ridges suggest exceptional ability for conceptualizing problems.

My interest in Einstein and his work took place in the decade following his move from Europe to America in 1933. He lived a simple life, played the violin, and sometimes found relaxation by occasional sailing in a small vessel of modest construction and condition. In his private life, Albert Einstein felt decided freedom from the religious atmosphere that always seemed to be associated with others, an atmosphere which he considered a "yoke" for uninformed man. On many occasions, he emphasized his special respect and appreciation for the cosmic laws and mechanistic power from which all men were derived and felt that all mankind should be required to acknowledge and understand them. Despite his depth of philosophy, Einstein was not a student of formal neuroscience or psychology for his life ended in 1955, well before the "decade of the brain" ended at the close of the twentieth century. The new science of brain mechanisms and the function of the human mind, such as those discussed earlier in this book, were not yet available but would certainly have provided him a better understanding both of his spiritual feelings and those around him. The rapid advances in neuroscience combined with more complete knowledge of the universe have indeed brought man to a new modern appreciation of his position in life. Since he could not bring himself to believe in the existence of any humanlike god or in the notion of communication with any spirit, Einstein had no expectations regarding prayer or any form of life after death. Although he exercised his great power of imagination with respect to the structure and function of the universe, perhaps more expansively than anyone before his time, it did not extend to imaginative spiritual matters. His imagination was fully based on rational extensions of recorded observations and other factual determinations, information that could be repeatedly subjected to the tests of reality and to mathematical definition. He was well satisfied with his life of acknowledgment and appreciation of nature and its transcendent mechanistic character. He occasionally expressed the thought that man might never find the answers to nature's transcendence. Even in his time of limited astronomical understanding, the mid-twentieth century, he sensed that the universe and its secrets were of infinite magnitude and consequence relative to the position and capabilities of man and, unlike many other great men of scientific accomplishment in that period, he refused to sell himself as a "speculative futurist."

**Spinoza's god:** Early in 1929, in the critical period of active religious concern and frequent examination of the Einstein beliefs, Cardinal O'Connell, the Archbishop of Boston, instructed New Englanders not to read anything about the theory of relativity because it was a "befogged speculation producing universal doubt about God and his creation." The archives show that Einstein subsequently received a cable from Rabbi Herbert Goldstein (175) of the institutional Synagogue of New York with the unambiguous and direct question, "Do you believe in God?" Einstein did not hesitate to reply that he believed in "Spinoza's god, who reveals himself in the orderly harmony of what exists, not in a god who concerns himself with fates and actions of human beings." Although his reply tells much of the story of his belief, many traditional believers doubted that Einstein saw any sort of "god" in his cosmic belief and simply called him an atheist. They claimed that an Einsteinian god would be totally useless to man, since prayer and houses of worship under his form of belief would have no purpose and there would be no "sense of immortality." One would expect inquiries by theologians of his time, but today concern is far more widespread. The forces of change are expected to continue to modify man's viewpoint as well as his culture as they always have in the past, century after century. The mind of Einstein was merely an early indication of the increased realization of that force.

**Cosmic religion:** In the half century or more since Einstein's death, interest in Einstein's specific terminology, "cosmic religion", has fallen from active discussion. However, since traditional religions are increasingly fading in various ways his concept of religion has continued to grow in more acceptable forms of secularism, humanism, and atheism. As more knowledgeable and learned society continues to question the imaginative basis of popular beliefs and as greater spiritual turmoil and aggression permeates the globe, the Einstein view is expected to remain the heart of secularism despite its impersonal concept of a god. Although pure atheism suggests the total absence of a god, a sense of recognition of a higher power behind the universe and its contents will no doubt remain in some form, however mechanistic.

In the opening chapters of this book, "wiring" of the brain was symbolically discussed in some detail. Neural circuitry was said to respond differently to different confidence levels of input information in the learning processes forming electrochemical records that largely remain in place and affect behavior throughout subsequent life. This book argued that such neural circuitry and related memory involve biological components not easily modified or erased and that any substantial changes in religious thought would likely be reflected only in subsequent progeny where new and different circuitry might be formed. Clearly, the wiring of the two minds discussed in this chapter was formed differently in reaching the age of reason and remained in effective control of their subsequent lives.

**The mind that gave us space flight:** The second famous mind I wish to describe is one that appeared prominently in the World War II era. In contrast to the Einstein philosophy, Wernher von Braun, a recognized forceful mind of superior technical talent, believed deeply in the existence of a personal god to which he frequently prayed. The Einstein philosophy was based on a comprehensive study of the physics of the universe whereas the von Braun background was more like that of an engineer and not a "scientist." Critics frequently make this distinction between scientist and engineer, feeling that most engineers do not look sufficiently into the underlying scientific basis for formation of belief to allow full analysis and exercise of rational judgment. Although much of their two lives were simultaneous, Einstein was a few years older. Von Braun was the son of a Prussian Junker and civil servant of Kaiser Wilhelm II, of a noble land-owning caste. As a brilliant engineer, his ambition was "space travel" even well before his maturity.

Von Braun became head of rocket development of Hitler's Third Reich, joining the Nazi party in the mid thirties and subsequently becoming an SS officer. He created the first major rocket propelled missile to be used by the Germans relatively late in World War II. No such advanced concept had ever been conceived and successfully realized. His famous World War II weapon was called the V-2, about fifty feet in length and six feet in diameter, fitted with a huge deadly warhead. Transported on a trailer called the Meiller Wagon, it could be fired from its portable platform most anywhere in northwest Germany and surrounding countries. It was responsible for destroying thousands of lives but came so late in the war that it never reached the intended point of domination. When the Allies became victorious in Europe, von Braun and much of his scientific staff were captured together with a huge amount of his rocket hardware and shipped to the United States. The code name of the program assigned by American authorities was "Project Paperclip." At an ordnance test site near El Paso, Texas, he and his group of scientists became engaged in an American V-2 research program in 1946, a portion of which I headed on behalf of Princeton University. The first American objective was to become familiar with the rocket vehicle itself, using it to make several important experiments in the upper atmosphere at high altitudes, heights that could now be reached directly for the first time. The assembly personnel and launching crew were largely former German scientists. Von Braun soon thereafter initiated a major visionary program with respect to the development of advanced missiles and space vehicles for the United States. His group was later responsible for the launch of the first American satellite to compete with the Russian "sputnik" and organized the manned space program eventually leading to the American moon landings.

Despite his spectacular accomplishments in American rocketry and numerous successful space projects, von Braun was periodically troubled by his wartime background in Germany. He was said to have supervised a military weapons program involving slave labor and was therefore responsible for the oppression and death of many Europeans. He was also accused of responsibility for the violent deaths of many innocent citizens of England and Belgium, victims of his V-2 missiles. He was simultaneously a war criminal and a great visionary whose chief life ambition was space travel.

**Death with human advancement:** Although his academic history as a student was brilliant in some areas and only passing in others, he graduated from one of the most prestigious German universities, the University of Berlin, completing a dissertation on the design theory and experimental basis of the liquid fuel rocket. His work was immediately classified as "secret" within the German ordnance system. He was said to be an unusual man "caught between his dreams of the heavens and the earthbound realities of life." He had subscribed to Lutheranism most of his early life but was "born again" in El Paso, Texas, near the White Sands Proving Grounds where he lived and worked on the German based missiles before being transferred to Alabama. He eventually became director of the NASA Marshall Space Flight Center in Huntsville, Alabama, responsible for incredible new technology and for sending over twenty astronauts to the moon. He died of cancer near Washington, DC in 1977 at the age of 65. Einstein had died in Princeton at the age of 76. As a Jew, he fled to the United States in the mid-1930s well before war broke out in Europe, whereas von Braun was captured and brought to America only after World War II hostilities ended in 1945.

The contrast in religious feeling between these two powerful men is particularly interesting, although difficult to analyze with any depth or precision. The differences were far more palpable in their personal presence than in their biographical accounts. Einstein was exposed to mandatory religious education in Munich, but his exceptional mind of reason could never accept the existence of a humanlike god. The religious training to which Braun was exposed, on the other hand, made a deep spiritual impression and the concept of a humanlike superior being and immortality of the soul penetrated his mind, never again to be questioned. It seems that the notion of an anthropomorphic god never seriously entered the Einstein mind, whereas it never left the von Braun mind despite his ability to exercise logic of the most complex technology. The Einstein mind was deep, quiet and unemotional, whereas the von Braun mind was highly charged with ambition and movement.

In a sense there was a connection between the scientists. Albert Einstein modified and clarified the principles of space-time relating to space vehicles and high-speed travel. He not only defined formerly obscure natural properties of the universe but also solved the "god problem," whereas Wernher von Braun, who in effect invented space travel, did not appreciably deviate from his "hurried" goal of vehicle development and space exploration. He clearly lived an emotional life of traditional belief, eminently satisfied with the status of his all powerful protective god who he felt provided him satisfaction, forgiveness, happiness, and potential immortality. Despite his deep religious belief, he seemed to adjust to the domination and arbitrary use of oppressed and starved workers to maximize his own status and that of Germany. These two men indeed had minds oppositely "wired" in terms of spiritual belief, and both of them openly and frankly expressed their private feelings as though there were no alternatives.

**Nonfactual thought is subject to change:** Consider a time in the future when imaginative spiritual thought will be far better understood and its lack of reality more widely recognized by much of society. Religion will be seen in a totally different light. Like Einstein, fully educated man will eventually understand the reality of his cosmic position independent of the vestigial purveyors of imaginative ideals and power. Advancing human knowledge and logical thought processes will eventually prevail, leading to general reinterpretation of past ages. Whatever is factually determined by science, society will remain firm in the growing "bank" of human knowledge. While religious history will continue to exist, factual history will increasingly become separated from it. Most history is subject to exaggeration and expansion as it is repeatedly written. Only the records of science will remain invariant, subject to updating and new discovery. Biblical history, as all history, will gradually be reinterpreted in terms of modern knowledge and meaning. There will be an increased tendency to move forward rather than backward as realism begins to build throughout the globalizing planet. Man will have respect and appreciation for the transcendent power of the natural universe as did Einstein and others, knowing that it will probably never be completely understood by man.

One can assume that the era of highly emotional religion such as that of von Braun will gradually disappear together with the many similar concepts of the distant past that have been retained and often distorted over time. It is not that life will return to the pre-Christian era of paganism and idol worship, but the social mind will have absorbed the facts of life and death as they represent reality. The subtle fear of disaster and death may linger but acceptance of scientifically determined fact will ultimately reign.

**New era of biology:** The era of physics is fading and a new era of emphasis on biology is well on its way. Freeman Dyson (176), a recognized world futurist, suggests that even children will play with elements of genetic science and will therefore be more fully informed and comfortable with reality. He predicted that an era of "molecular biology" will replace the era of the "computer." Future children will mature to better understand the workings of nature and they will gradually lose much of their inherited animal fears. They will understand and appreciate both life and death and will become more knowledgeable about origins and evolution. It is ironical that human progress may well have been deterred by long retained and promoted religious notions, by limiting the comprehension and acceptance of reality when its underlying rational structure is "designed" to know better. But it is well known that even the present somewhat capricious and unsure mind of man has frequently allowed unacceptable behavior while being fully aware of possible dire consequences.

Einsteinian religion has not proliferated in society by its name, despite its compatibility with advancing human knowledge. First, imaginative religion is taught to the young organized believers as a priority and deeply absorbed. Second, numerous institutions and edifices have arisen in connection with imaginative religions that continue to draw people to participate in stimulating rites. Third, no organized framework of celebration was offered by those who subscribe to realism with the possible exception of Humanism described earlier. As faith in imaginative belief is slowly replaced by secularism more organization and formality may be introduced to more fully satisfy human emotions and interests more like that to which people are accustomed in existing organized religions. In formal Humanism, life events such as births, marriage, and deaths are sometimes celebrated with appropriate ceremonies, and appropriate music as for traditional religions.

**Causal basis of thought and behavior:** Science and religion were described earlier as separate disciplines, but the methods of science are effectively applied to the understanding of the surrounding world and to the complex mind of man, and even to his belief systems. Mechanisms of interpreting religious thought were included in the introductory chapter to set the principal course of this book. Scientists and philosophers like Einstein have frequently suggested that there is a high degree of determinism in the human mind structure since every thought is the product of another neural action derived from memory or from real time. Many therefore believe that there is a causal property in all thought and no overt exercise of free will. This idea has been argued extensively over the centuries and only recently have neuroscientists come to the conclusion that pure free will does not exist (177). Humans are thought to be subject to causal laws as in the rest of nature. Laboratory measurement made of brain function recently showed that neural electrical activity takes place well before a conscious decision is made to move an arm. However scientists argue the case for determinism, others find human actions to have a high degree of freedom and responsibility. The human brain is very complex and involves a great many interconnections. The information it contains as the result of learning and experience clearly influences belief, decision making, and behavior. The degree of "freedom" therefore becomes specific to each brain. This book has suggested that the powerful indefinable transcendent forces of nature constantly act on the universe and everything in it. Among its effects is the changing and developing character of man who retains many primitive emotions. This study has emphasized his greater pursuit of factual knowledge and interest in more peaceful coexistence as significant factors inherent in behavioral objectives and will therefore be implicated in his ultimate direction. Enhanced power of reason and behavioral concerns are expected to result in widespread religious indifference and secularism as well as gradually improved conditions of coexistence, all influenced by the observed trickle-down application of those natural forces.

**Judging new knowledge:** Over the past century, as understanding of theoretical physics was substantially advanced by major contributions of extraordinary scientists such as Einstein, Bohr, Heisenberg, Born, Wheeler, Feynman, Hawking, Mach, Millikan, Pauli, Sagan, Weyl, and Schrodinger, knowledge about his world took several giant steps. Interpretation of religion more recently took an equally giant step. Physicists and theologians entered a period of interaction that could not be matched by any prior period. Two particular phases of science entered the picture. They were the theory of the big bang and the theory of quantum mechanics. The Einstein theory of relativity also played an important connecting role. Up to now, most of the discussions were carried on among scientists and intellectuals. However, the general world population is now beginning to feel the effects of improved education and more of the population is capable of subtle understanding of the physical principles and conclusions involved. The pursuit of scientific principles has in turn improved reason and judgment, and has clarified the bases of both religious belief and lack thereof. Advancing secularism is largely the consequence of that expanding judgmental process.

**Einstein spiritual views:** Well before the decade of the brain prior to the end of the twentieth century, scientists and intellectuals were already anticipating changes in religious thought based on scientific advances and new knowledge. Einstein summarized the origin and practice of religion as he saw it, even in the mid-twentieth century, having become more deeply involved in response to streams of questions from prominent religious leaders. His frequent talks and publications appeared to force him to establish a firm public expression of his own belief and make his concept of cosmic religion quite transparent. He often reviewed the processes of religious development and frequently talked about the purpose of life, referring to the ideals of kindness, beauty, and truth, about political ideals and democratic principles without which he felt his life would be "empty." He did not believe in absolute human freedom in the philosophical sense, saying that man acts largely under external compulsion and inner necessity with "knowledge of the existence of something that one cannot penetrate, perceptions of profound reason and radiant beauty in only their most primitive forms accessible to our minds." He felt that it was this important emotion that constituted true religiosity. Again, it is what many people mean when they now say they are "not religious but spiritual." In that sense, he claimed that he was deeply spiritual but he could not conceive of a god who rewards and punishes his creatures or one who has a will of the kind that man experiences himself.

His concept of God did not involve any humanlike image, suggesting that a god could only be conceived "through rationality and intelligibility of the world that lies behind science." He confirmed the notion that it was human emotion that originally led to religious thought and beliefs and that it was primarily fear that initially evoked religious feeling, fear of hunger, fear of wild beasts, fear of sickness and fear of death. He felt that understanding of causal connections was poorly developed in early man and therefore, the mind created imaginary images on which his life would depend. As is shown in history, to appease these god-like beings they were frequently offered sacrifices.

**Fear and desire but retaining reality:** Professor Einstein accordingly felt that the "first stage" of religiosity was deep, primitive fear, the most basic of inherited emotions. His "second stage" was the social and moral concept of god that arises from the desire for guidance, love, and support. Man's god became an imaginative figure who rewarded and punished, who comforted in times of distress and who preserved the souls of the dead. Man chose the concept of an anthropomorphic, a humanlike god, a most natural form for he could best deal with the human characteristic he knew. The "third stage" of Einstein's explanation of religion was the development of a cosmic religious feeling that is very difficult to impart to anyone else without basically feeling it oneself. He spoke of the sublimity and order that were revealed in nature and in the world of thought. He spoke of individuals feeling the futility of human desires and aims. He spoke of existence as a sort of prison and the desire for an expression of outside experiences. He recognized that people with cosmic religious ideology may be regarded by contemporaries as atheists, but on the contrary, were filled with the highest kind of religious feeling, despite having no specific church whose teachings are based on it.

If cosmic religion lacks any organized authority and visual imaginaries, how can it be propagated and taught? He felt that it was there that religion and science must meet. He suggested that the relationship between religion and science, if properly conceived and defined in the form of cosmic religion, would no longer be that of irreconcilable antagonism. He further suggested that cosmic religion is the strongest and noblest driving force behind scientific research and human progress. He went so far as to state that, in principle, a man's actions are determined by necessity so that, in his god's eyes, he cannot be responsible any more than an inanimate object is responsible for the motion it undergoes.

If one has not had a complete education that included an understanding of natural laws of the universe, the formation of life on planet Earth and the emergence of human form from animal ancestors, information contained in this book, may not be understood or appreciated. Human knowledge has finally reached a point at which general understanding of the human position is accompanied by few doubts. Persons who resist true understanding of the universe, the origin of life, and its development as determined by measurement and interpolation may well become permanently vulnerable to imaginative concepts.

**Clashing views:** Differences in belief often lead to a violent clash of personalities. As an example, the Dawkins book, *The God Delusion* previously mentioned, was followed by a counter book written by Keith Ward titled *Why There Almost Certainly is a God* (178). Ward claimed that he had been appointed to a most highly respected position of senior professor at Oxford, the Regius Professor, a theological position which Dawkins refused to acknowledge as being meaningful. In effect,

Dawkins seemed to feel that Ward knew nothing of theology since it was a subject which he did not even recognize to exist and certainly nothing about evolutionary biology, much of the basis for his own science. These two opposing minds appeared to become entangled by ego and politics and the fundamental question of the existence of a god was somewhat lost in the clash. The Ward book contained no convincing argument that a traditional humanlike god existed, not even "almost certainly." The true basis for the claimed existence of such god by man was clearly explained in the early chapters of this book, leaving little logical basis for an effective counter argument.

Humanism similar to cosmic and natural religion: It was previously indicated that the natural or cosmic religion of Einstein was essentially the same as that now being promoted under Humanism mentioned in earlier chapters. The principle of non-belief in the existence of a humanlike god is certainly not confined to persons of high intellect, philosophy and science. There are now millions of persons represented in all walks of life whose knowledge and reason simply turn them away from imaginative belief. They therefore regard themselves as atheist or agnostics. Unlike the publicity involved in private feelings of prominent personalities, most nonbelievers simply conduct quiet lives without public membership to any religious institution. This book has described the "discomfort" of atheists in the form of stigma in a very religious society as might be expected. It is for that reason that several books and articles have appeared in recent times declaring a more extensive "age of new atheism" than was earlier apparent to the average public. The increased formalization of humanism is taking place for similar reasons.

The American Humanist Association points out that there are several different categories of humanism. There is Literary Humanism, devoted to the humanities. There is Cultural Humanism, which refers to the tradition of Rome and Greece that evolved through Europe and now constitutes the Western approach to science, politics, ethics and law. Philosophical Humanism is a way of life based on human needs and interest that includes Christian Humanism, advocates Christian Principles, and Modern Humanism, which rejects all supernaturalism and relies on reason and science. Subgroups of Modern Humanism are Secular Humanism and Religious Humanism. Secular groups such as the Council for Democratic and Secular Humanism and the American Rationalist Federation include many academics, scientists, and philosophers. Accordingly they are known as Secular Humanists. Religious Humanists emerged out of Ethical Culture and include the church of Unitarianism and Universalism. Hopefully, the humanistic principle is basic and will not fall victim to a multitude of unnecessary conflicting interpretations such as developed over earlier centuries in traditional world religions. Despite the several descriptions, the principle of humanism tends to encourage secular unity in accordance with the expectations of this study.

**Freedom from belief:** At Harvard University, Chaplain Greg Epstein (10) is currently a member of the executive committee of the thirty-eight-member interfaith corps of Harvard Chaplains. He said, "I have zero belief in God, goddesses, or any other manner of supernatural spirits. My conviction that this life is all I have, however, is precisely why I don't want to spend my days focused on the worst in religion. I prefer seeking the best in each of us. I am not an antitheist or simply an atheist, but a Humanist." There are currently fledgling humanist Chaplaincies at Columbia and Cornell and probably at other universities. The Harvard Chaplaincy was founded by a former Catholic priest and is said to be dedicated to building, educating, and nurturing a diverse community of humanists, agnostics, atheists, and the nonreligious at Harvard and beyond, seeking to advance dialogue among religions, cultures, and civilizations.

Although there are many members of society whose beliefs may be classified as humanist, atheist, nontheist, cosmic and naturalist (219), it is still a relatively small number worldwide compared with believers associated with traditional religious groups. Because of the principles set forth in this writing, there is little doubt about the continuous growth of the nontraditional beliefs. There is no currently valid poll of how the world stands, nor the rate of conversion. To my knowledge there is no active proselytizing effort by humanists equivalent of the missionaries of traditional religion, for it is generally understood that the imagined gods of traditional religions have ordered membership to "go forth and teach." Conversion to realism, however, must depend largely on the results of factual education and rational decision. There is little doubt that that is the direction of the forces that drive human life as described in chapter one. In my fictional story The Man Who Created God (15), the protagonist sent his students forth following an agenda typical of that of traditional religions in certain sectors of the world and thereby accomplished his objective in a relatively short time. The natural drift to secularization in the real world however is related to cultural and evolutionary processes, processes that depend on the effects of complex stages of interaction well known by psychologists, anthropologists, and neuroscientists familiar with mammalian development and function. Belief must gradually be separated from cultures and separate cultures must make an effort to understand the need to become compatible with other cultures based on common objectives.

Although there are differences in humanist titles and definitions, all humanists appear to share the same worldview and the same basic principles. It makes little difference whether humanism is called a religion or secular. They all reject the concept of a humanlike god and the supernatural as well as an afterlife. All embrace the methods of modern science including the theory of evolution. Although most traditional religions claim to believe in the teachings of science, they relate them to an anthropomorphic god and to notions associated with imaginism. Over time, the detail of such beliefs has been subject to changing interpretations by changing leaders. For purposes of simplicity in this book, it was necessary to define religious belief as belonging to one of only two clear-cut forms of belief systems.

Humanism is described in much more detail on the Internet and in applicable literature published by its various groups. A general description of humanistic philosophy claims that humanism is for the "here and now." It is a philosophy focused upon means for comprehending reality, a philosophy of the pursuit of knowledge through reason and science. It is a philosophy of compassion and human ethics committed to civil liberties, human rights, separation of church and state, the pursuit of democracy in government, in the workplace, and in education. As earlier indicated, some humanist organizations encourage the same rites of passage as are found in other established religions and its descriptions are consistent with the theme of this book.

**The same except learning:** In closing this chapter on famous contrasting minds, emphasis might be directed, not so much to "differences" but to "likenesses" in human life. All human structures have a common form of makeup and all are commonly subject either to a god of imaginism or to a mechanistic natural force of realism. All are emerged from the same ancestral animals. There are also many people on the globe who are just too concerned with basic survival to contemplate the formality of any belief system. They have no material things and no hope or power of progression. However, the beginning and end of human life is fundamentally the same for everyone as often emphasized in these pages. Man is currently passing through an age in which he is still uncertain about many aspects of his existence, about his origin and about his larger world. Accordingly, he is inclined to be taught differently, his mind can therefore react quite differently, and his learning and memory can be oppositely registered. It is not that his god or his world is different, but the information acquired is different and inexorably mixed with many other practical inputs. The basic belief and behavior is similar for all, but in a sense have been randomly and involuntarily distributed. Although religious difference is just another difference among man, it has been one of the most troublesome as has been evidenced over past millennia. It is a difference that cannot reduce or eliminate evil or misbehavior and he must look elsewhere to a very much broader unity for its conquest.

Life clearly has only one interpretation and, in the course of time, it will be interpreted very much the same by all. It will be based on future more complete and more accurate discoveries and determinations, on the sound convictions of new minds. It will be based, not on the idealism that the mind desires and can imagine but on what the mind can detect and measure with its given senses. It is therefore most important that the best informed and visionary minds of man continue to contribute to discovery and interpretation of nature, to factual understanding and to expanded education throughout the world population.

**The mechanistic universe is not humanlike:** In closing this chapter I must again emphasize the fondest hope of our greatest contributing men of history, that it is indeed time that society is freed from emotional and imaginative spiritual uncertainty, from the induced spell inflicted by the desire for a different kind of more ideal world, what it can readily imagine, quite different from what it measures and observes. Man got started on the wrong spiritual track. It is time for man to find his interpretation of existence in reality, to responsibly consider the plight of fellow man on Earth, to adopt for his own use the benevolent character now attributed to his imagined gods in order to finally effect settlement of differences and conflicts. Man must gradually become aware that life can and does exist perfectly well without imagined gods. It is precisely the same. He must become aware of the magnitude of his natural privilege and responsibility to serve among fellow man as a temporary occupant of the planet. The observed mechanistic universe cannot represent humanlike character for it has no capability of emotion, mind, judgment, or response as are attributed to a humanlike god. Despite the existence of many belief systems, this underlying fact has long been generally accepted and understood by the most educated humans. The character of human behavior must be disconnected from an imagined god and responsibly applied to the satisfaction of human needs.

# Chapter Seven

## Vulnerability and Behavior

"Circumstantial evidence is a very tricky thing," answered Holmes thoughtfully, "It may seem to point very straight to one thing, but if you shift your own point of view a little, you may find it pointing in an equally uncompromising manner to something entirely different."

—Sir Arthur Conan Doyle

It is fate that there are wars and that one part of mankind has to be ruled by another. It is fate that the amount of suffering can never be less than it always has been. Fate may be rationalized philosophically as natural law or as destiny of man, religiously as the will of the Lord, ethically as duty—for the authoritarian character it is always a higher power outside of the individual, toward which the individual can do nothing but submit. The authoritarian character worships the past. What has been will always be. To wish or to work for something that has not yet been before is crime or madness.

—Erich Fromm, *Escape From Freedom*, Rinehart

Man was made for joy and woe,
And when this we rightly know
Through the world we safely go.
—William Blake, mystic and visionary, 1757-1827

**Behavior a function of the educated mind:** The human mind, exclusively and directly responsible for personal behavior, still has many inherent weaknesses, but the concept of a behavioral relationship with an imaginative monitoring god is beyond the biological sense of modern human reasoning. The attempt to control behavior through spiritual fear is becoming increasingly unsuccessful and it is time for society to take a giant step in its redevelopment with cultural advancement. Education and encouragement of moral responsibility at an early age are essential, but it is equally important to teach the elements of alternative realism before the mind is otherwise exclusively wired. Misbehavior with respect to such matters as drugs, material riches, sexual desire, political power, and corruption and conquest is rampant and largely out of control worldwide, but the spread of religious fear to curb such behavior has been generally ineffective and by its various diversions and deflections might have actually contributed to increased world problems. The excesses and extremes in religious behavior itself now must also be added to the list of behavioral problems. Human behavior is an exceedingly complex issue and must be subjected to proper education relating to more widespread established rules, laws, and accepted practices of society. The discipline of proper behavior should relate to inherent human makeup and not necessarily to the teachings of traditional religion and idealism. It should become an essential part of mandated formal education worldwide.

**God as a drug:** Among the discussions of diverse behavior in this book, little or nothing has been said about abnormal religious behavior and extremism. As explained at the outset, religious belief is a function of the innermost mind and is therefore intrinsically related to many other mind functions. Without raising profound psychiatric issues at this point, it is important to recognize the existence of "religious addiction," a little publicized problem that has appeared from time to time during periods of active religious development. It is a consequence of over-stimulation or over-reaction of the religious mind that can affect individuals or groups in society and might well account for some of the human behavioral problems mentioned in earlier chapters. An entire religion or cult can readily be thrust in some antisocial direction. A new book, *When God Becomes a Drug*, by Father Leo Booth (77), provides an outstanding review of how religious addiction occurs and how it may be dealt with.

The human brain is not only biological in structure, it is also chemical in function. It is further, a form of "electrical" system as are computers and data systems and it can become subject to "overload" or to circuit malfunction. There are many known addictive influences that directly or indirectly modify the human mind. Prime sources can be verbal, sensory, or chemical in origin, directly or indirectly for all emotions have a characteristic chemical component. Imaginative belief by its very nature can reach well beyond normal limits, resulting in extremes of behavior and in abnormal interaction with other members of society. Drugs and alcohol often stimulate or alter the sensitive and vulnerable mind to the extent that its function may become dangerous and cause unpredictable and uncontrollable behavior. Since numerous stimulants tend to act on the "pleasure center" and other behavioral centers of the mind, they can create a strong desire for repetitive continuity. This process is confirmed in animal experiments as witnessed by most students of modern psychology. Substances such as nicotine, amphetamines, cocaine, heroin, coffee, marijuana, and various opiates are particularly well known for their common usage and for their biological influences on critical areas of the brain. Abnormal dependence and expectations with respect to religious belief are sometimes related to such chemical stimulus. The relationship between drugs and religious belief is discussed more specifically by John McGraw in his book, Brain and Belief (44).

**Man seeks something more:** A February 2009 New York Times article emphasized the high cost to American society of the use of methamphetamines alone. The cost of abuse was said to approach $24 billion, not counting associated intangibles such as crime, criminal justice, children put in foster care, reduced productivity; drug treatment, health care, injuries and fatalities, and on and on. The rate of use was said to be higher than that of heroin but still only half that of cocaine. Worldwide, the magnitude of proliferated drug use can only be imagined. Whatever the quantitative usage of such chemical substances intended to escape reality; society should not knowingly, add imaginative spiritual stimulation as another major behavioral factor. Although there are few clear-cut treatments for spiritual excesses, religious education should instead emphasize factual reality without pretension.

**Forcing belief on others:** Man has learned a great deal in the psychology laboratories in recent years but his knowledge has not yet sufficiently reached the public mind. If effort to teach behavior based on religious fear were, instead, directed toward teaching its full physiological meaning, there would be noticeable returns. In addition to the effects of ingested chemistry; people also become addicted to common forms of social stimulus, including gambling, sex, shopping, and dieting, all of which can cause uncontrolled obsessive desire for increased satisfaction (179) and further sources of problems. With respect to "religious addiction" however, the sensitive and vulnerable mind can develop a deep unnatural response related to one's imaginative god and its idealized surrounding framework. Human imagination knows no limits and can readily exceed reason and good judgment. It becomes a more serious problem when extreme elements of religious organizations and sects take it upon themselves to force their beliefs on other members of society, often with violence.

History has recorded many cases of abnormal excesses in beliefs and has sometimes associated them with "possession" by various spirits and with attempted remedial processes called exorcism. Society has not yet progressed to the point where belief in such spirits is effectively explained and fully eliminated. Ironically, many persons afflicted with religious addiction are not criticized or offered appropriate therapy, but are often admired for their "religious zeal." Society as a whole, is still reluctant to respond effectively to protect victims of extreme religious behavior. Despite observed suffering, agencies cite individual rights, lack of authority, violation of sovereign nations, and religious rights and little is accomplished. This book has frequently suggested that long-term preventive measures are best instituted in the form of factual education and in the publication of more rational critical analysis of imaginative belief throughout all of society. The intervention by central world authorities should be considered.

Although the writings of Father Booth have made a worthwhile contribution to many victims of religious addiction, he made it clear that he remains a firm traditional believer. But it is not exactly clear just how he views these activities in relationship to his concept of a benevolent god. Earlier chapters of this book explained how deeply religious promises can penetrate the emotional mind without limit and with little likelihood of modification or neural escape. If society were not so frequently, so dogmatically, and so strongly subjected to ideal spiritual promises without appropriate disclaimers, problems of this type might not so easily reach a level of serious social threat. It is not clear if excesses in religious teaching assume the role of stimulant or placebo. The modern trend toward secularism is however, expected to gradually lessen instances of such addictive and extreme behavior.

**The human drug issue:** Today problems begin at an increasingly younger age unprotected by maturity and experience. The Associated Press reported a case of freedom of speech (180) that arose because a high school student in Alaska apparently invoked the "power of Jesus" to support the use of drugs displaying a large banner with the coded words "Bong Hits 4 Jesus." In effect, it suggested that the omniscient and benevolent god, about which he learned in his early training, would condone the use of drugs. Action of the school principal, prohibiting its display, was eventually upheld by the Supreme Court with Chief Justice Roberts writing a 5-4 ruling. Apart from the drug issue itself, it is clear that young people can become grossly mislead by the imaginative concept of a powerful humanlike god, leading to deflection of essential balancing information in the early learning processes. The lack of spiritual disclaimers and unnecessary high court involvement are indications of social disorganization that should be given appropriate attention.

Last year, Washington announced a $1.4 billion assistance package for Mexico and Central America. The budget proposal for 2009 for the prevention and treatment of abuse was $4.9 billion. The Colombian government has been given more than $5 billion since the year 2000. Thousands of police and civilians have died in Latin America fighting traffickers yet, with all the blood, tears, and cash, there has been virtually no impact on the amount of drugs used in the United States. The report warned that there was little hope of ever defeating the traffickers abroad if government is not effective in reducing demand at home. As implied earlier in this writing, the human desire for stimulus seems to be insatiable and routine education must offer means of appreciating life without resorting to imaginative religions that are increasingly becoming ineffective. Drugs and alcohol are very "available" and bars are well attended most every day in most developed countries. Indeed, religion has not contributed much to resolve the inherent problems of human behavior and may even be an errant stimulus.

**Drugs, emotions, and gods are all chemistry:** The illegal poppy growers in Afghanistan do not seem to be able to keep up with the world demand. Is the demand for mind-altering substances related to the demand for imaginative beliefs? Will alcohol and drug abuse eventually get completely out of hand in society? Is there an addiction mechanism in the human mind that feeds the demand for imaginative belief as it does for offending substances? Will fading religious belief lead to more alternative drug use as the imaginative nature of religions is increasingly discovered and acknowledged? Will the opiate effect of the emotional chemistry be completely replaced by medicinal chemistry? Drugs have been the subject of concern for centuries, but how will usage change with further human development? (181) As one surveys "Christian America," one wonders why drugs are pouring non-stop into the country and why so much money is available for their purchase in preference to other needs. It may not be a phenomenon related to unsatisfied spiritual expectations but completely biological in nature. Man apparently requires more than religion at times to escape from reality. Is it not time for modern man to employ his full capability to better understand and adequately resolve both his religious and drug needs?

As the findings of this study are more thoroughly digested and better understood by more widespread society, the problems of addiction might eventually wane on the path toward secularism. Although they might be replaced by other problems that involve similar mechanisms, mind function is becoming increasingly understood and appropriate remedial therapy is more generally available. The human mind is an amazingly sensitive mechanism in its response to stimulus which, in the final analysis, we have already agreed is a matter of mind chemistry no matter how it is induced, by learning or by drugs. Again, appropriate education and training remain as the most effective means for social improvement. Perhaps education should more effectively consume the time and effort now applied to religion.

Psychologists have extensively studied animal behavior in order to ascertain mechanisms of parallel human behavior under specific conditions. The last century has accordingly produced far greater understanding of human behavior than in any prior century. Using standard test procedures involving sound and electrical shock for various tasks, response to stimulus has been placed under increasingly more accurate observation. In general, an animal will respond repeatedly and reliably only if it can reason with respect to cause and effect (182). On the other hand, if it becomes confused and disabled, it may revert to a state of "helplessness." This is simply a way of nature, the mechanism for which is contained in the present neural system, a mechanism that may be subject to long-term change with adaptive encouragement of improved understanding and behavior. Humans can also become confused when cause and effect are not understood. Belief in a controlling god has been an easy route for avoiding helplessness with its overly simplistic explanation of origin and purpose. However, the increasingly recognized question of reality is expected to contribute to long-term ineffectiveness.

**Behavior in the next millennium:** Whatever the nature of belief, the behavior of society is a basic and fundamental issue in the globalizing world. Neither imaginism nor realism can, on their own, convert the "animal derived" human mind to perfect behavior within present view. "Existentialism," a movement of the twentieth century, was centered on the analysis of "individual existence in an unfathomable universe," existence that represents responsibility for assumed acts of free will without certain knowledge of what is right or wrong, good or bad. It attempted to define the newly determined position of man in simple language. However, the matter of right, wrong, good, or bad is indeed now determinable in society by the rational application of judgment of the further developed mind with respect to experience and observation. Behavioral judgment, however, is constantly changing with time and with ongoing cultural development. Despite knowledge and experience, imaginism has only moderated behavior because of the fear of "sinning" against the wishes of an assumed god (153). Unfortunately it tends to be temporary and intermittent and more recently less recognized and therefore more unreliable. Modern views of moral and ethical behavior have also been increasingly integrated among peoples of the planet. Hopefully, the current processes of greater communication and globalization will tend to normalize social practices and eventually result in a more acceptable common basis for everyday rules of behavior and for more uniform and formal terms of worldwide law and order. Again, this book claims that behavior must first be removed from the imaginative concepts of religion and brought to a common, more realistic understanding.

**The separation issue:** The preceding chapter dealt with the contrast in the neural wiring of two strong minds. An article of related interest concerned the mixture of "piety and politics" exhibited by a member of the Kentucky legislature. It described a Democratic representative who served for over a quarter of a century, recognizing no barrier between church and state. He was said to look for God everywhere, and in places where he did not find him, he tried "to put him there." He was determined to preserve the historical perception of "a nation under God," claiming that the "safety and security of the Commonwealth cannot be achieved apart from reliance upon almighty God." Constitutionality seemed to be in question here but, convinced of his sincerity, other politicians were afraid of using attack ads that would suggest they voted "against God" if they voted against his religious measures. This vignette is the epitome of current reality with respect to the depth of imaginative belief that remains as the result of past missionary efforts. It is also a revelation of how legislative bodies can sometimes function in practice. Spiritual calling, a frequent reference to an imagined direct influence by God, is nothing more than the response to an emotion as described in earlier chapters. It simply confirms the suggestion that the process of eventual spiritual unity will take a large number of generations, even under carefully established conditions of factual education.

**Change begets change:** Changes in all living things are increasingly under constant close scrutiny and much has been determined about the processes of life only in the past half century. The publications of science frequently record the disappearances of species, never to return to earthly life again despite natural adaptability. It may well happen to human life. This thought again emphasizes the fragility of biological life and its sensitivity to fundamental natural mechanistic forces. In the long-term process of continuing evolution, behavioral relationships can therefore be most important. Man is now reaching a point in his development at which he can influence his own natural evolution to a greater degree than was possible in the past. He knows and understands more about his factual existence and his vulnerability to imagination and unrealistic interpretations of his natural environment. The causal basis of changes is far better understood and vulnerability should gradually lessen in intensity.

**Calvinism another theory:** In terms of "modern" trends in religious behavior, a recent news article (183) described the antics of an unusual preacher whose sermons were even "too racy to post on godtube." This story not only demonstrates the wide extremes of information that may be presented to a church congregation today, but also suggests a tendency toward radical and unpredictable social change. The ministry involved is said to have grown from a Bible study room to a megachurch of seven to eight thousand members, primarily because of the openness and perhaps, "looseness" of its sermons. The mainstream media calls it an "evangelical crackup," an erosion of the religious right. The church's success is said to be based on a new sensitivity to the "body-pierced and latte-drinking" seekers. It suggests that human beings are totally corrupted by original sin and predestined for heaven or hell no matter what their earthly conduct. Its philosophy is derived from the roots of John Calvin's sixteenth century doctrines. The church message is said to be raw, disconcerting and insensitive. The Calvinist theory holds that God has mapped out human lives and actions but that people are still to blame for their sins. Considered totally depraved, they are said to be held to the "impossible standard of divine law."

**Depravity is cool:** Critics predict that this unusual church, which thrives on paradox and on the delicate balance of opposites, cannot last very long. It is too bent on staying perpetually "hip." However, the pastor continues preaching to his seven congregations on most Sundays, and his sermons are said to be broadcast from its main campus to jumbo-size projection screens around the city. He continues to preach the doctrine of total human depravity in the special language of his "cool" congregation. This is an important example of vulnerability of current minds, how easily man can be influenced by a strong-willed preacher employing dramatic but relatively crude and vulgar language backed by his own interpretation of the "word of God." It is perhaps what some people want to hear, especially if it becomes available to them from the respected pulpit.

In today's social structure many very religious people are considered "good people" for, in terms of behavior, they recognize, if not adhere to principles of the Bible and to those principles taught by various churches. This does not mean that nonreligious people are not good. Many nonreligious people may in fact be more righteous and do more good than many who claim to be religious. However, a false association has unfortunately become part of the social pattern, as discussed in greater detail in chapter four on religious stigma. The natural instincts that govern behavior are not necessarily related to "good" people. Many people, in fact, want to be "bad" and some are bad much of the time.

**Righteousness its own reward:** The question of reward for being good is often raised. A book by authors Stephen Post and Jill Neimark (184), titled *Why Good Things Happen to Good People*, discusses why good people who are generous, loving and contributory in their nature, may live longer, healthier, and happier lives. Commenting on this book, psychologist Martin Seligman, who has long studied behavior, called it a "new science of genuine love" suggesting that its dictates would make the world a better place. The concept of extended longevity for religious and good persons has long been a subject of discussion. It seems to be a temporary psychological factor that may well be related to less stress and particularly less concern about death, looking forward optimistically to a beneficial afterlife. However, not all expectations of life after death turn out successfully. In more extreme belief resulting in suicide bombing, life and its potential vanish forever from the universe and from the imagined world beyond. The relationship between mind function and physiological effects is continually studied and is better understood decade by decade. An early paper at the National Institute of Health demonstrated this previously obscure effect very clearly. This book assumes that human life will eventually become fully adjusted to secular belief, that happiness and satisfaction are based on settlement of the mind to the true character and basis of life, and to the permanent elimination of uncertainty. It is in line with the positive direction of evolving man as a result of the continuous forces that change life mentioned earlier. Longevity is also said to be increasing with better medical understanding and care of human life and is expected to continue to gradually improve on that basis.

Despite more advanced scientific knowledge and understanding about the vast and somewhat frightful natural forces of the universe, man is indeed subject to its uncertain cosmic conditions and to the unpredictable secondary conditions that affect life on his own planet. All such natural effects remain the same irrespective of human belief. No supplication or wishes of man can possibly have any effect on its larger mechanistic function. There are many levels of human behavior in the course of life—personal, marital, community, national and global. All levels are influenced by nature but are controlled by man himself, and not by any humanlike god. As time moves onward, there will be far more human interaction, more common language, greater knowledge, and eventual common belief, and indeed, more human understanding of satisfaction and happiness than ever before. Its change is sorely needed in view of the increasing unsettled condition of clashing societies. There has been a major recognized change in human interaction within only one century because of increased globalization, travel and new electronic systems development. Languages, previously unintelligible, have already achieved a new level of interpretive understanding. Advances have been particularly notable in international negotiation and in trade. Knowledge is being more widely disseminated through daily exchanges, including scientific and medical findings. The significant changes in the more rapidly globalizing world are discussed more fully in chapter nine.

**Man determines omnipotence and omniscience:** With respect to international behavior, the concept of the United Nations, with its variety of subcommittees, has been a boon to advancing globalization. However, it represents only the beginning of much needed unity and world organization. It still seems relatively weak and powerless for the total mission at hand. It requires at least another order of magnitude of organization, support and cooperation to be effective in neutralizing the visible modern world. There is much to be done with respect to greater unity of functional civilizations and more fully effective peaceful existence of nations and, indeed, far more widespread correction of poverty, homelessness, and disease as the world continues to shrink. Because of its supreme importance, this subject was broached in chapter five and is again discussed in chapter nine. The forces of human advancement are strong and relentless and will in time force man to better serve man, hopefully through democratic world organization to better guide the processes that determine behavior, equality, and freedom. He must penetrate the weaknesses and unworkability of separatism of greed and questionable competitive governments, especially those that have no genuine sovereign or democratic basis. He must do so without the help of an imaginative humanlike god, without the expectation of a greater governing power higher than that of man himself.

**Nobody has ever been watching:** The majority of believers have been sold on the idea that their humanlike god is omnipotent, omniscient and omnipresent and is closely associated with goodness and righteousness. They are told that their conduct and behavior will be judged by such a god at some point and is a key factor in their entry into the immortal kingdom of heaven. This means that they must sense the existence of an invisible spiritual watchdog that functions in some indefinable manner from whose eyes they can never escape. It is a notion widely recognized by religious adherents without any depth of thought or analysis, one that has little rational basis. Surveys and analyses have generally shown that religious and nonreligious cross sections of society act about the same with respect to fundamental behavior and experience exactly the same final form of death. To assume there is anything beyond that is a primitive expectation.

Earlier chapters described how spiritual visions can develop so strongly that they become reality for the mind and resultant behavior may vary widely from careful compliance, indifference, and rebellion to unmanageable violence. Man is frequently determined to lay down his own life in defense of an imagined spirit. With respect to traditional religion, no more realistic example can be noted than the behavior of the crusades as "intolerance and mayhem" in the name of an imagined god. The sack of Constantinople was said to be an "unsurpassed crime against humanity." The confrontation over possession of the holy city of Jerusalem has also been suggested as legitimacy of violence in the service of religion. It was claimed that the crusade had anti-pagan interests as well as objection to universal Muslim conversion. The imaginative nature of traditional religions renders these struggles quite useless and without purpose. This book admonished man to consider modern events in these terms in order to rethink purpose and value to developing mankind. Man is once more facing a challenge marked by the sharp differences between Moslem and Christian beliefs as they penetrate cultures of the modern world. Hopefully, these chapters will influence the establishment of new viewpoints and different behavior, especially that involving the younger generations, the up and coming leaders and guiding management of the future world.

The world is replete with humans who break laws and traditions or make strenuous effort to avert them to their advantage. World function falls far from ideal behavior. Tighter control or more effective education and training is required. Voluntary adoption of affair and equitable human exchanges cannot be expected. In many cases those that preach goodness and ideal behavior in public behave immorally and greedily in private. They might even commit the most despicable crimes if it is to their advantage and if there is a high probability of doing so with impunity. Surprising behavior involving some of the most respected persons is disclosed every day. Even members of some of the well known crime organizations claim to be very religious church goers. As discussed earlier, some highly regarded members of religious organizations have been found to engage in unbecoming moral conduct and in illegal financial transactions.

**Behavior needs more than religion:** Business is rarely accomplished in international circles without substantial levels of bribery or beneficial exchanges. Even in the American Congress pork and similar allotments for private interests are understood as the way of life. Lobbyists have reached an unprecedented level of interchange with government in America resulting in frequent incidents of poor or unintended legislation. Similarly much legislation is voted along party lines to achieve maximum political benefit. The system is so well organized and protected that it is relatively immune to taxpayers and sometimes even to the High Court. Without a real god to protect them, it is easy to fall victim to natural instincts that tend to work against the most beneficial social standards.

**A blood disease:** Negative behavior has frequently involved a mixture of religious conflict and cultural prejudice. There has been no indication of authoritative divine leadership or inspiration that has contributed effectively to failed relations such as exist in the Middle Fast. Conflict seems to continue year after year and only intense diplomacy and difficult compromise can bring even temporary resolution. It is sometimes said to be "in the blood". Generations are evidently taught to carry on the conflict based on retaliation and vengeance. The Palestinian/Israeli conflict has been long and has deeply penetrated their respective minds. Solution is not easily available in the highly charged emotional minds of the region. A satisfactory solution may not even be possible under the circumstances and may require forceful cultural separation.

A recent news article (185) emphasized the difficult nature of the conflict when the last sacred annual pilgrimage to Mecca was in process. Movement was limited by the Israeli government. The Gazans were being deprived of performing their most basic religious duty in joining the religious pilgrimage sponsored by Saudi Arabia. Disorganization and conflict reigned throughout the area. Rockets were being lobbed, tunnels from Gaza into Israel and Egypt were discovered, random terrorism and raids were completely out of control. The sacred hajj was destined to be lacking many Palestinian Muslims. Alas, if only a small portion of the benevolence and omniscience that they attribute to their god had been effectively made available to his believers.

Despite the prediction of gradual drift to world secularism it was reported (186) that Tajikistan, recently becoming independent, has had a resurgence of Islamic religion. It had been part of the Soviet Union where religion and its public expression were banned under penalty of harsh punishment. It now might well be a backlash effect and an important new expression of freedom for a national society. Mosques have sprung up and there was a tenfold increase in the most recent hajj, the annual Muslim pilgrimage to Mecca. With the collapse of the state-run school system, families are expected to turn to religious schools for their children. Experts predict that religious leaders will gain disproportionate power in Tajikistan society and, with its economic crisis, could lead to a surge of fundamentalism or militancy. This is an example of the power and pace of cultural evolution, the spread of contagious religion in an undeveloped and vulnerable area. Like for many other areas of the developing world, this book predicts that the Tajikistan will enjoy improved education and more modern development but will be totally secular by the end of this century or sometime in the next.

Conflict among men has many sources and levels, but serious conflict is most generally due to an expression of strong feeling or intense belief (187). It may be rooted in culture, race, religion, mistreatment, nationalism or leadership. It has been said by many nonbelievers that religious organizations are often hypocritical with respect to behavior, that the ministry for some is a form of business. Money seems to be as important as saving souls among some religious groups. At times huge contributory congregations have been amassed and leaders made wealthy. Some churches have broadened their mission to include large real estate developments (188). The evangelical Oral Roberts University sported its own jet. The use of funds for private purposes had even led to the resignation of its president (189). The Cleveland dioceses were accused of embezzlement (190). The Internet is now being used for more effective collection of funds. There is a Christian Debt Trust that advertises loans specifically "to Christians". A federal faith-based medical mission was reported to have had problems of clearing its alleged misuse of assets (191). This list is various and endless, but sincerity in service to fellow man must always be highly respected. The achievement of secularism would tend to "purify" much of the behavior of man in these respects.

**Spinoza finally became an atheist:** In the religious world there have always been attempts to relate human behavior to the assumed desire of its god. Man seems to be able to establish the essence of divine thinking with a high degree of certainty. Because he created a humanlike god, its behavior is well understood. However, the humanlike notion has now outlived its purpose and usefulness in the more knowledgeable and rational modern world. In the seventeenth century, Spinoza attempted, through logic, to describe the properties of God in terms of human nature and emotion. He tried to define the way by which man should lead his life. Perfection was thought to be obtained when reason, in striving to understand, attains knowledge of nature which Spinoza related to the knowledge of God. But his god was essentially a god of nature as viewed by Einstein, who carefully followed the Spinoza philosophy. Torn by doubts and fears, Spinoza eventually became an atheist and was excommunicated. Despite having no god for behavioral guidance, he lived an exemplary life by seventeenth-century moral standards. As a philosopher who lived by reason, he was constantly concerned about the masses, who he felt reflected "erratic passions and fervent imagination" unchecked by reason. He looked to a civil religion that would guide them to rational behavior and thought that such civil religion should be under the control of the state rather than a church.

As described in an earlier chapter, Spinoza became fully committed to a life of contemplation after describing his concept of ethics, his effort to establish principles for achieving a perfect life through reason. His pious contemporaries found it difficult to reconcile his saintly life with his atheist beliefs. For Christian thinkers of the time, being a godless Jew and leading a morally excellent life was a dreadful contradiction. It was said that if there were no god in one's life, anything bad was possible. By seventeenth-century standards of morality, his only indulgences were said to be "smoking and wearing silver shoe buckles."

**Sex issues:** Many modern believers are convinced that, at some point in the past, God had made his desire clear to man. The Bible is held out as an expression of unverified evidence. A framework of behavioral standards had been created in the name of an imaginative god and presented to believers by various religious leaders. In the Catholic Church transgressions are called sins and there are different classes of sins, such as mortal and venial, with appropriate consequences. There are many differences between the nominal laws of churches and those of states and little effort to bring them together in reconciliation. The words of some editions of the Bible are often quoted in defense of a religious position. Laws of the state are based on modern interpretation of morality. The issue of homosexuality has become a major difference not only between church and state but among members of religious organizations themselves (192). This issue is gradually working its way through modern society and will surely reach a position of public understanding. Religious leaders have frequently become the victim of their own emotions and have themselves periodically engaged in illegal sexual relationships. Principles of abstinence have been promoted by many churches but have never worked effectively in practice and are considered a failure (193).

Every major city in the world contains areas in which prostitution is rampant. This has been the case throughout known human history. Love, marriage, and divorce are very personal matters, but a successful society requires a set of applicable rules and laws for adoption with great care and sensitivity. The latter part of the twentieth century has shown substantial change in attitudes about sex based on greater understanding and frank education. In more recent decades, new methods of contraception have provided some relief. Sex and money are often related. Money allows people to be more independent, adventurous, and more subject to experimentation. Because of its inherent emotional stimulation, the sale of most products to the public utilize as much sex-related advertising as is acceptable or tolerable. Moving pictures and stage plays follow those same tendencies. Words and conversation heard today would not be tolerated only a few decades ago. Sex and pornographic moving pictures are readily available everywhere, including the public Internet. Although prostitution is constantly and vigorously opposed by many segments of society, it is acceptable to others but legalized and effectively controlled in only a very limited number of countries. Although some increased understanding and spiritual concern may be anticipated with the advent of secular unity, it will probably long remain a social issue because of its very basic nature.

As the world becomes more and more global, the whole issue of sex is expected to become more normalized. The world is made up of many different religious groups and many different governments, but the issues of human sexual behavior are singular. Improvement in common standards of behavior must await greater unification of both. This book has suggested that the forces of change are constantly acting in a direction that favors more uniform understanding and man's best long-term interest. Beliefs in the power and control of a humanlike god that is thought to understand all the issues of man and to prescribe his behavior are still so strong and pervasive that several generations will be required for even reasonable resolution.

**Sex is for the educated and responsible:** The issue of sex has been both a social and religious issue for thousands of years and it is therefore truly surprising that it is not under more satisfactory management and control worldwide (47). The first lesson in sexual matters must be taught at the early preteen level. Until the time of the famous Kinsey report, sex was kept very much in the closet and little was said and known about its methods, practices and treatment. Today, we are living in a world of relative sexual freedom, making even more important to have frank education and established norms.

The first step in sexual education of children is to eliminate references to God and the Bible and to other religious documents and talk straight to children on a medical basis at the earliest possible time. A recent article (194) asks a question at what age to start talking to children about sex and AIDS. It answered, "How about, oh, 4?" It spoke of two precocious sisters ages 4 and 6, who accompanied their parents to an AIDS conference in Toronto. The interview involved AIDS experts, gay activists, condom distributors, a sex toy saleswoman, a cross-dresser and an Indian transgender. They asked pointed questions and even one of the world's leading scientists began to squirm when grilled by a child. The story is an interesting example of how difficult it is to get information about sex and sexually transmitted diseases across to children who have little knowledge or experience and yet need to know the realities of life. A comment made in the article was especially interesting. "I wouldn't tell a child there is no Santa Claus or why I am an atheist without a parent's permission."

Churches still play an important indirect role in sexual behavior since, although not specifically taught or controlled, there are some related rules and observations and it is encountered in specific religious education and in biblical study. Because of the uncertainty and limited descriptions and the personal complexity of the issue, it becomes highly confusing. In any case religious rules and prohibitions are just not observed in practice. All persons must be properly informed, beginning at an early age, that sexual desire is built into the human body very much as in animals and that certain respect and consideration must be observed in the practices of society. Different cultures have different views but hopefully, in the course of time, with accelerated globalization and new methods of communication, the entire matter will become more unified and resolved within widely accepted standards.

**Life is not the property of any god:** The Vatican recently issued new instruction for its Catholic membership (195) regarding bioethical issues because human understanding of biological processes has advanced appreciably in recent years. It essentially reinforced its opposition to "in vitro" fertilization, human cloning, genetic testing of embryos before implantation, and embryonic stem cell research. Its multi-page document, The Dignity of the Person, was issued by the Congregation for the Doctrine of the Faith and approved by Pope Benedict XVI. The new document, claiming that every human life is "sacred," bans the "morning after" pill, intrauterine devices, and specifically RU-486 medication, claiming that these techniques result in the equivalent of abortion. The instructions also ban surrogate motherhood as well as the freezing of embryos, since it exposes them to manipulation and raises the question of eventual disposal. They are claimed to give voice to those who have no voice but also that they protect life, claimed by Catholic doctrine to be the "property" of God. A uniform understanding based on medical knowledge in future worldwide secularism will eliminate confusion and misunderstanding. Sex and pregnancy are far more complex matters than is generally acknowledged and applicable advice should be based on capable professional training. They are human issues that are just too critical to be based on an imaginative deity. Life is indeed the property of man and man must handle it with the greatest care and understanding.

The position of the Catholic Church that babies should only be created by "intercourse of a married couple" is indeed, clear and suggests great oversimplification. However, realistically, the issue does not end there. However common is the processes of human reproduction in expanding society it must be subjected to careful medical considerations. There are frequent complications in fertility and serious intervention is often required. Sexual intercourse is a very common worldwide practice and cannot be made subject to varying narrow laws of prohibition. A recent editorial (196) spoke of the abstinence-only concept that has been widely promoted which is said to be at best a "delusion." It just does not work. As already claimed in these chapters, life is a very complex process that must be regulated by good biological and medical advice and understanding and, like the basis of belief, must be considered a private matter based on full and adequate education. Specific guiding rules and practices and reasonably established laws, independent of religious belief, should be formulated and published worldwide with respect to sexual behavior compatible with prevailing social norms.

**Embryo is parental responsibility:** There has been wide speculation as to the rights of an embryo. One must bear in mind that the nervous system that makes feelings and consciousness possible does not even exist at the time of fertilization and may not become active and operational until well into the second trimester, if then. Whatever the biologic structure, it cannot be successfully argued that a new life is truly in existence before that time. It is only potential life, exactly like the haploid elements of its makeup. The separate cells of sperm and ovum, the actual building blocks that make up a new person represent only potential life and are constantly subject to separate disposal in large quantities without special religious consideration. Fetal third trimester growth involves another consideration. The entire matter is indeed personal and should never be subject to church or state in primary consideration. Subsequent growth of the chemical combination to create a full life is little different than it is in plants or in other animals.

**The regular disposition of life:** In males, hundreds of billions of living sperm cells are produced at puberty and regularly disposed of. In females, living reproductive egg cells, present in much lesser quantity even at their birth, are also regularly subject to disposition. They are the major components of life. These cell forms contain only half of the normal number of chromosomes and genes, a normal quantity of 23 representing each parent. The fertilized egg combination therefore, involves a complement of 46. Although the normal body cells called "somatic cells" are freely manipulated, manipulation of the "germ cells" is considered to be the concern of all mankind. They have therefore become a religious issue, thought to be the creation of God apart from man. By a process of cell mitosis the fertilized cells eventually create about 100 trillion cells of the body, giving rise to brain cells, muscle cells, red blood cells, liver cells, bone cells, and so forth, first producing an embryo usually about eight weeks, and then a fetus achieving its structural plan usually from about three months to birth. Surely, the concern with life structure must not be left to an imaginative god.

**A fetus is not public property:** The claims of human interpretation of the mind of an imaginary God can lead to dangerous extremism. Some ardent believers tend to interfere with the lives and rights of others by actively opposing abortion. Again, pregnancy is a very serious medical matter and can result in various malformations of a fetus (197). In many cases the life of the mother is also at stake. One should bear in mind that because of evolutionary changes that the modern baby is large relative to the normal birth canal and a Caesarian section is often required. The mother in consultation with her medical adviser must be given the right of decision as to birth or abortion. Those that violently oppose abortion should be required to take a course on pregnancy and reproduction in order to become aware of its many serious complications. They should also be required to support any malformed or disabled product. Pregnancy is a totally private matter and not the concern of a third party. Upon due medical consideration the mother must have the undisputed right of choice in all these circumstances. Claims of the defense or protection of life of the unborn by some outside third party agent should be given no consideration whatsoever in law or in practice unless there is no other alternative. Overt interference with this inalienable right of parental control of human reproduction should be prosecuted to the fullest extent.

**Same-sex unions:** In recent years there has been renewed conversation about gays and lesbians and about same-sex marriage. Since such practices have been looked upon by many citizens with a jaundiced public eye, they have nevertheless continued in private. To use the vernacular, it had all been kept "in the closet." But that was the last century and times have changed. It has more recently found its place "out of the closet." Gays and lesbians are now part of accepting society, they have become organized at schools and in some churches and, with acceptability reasonably achieved, they have considerably expanded their rights. There are increasing numbers of same-sex couples who have chosen to live together openly and have mounted considerable pressure on states to recognize their "marriage" as one sanctioned by the state in a civil ceremony, accruing the same state benefits as afforded a heterogeneous marriage. The states are gradually succumbing to these natural forces. Whether such combinations are condoned or prohibited with reference to the Bible is no longer a matter of serious concern. It is simply the force of human instinct being expressed and it has become a force level that is too strong to be barred from accepted social practice any longer. Some distinction should be made however, between union of opposite sexes and same sexes. In an earlier writing I suggested that the definition of marriage of man and woman should be as always defined. However, same-sex unions affording similar civil rights might be called "marriaget or marriagette" in the case of women and "marrigent or marriagente" for men in order to more appropriately identify the form of relationships as distinct from traditional marriage. The term "same sex" has unfavorable connotations.

**The tolerance of difference:** Gay behavior has become an even more public consideration as it concerns the officers of the church. One of the most publicized stories at issue has been the pressure by leaders of the Anglican Communion (192) to refrain from establishing openly gay and lesbian bishops and to ban blessings of same-sex unions. As the world's third largest Christian denomination, made up of varying opinions, it has indeed, been a threat of great division. Dioceses have threatened to leave the Episcopal Church and proposals have been made to have parallel leadership structures. Of some 77 million Anglicans about two and a half million are member of the Episcopalian church. In the threats to alternatively align themselves with various sections of the faith there are many legal questions of authority and particularly questions of property and other asset ownership. This entire matter represents change and development of civilization as globalization brings different expanding modern minds together. The problems of human society are reflected in behavior subject to laws of all churches. The imaginative belief that a humanlike god has preordained behavior is gradually taking its toll. There is no compelling reason why the church hierarchy cannot be of the same form and have the same rights as its adherents. It is only a matter of how it desires to be organized. Most organizations are bound by a set of prepared by-laws in some agreeable manner. The Bible has taken a back seat on many practical issues in which inherent human behavior is naturally dominant and society continues to pursue the many sexual concerns on a modern medical and emotional basis rather than on a biblical basis.

The unification of thought in a future secularist world will finally disengage human behavior from biblical references and from the attempt by religious leaders to interpret them as they and their imagined gods see fit. This will allow man to pursue more important and more urgent current areas of society such as poverty, human rights, happiness and satisfaction in greater world freedom and equality.

**Responsible to self and society:** There are many basics that must still be taught universally, including the mechanisms of contraception, and abstinence where and when it is applicable and workable, recognizing the importance of avoiding sexually transmitted diseases and unwanted pregnancy. It is also important to have a good understanding of modern methods of fertilization and contraception that are now available. Lastly, for procreation good advice and experienced counseling is always highly desirable. Pastor Paul Wirth of the Relevant Church issued a challenge to his married congregation to engage in sex "every day for one month" to counter the increasing instances of incompatibility and divorce. The related article suggested a decline in divorce rate attributed to a rise in cohabitation, older ages of marriage, and a decline in marriage rate by thirty percent since the mid-80s.

**The celibacy ritual:** In his most recent visit to the United States, the Pope took the opportunity to comment on the sexual behavior of pedophile Catholic priests. Despite the teachings of the church and despite the concept of a monitoring god in heaven, Catholic priests have frequently been accused of sexually abusing young people who have acquired their trust and confidence in the course of frequent spiritual contact. Although the Pontiff expressed shame, the Church has admittedly spent billions in restitution and in legal fees throughout the country. It is certainly not a minor issue for it violates and renders valueless the declared principle of righteousness and propriety by trained and experienced members of the clergy sworn to the will of their god. It represents private and personal indifference to the god in which they believe. The National Review Board of Laity seemed to indicate that more than seven hundred priests were dismissed and prelates were admonished for protecting their charges. Five dioceses went bankrupt and payouts to victims totaled almost two and a half billion dollars. The apology has not been good enough for many victims (103).

**Celibacy is a fallible idea:** One woman, now fifty years old, said she was abused by her parish priest in Virginia from age nine to age eleven.

Her group exhibited a banner representing more than sixty children who were abused by priests explaining that fifteen faces were those that committed suicide. Such suicide would not presumably occur without a religious involvement. Irrespective of the psychological explanation of behavior and any defense that can be mustered under the circumstances, it is clear that it is certainly not the work of any benevolent God "testing his flock", but an attempt by the church to explain chastity and celibacy in its declared spiritual representation by priests. The issue of celibacy is an unnecessary ritual for any church in the modern world and a high risk to society because of strong natural temptation and the tendency to capture the hearts and minds of innocent young people. There is no more heinous crime than one that shuts out the trust of the young before life begins. The claim of a protective or benevolent god is certainly rendered false in the modern view. Church appreciation of human makeup and exercise of public responsibility is in serious question and the risk to society must be subject to legal consideration. There is public outcry about action to at least add pedophilia to the Code of Canon Law that specifies acts disqualifying priesthood. The Catholic Church must eliminate the threat of sexual abuse of children in order to support its imaginative framework of organized religion to which so many people have subscribed. The Pope said the church must "address the sin of abuse within a wider context. What does it mean to speak of child protection when pornography and violence can be viewed by so many homes through media widely available today?" Sexual missteps and pornography, like many other vices, are inherent in the characteristic mind of man and must be made part of the behavior standard of society with appropriate chastisement, penalties and punishment. The entire world must acquire firm understanding about the inherent character of sexual emotions and the constitution of reasonable human behavior and practices limited by rules and laws of modern society.

**A process of self-destruction:** In the March 22, 2009 issue of the New York Times Cardinal Edward Egan was said to have breathed life into the celibacy issue. Pope John Paul permitted the ordination of married former Episcopal priests who wished to convert to Catholicism, in 1980. About two hundred joined the Western church in America most of them married with families. At an earlier time the church had added about fifteen thousand deacons, ordained men who could be married and empowered to perform almost all the functions of priests except confessions and consecrating the Eucharist. It is said that the tradition of celibacy had its origin in the biblical portrayal of Jesus Christ as celibate and that the role of spiritual teacher required a single-minded dedication to the community. It was also said that as a practical matter marriage could result in claims on church property by priest's offspring. In the Eastern Rite churches, the Ukrainian and Melkite denominations recognized as fully Catholic by the Vatican, celibacy was never required. Married men can be ordained, although priests, once ordained cannot marry. In 2003 almost two hundred priests in a Midwestern Archdiocese petitioned the church to open discussions on the matter but they were rebuffed. Church closings because of priest shortages were clearly at issue. It seems to many that the matter of celibacy has become a somewhat arbitrary rule without meaning in the bigger picture and an issue that will eventually contribute to the process of fading religions as claimed in this book.

**Something amiss in the system:** A recent news article told the story of members of the church of St Francis Xavier Cabrini in Scituate, Massachusetts, guarding the building around the clock to prevent closure by the Catholic Church. Parishioners take turns sitting in the church for hours at a time, including overnight shifts in the sacristy. The vestibule serves as their living room. The church closing was part of a wave of closings around the country brought on by a shortage of priests and dwindling attendance and money problems. The archdiocese announced an eighty-five-million-dollar settlement with victims of abuse by priests in 2005 but proceeds from the sale of churches have fallen far short. Some parishioners have grown so disenchanted with the church hierarchy that they cannot imagine returning to tradition. One member indicated that she would always be a Catholic but may not be able to worship in the "mainstream Catholic Church." The Church is clearly going through a major metamorphosis in a mixture of misbehavior and fading religion and it is difficult to predict how and when its eventuality might fit the concept of developing secularism.

The Pope also made an interesting comment at the time of his visit. When asked if America would serve as a model for more secularized Europe he replied that America started with the positive idea of secularism. He implied that people fled communities to escape purges and wanted to have a secular state that opened the possibility for all forms of free religious exercise. He said that America was intentionally secular and the exact opposite of state religions but it was secular out of "love for religion" and for an authenticity that can only be lived freely. It is this form of "freedom" that is obtained in a purely secular society without the artificial restrictions and added complexity of a framework of an imagined ideal life outside of reality.

**Inherent procreation will win out:** A news article (199) indicated that the "Voice of the Faithful", a lay group, is calling for the Vatican to review the requirement that priests be celibate. It is attempting to apply new leadership and energy to convince the average Catholic that there is strong need for an independent lay voice in the Church governance. The issue of pedophile activity and sexual scandal has been much more widespread than ever realized as case after case was exposed. It makes one wonder how many cases have been left unexposed and uncorrected over the world. Sister Mary Ann Walsh, spokeswoman for the "United States Conference of Catholic Bishops", was quoted as saying "There are lots of other groups that are talking about celibacy. Don't waste the bishop's time on it—they can't do anything about it. You might as well have a great discussion on what goes on on Mars." In this way the carry over of ancient times as to religious ideas has influenced the behavior of men in modern society. Man must indeed, await the process of further expert development to apply modern reason to behavior and to appropriate adjustments in church tenets as variously suggested in this text. Indeed, one can see that conflicting and troubling religious institutions are replete with unsolvable problems that stem from their very imaginative base and must evolve toward the invariance of secularism.

Anthropologists have constantly made man aware of inherited sexual instincts. Man is simply a procreative mammal. The problem is deep and complex and should be made more subject to logical and reasonable standards. Sexual desire is without a doubt the strongest natural instinct that reflects constantly in man's behavior. It is frequently implicated in rape, not only of adult females, but also of children, often of children in one's own family household. When invading armies are on the move they often assume a free hand in raping the women of the area being invaded. It became a serious problem in the Serbia/Bosnia and Darfur conflicts. It is indeed, an animal-level act relatively uncontrolled by reason. The mind of reason is just not yet strong enough to overcome all emotion. This book claims that the mind of reason is still below the level that can properly resolve such issues including the ideal imaginative promises of religions. Although the mind of reason and judgment is naturally developing through processes of both evolution and cultural evolution earlier defined, it still has not reached a point of general acceptance. The neural capability of judgmental control must be given more time, study, and serious attention.

Education is slowly evolving and expanding all over the world. The days of referring social and medical problems to biblical writings are also fading. There has always been an implied connection with biblical writings and religious teachings of the churches for most of the activities of life. Originally intended as a guide for morality based on the teachings of Christ and others, circumstances and knowledge have long since rendered much of it useless, ineffective or simply incorrect. Religious instruction, having deeply penetrated many human minds as described in the first chapter, causes much delay and reticence with respect to following a proper modern course of action.

Not many years ago, there was public reluctance of many mothers to allow the marriage of their daughters to Catholic men. They feared that the mandated prohibition of contraceptives might produce too many children too quickly. In sophisticated society that has all changed and reasonable family planning often takes place in violation of that rule. There are also many family planning organizations that work with women in direct opposition to teachings of the Churches. Modern education and diminishing recognition of unreasonable church rules has brought much relief. However, where education is incomplete or in trouble the forces of inherent animal instincts continue to function and in many areas of the world reflect the more primitive function of the emotional mind. The widely publicized erectile drugs and the pervasive product sales with increasingly unabridged sex-related advertising has few limits in modern society.

**If God made sex:** The Catholic Church continues to oppose contraception relentlessly. Sexual intercourse is considered easier and more enjoyable without its concern and control by reason. It is therefore, often obscured by the overriding emotions of the moment. However, both overpopulation and sexually transmitted diseases are in question. Pope Benedict XVI as supreme head of the Catholic Church and claimed to have infallibility with respect to spiritual matters and church doctrine recently made a flight to sub-Saharan Africa. As usual he was met by millions of Catholics and others who considered him to be a great leader with divine influence and implied heavenly connections. Many Africans have suffered, not only from lack of education and hunger but also from the uncontrolled spread of the AIDS virus. Regions of the continent are devastated by HIV, the virus that causes AIDS. The disease is simply not understood. Governmental, religious workers and health organizations have made major efforts to conduct campaigns to reduce the number of sex partners and engage in safer sex practices with little improvement. As in the case of relative emotional and judgmental strength discussed in earlier chapters with respect to religious belief, the emotional desire for sex satisfaction clearly wins out in the presently structured mind.

The Centers for Disease Control have determined that proper use of a latex condom will prevent transmission of disease. The alternatives are to abstain from sex or to assure long-term mutually monogamous relationship with an unaffected person. This is clearly not a realistic direction in terms of expectations and experience has been negative. Although the Pope told reporters on his African flight that distribution of condoms would increase, not resolve the AIDS problem at hand, a respected organization called Cochrane Collaboration claimed that it would reduce transmission of the virus by eighty percent. Surely, the Pope must have anticipated that such distribution would encourage irresponsible sexual activity. However fallible, he is obliged to further the precepts of the Church irrespective of modern factual knowledge. The world is moving through another era of change under the fundamental universal forces that govern life on Earth and resolution of these types of problems by modern means is bound to accompany the slow movement toward secularism.

**The divine power of snakes and priests:** Perhaps not the best example of how deeply entrenched ancient beliefs interfere with proper medical attention even in this modern age, is in the area of the Cobra-infested town of Mushari, a region north of Calcutta (200). It is an area where five foot Cobras constantly slither through throngs of barefooted children, where people sleep with them and live with them without fear. They seem oblivious to the fact that as many as fifty thousand Indians are killed each year by serpent attacks. However, the belief that the legendary snakes have divine powers and the resulting religious prohibition in harming them becomes a major problem. They are told by the priest that if they do not go to a doctor but come to him, the bite will be cured in two or three days but if they go to the hospital, their limbs will swell and there will be major complications.

However, faith in the priests and their special powers are only now slowly beginning to crumble. One victim spent the first few hours with the chanting priest but became increasingly sicker and asked to be rushed to the nearest hospital twenty-five miles distant. She barely arrived in time to be saved. Near-death experiences have finally convinced most residents to obtain modern hospital treatment and no longer maintain rigid belief in the priests but many are still unconvinced and feel that the rituals are the only cure. This story emphasizes the great power of belief solidly fixed in an uninformed mind as I earlier discussed in the first chapter. However, education and globalization is gradually penetrating many remote regions around the world. Recent stories described the introduction of medical doctors to replace the primitive treatment by "witch doctors" in the mountains of Kashmir. But the principle is not confined to minds in undeveloped areas. It still applies as well to those in New York, London and in Paris where primitive beliefs are still occasionally found. Each step represents change toward greater realism based on advancing human knowledge, change that very slowly tends to resolve the enigma and stigma of the planet.

The biblical distinction between good and evil is insufficient for modern society. The definition of sin is passé. The churches have long published documents on behavior required for its members, documents that have or should be updated for the modern world. As discussed in earlier chapters many people still believe today that persons who do not subscribe to a traditional religion and a humanlike god have no standard of reference for righteousness and must certainly be immoral. Simple reason tells us that human ethics and morality do not have to be deduced from traditional religious beliefs. The most reasonable objective of man is to make individuals responsible for their behavior. The action of churches imposing their views of moral value and sin, sexual conduct, marriage, divorce, birth control, abortion, etc. is obsolete. These are all views that must be agreed in the social complex based on the best interest of man. The underlying basis must be medical and scientific coupled with the best established social behavior. Living under the rule of law as established with democratic procedures it is possible for all human beings to live a meaningful and happy life. Proper ethical judgments can be formulated by society without reference to the dogmatic teachings of traditional religion. It is not necessary to create fear and hypothetical punishment in the afterlife. This is a practice that must be discontinued at once in the interest of human reason and progress.

**The rules of medicine, science and the law:** The study of morality and ethics is becoming far more prevalent at all levels of education. There are more publications on the subject than ever in the past (201). In his book *Bending of the Rules*, Robert Hinde addresses the subject of how the behavioral sciences apply to the study of morality. The propensity of people to divide their world into "in-groups and out-groups" he argues, leads to moral behavior within the group but a bending of that code when dealing with the other. Morality is a product of biological and cultural evolution between the two opposing human potentials of selfish assertiveness to win human competition and pro-sociality to facilitate group cohesion, in part to win competition against other groups. Hinde concludes that what is seen as right may actually differ with the situation. Thus, there is no single moral answer and there are no general objective tests for morality except that most actions, to be moral must be conducive to group harmony.

In an attempt to introduce his normative views in three ways, first, he submits that using scientific methods from biology, ethnology, psychology, sociology and anthropology can help us understand the other in different groups and cultures and understanding can dampen conflict. Second, the author argues that there are certain moral principles found in all cultures. These pan-cultural principles contrast with precepts, rules, and values for guiding behavior that are found in different cultures. The golden rule is emphasized as an example. Third, Hinde offers a set of personal, but quite general ethical perspectives on law, personal relationships, physical sciences, medicine, politics, business and war. Most of the larger universities now have departments tuned to the fields of morality and ethics and all students must do their best to participate.

**Why terror?** Behavior with respect to violence and terrorism has also become a matter of great concern in recent decades. Mark Juergensmeyer produced a book entitled *Terror in the Mind of God* (159). It is a wonderful summary of how religion can become militaristic and produce terrifying violence whereas it claims to be the source of peace and end-all for man. Its first part describes the endless number of global incidents by various religious extremists. It explores the use of violence in recent times by groups or individuals within the five major religious traditions. For Christianity, it cited the abortion clinics, the Oklahoma

City bombing, and Northern Ireland, for Judaism, the assassination of Rabin and Kahane, for Islam, the World Trade Center and suicide missions, for Sikhism, the assassination of Gandhi and Singh and for Buddhism, the Japanese subway attacks. Using powerful new explosives, the lowest form of violent behavior is frequently tested by man. The many attacks on embassies, the countless suicide bombings, bombing of abortion clinics and on and on by religious extremists constitute a turn to mental disorder from religious emotion. The author claims that it has much to do with "religious imagination" the very subject of this book. These forms of distorted minds must be modified or in some manner eliminated from normal and peaceful society.

In the second portion of his book, the author examines common themes and patterns. He claims that the choice of targets and dates exhibit symbolism and connections with religious ritual. Opponents are demonized and the perpetrators become martyrs. Savage attacks have become a way of life throughout the world, in shopping centers, government buildings, Olympics, subways, coffee houses, riverboats, restaurants, and military establishments. It is not confined to the Middle East, to Sri Lanka, Northern Ireland, Kashmir and Pakistan but has taken a huge toll in New York City and in other areas of American soil. There are also extreme elements of religious belief that feel bound to correct the behavior of other members of society by forceful beatings and by other means. A balancing effort by the establishment of reasonable common rules of behavior around the world independent of religious belief is most essential. Human suffering due to the narrow interpretation of proper behavior by a few religious extremists cannot be allowed to continue in any area of civilized society. The planet is far too small for tolerating such a high degree of abnormality.

**Imagination or reality?** The imaginative effect of belief, lives on with amazing reality even in the twenty-first century. A recent story reveals a great deal about the mechanisms and reasoning of the uninformed and eager minds of belief. In a village in northern New Mexico known as the Lourdes of America (202) Reverend Roca remarked, "It's not the dirt that makes the miracles!" Tens of thousands of pilgrims walk eight or more miles to his shrine on Good Friday and at other times of the year, some bearing heavy crosses and others on knees, hoping to cure diseases or disabilities with prayer, holy water, and most famously the healing dirt that visitors collect from a hole in the floor of the church. It has become a stopping place on tours en-route to Taos. Declared a national landmark, visitors bring their own baggies or containers or buy plastic containers marked "blessed dirt" at the church gift shop. The reddish dirt is scooped from an eighteen inch wide well that is replenished by a caretaker from unknown sources.

Father Roca was careful to explain that the pit is not refilled by "divine intervention" but few leave without it. They eat it, brew it, or brew it for tea, or rub it on the body. On one occasion rubbing the dirt on a man's chest seemed to improve his cancer condition. The lore of miracles on the spot of the hole accounts for its supposed power and the continued faith of visitors. However, the padre remains vexed about the continual loss of his dirt and his concern about the unholy nature of its source. To a rational and logical mind this sort of behavior sounds absurd and foolish but it is typical of behavior that can often be found over the globe. Although each century sees less and less of such ardent dependence on an imaginative god, it is an indication of the still undeveloped mind of man. Do these practices get started by imaginative religious leaders? Does anyone ever intercede on behalf of the pilgrims? Will they ever become educated and enlightened? Or do we simply say, that is the way it must be, that is the way things are. Who is responsible for teaching reality to the overly spiritual and naïve world? Are the religious publications without disclaimers the root cause of these beliefs? Is the modern human mind still not quite sure of its spiritual position and of what constitutes reality?

**Faith hope and reality:** This account of spirituality is not intended as a mockery but as a serious indication of how modern minds can still be "lost" in emotional neural loops of feeling and thought, indeed in a strong desire to be cured under any consideration of reason. This is an indication of how uninformed minds, sensitive and vulnerable, react to information they obtain about the miraculous works of God. The collectors of miraculous dirt believe it has some power, though it might not be understood or proven. It is a limited expression of logic and reason and in its extremes can result in serious mistakes. However, it is in sharp contrast with the knowledge and understanding of the philosophers, intellectuals and scientists who are attempting to analyze knowledge in modern terms and broader human understanding with the hope of achieving greater efficiency of life in the direction of unity. Religious expectations are often expressed outwardly. Athletes are seen to bless themselves prior to performance, presumably to obtain the help and support of a superior being in the execution of a task. Some teams pray for help to win a critical game. In recent years religious faith was sincerely expressed as an important factor in winning the world series of baseball (203).

**Religion requires violence?** Religion has supplied the motivation and perceived justification for public acts of violence. In many cases religious violence is converted to related emotions, to ethnic interests or separatism. One of the haunting questions asked by great scholars is why religion needs violence and violence religion and why there seems to be a divine mandate for destruction accepted with such certainty by some believers. The author of a previously referenced publication expressed puzzlement, not about why bad things are done by bad people but rather why bad things are done by people who otherwise appear to be good, by pious people dedicated to a moral vision of the world. He questioned why such violence appeared with such frequency at this juncture in human history. This book suggests that the continued marketing of the promises of traditional religions is partially responsible. It also suggests that the world is passing through a transitional era, a transition in which religion is no longer the opiate it was in the past. The opiate is neither strong enough nor too weak to exist. The continued state of mind, balancing on the religious and secular fence is dangerous and must be treated as a disease that can only find resolution in all levels of rational education. Contrary to the principles pursued in this book the author suggests that the cure for religious violence might ultimately lie in a renewed appreciation for religion in society. In the theme of this book that view suggests human retrogression and defeat.

Faith and wisdom may be incongruous: In Runciman's comprehensive history of the Crusades in 1951, he wrote that faith without wisdom is a dangerous thing. In this book, wisdom is synonymous with the careful practice of logic and reason. He reminded man that high ideals were besmirched by cruelty and greed, enterprise and endurance by a blind and narrow self-righteousness, and the Holy War itself, was nothing more than a long act of intolerance in the name of God, "a sin against the Holy Ghost." It is generally felt that Christians had more difficulty than Islam in accommodating the notion of the holy war. The struggle was intrinsic to Islam enjoined on all practicing Muslims, often described as the sixth pillar of Islam. Jihad takes two forms, the greater being internal struggle for purity and the lesser the military struggle against infidels outside Islam, a struggle that is sworn to continue until the whole human race accepts Islam. It is said that Christian justification of holy violence was roundabout and that the pacifism of the beatitudes could not be sustained once it became the religious war of the Roman Empire. The Christian Holy War, a war of "sanctified slaughter", a war brutally uncivilized and basely motivated, was basically contradictory. It was said to have protected the religion of Christ who commanded followers to love their enemies, to turn the other cheek, and warned that all who took up the sword would perish by the sword. Although armed confrontation over possession of the holy city of Jerusalem has affected the thinking of both religions about the legitimacy of religious violence, the underlying emotions continue to this day. It again raises the question of what is holy in the imaginative framework, and therefore what level of action might be justified. This book strongly suggests that the objectives of both Islam and Christianity are empty objectives with respect to the current assumed direction of human life.

**The human mind of God:** Freeman Dyson in an Ekeland book review reminds us that the French scientist, Maupertuis, who investigated the principle of "least action of nature", explained how satisfying it is for the human spirit to contemplate these laws, so beautiful and simple, which may be the only ones that the Creator and Ordainer of things had established in matter to sustain all phenomenon of this visible world. He identified action with evil behavior, so that the principle of least action became a principle of maximum goodness. He concluded that God had ordered the universe so as to maximize goodness. The world that we live in is the best of all possible worlds that God might have created. It is said that this principle unites science with history and morality.

Many of the great scientists of the seventeenth and eighteenth centuries struggled with the concepts of god, nature, mathematics, science and morality. The times were rich with publications of the most enlightening nature. It was a period of reason and logic in which many of the scientific principles were born. Although some were subsequently modified, altered and supplemented, it led to the more pronounced split between science and religion in the nineteenth and twentieth centuries and will continue in the movement toward secularism in subsequent centuries. Today the discovered principle has little to do with the spirituality of man. Discoveries of intimate correspondence of man and animal gave rise to the more modern concepts of ethics, morality and behavior in which religious belief began to play a lesser role. Today psychology and neuroscience depends largely on animal models, not on belief in a god, to analyze human behavior.

Human behavior concerns not only the individual, but also major groups and nations. For some years the question of Muslim, Christian and Jewish coexistence has been in the headlines. In more recent times, terrorist action and responsive military action by the United States in the Middle East has momentarily modified the picture. It is not yet known where the current antagonism, mistrust and outright war action will end. It is another case where religion is intrinsically mixed with politics, terrorism, and perhaps with Mideast oil interest. The jury is still out and Mideast religion remains strong. Because of the restricted level of factual education, the imaginative nature of religion will take some long time to penetrate and be duly recognized by the holy regions of highly emotional believers. When, and if, the area is swept by the predicted wave of contrasting modern reason and logic of the intellectuals and scientists the fray will certainly be appropriately intensified.

The subsequent generations are not expected to wallow in tradition and in opposition to obvious change. New generations of logic will breed increasingly greater impatience for appropriate resolution and change. The traditional meaning and purpose of religion will begin to change the nature of the entire conflict. It will become a conflict of culture and race before it subsides in more widespread unity, cooperation, and peaceful coexistence. In the Middle East the principles of economics will continue to rule for a time because of richness in oil in the region but the local differences will tend to be absorbed in the larger process of advancing technology and globalization. However, the new world effort toward independence in more environmentally favorable fuel, will certainly produce major changes in both Middle East religion and culture.

The conventional thinking is that human beings are bestial by nature and ethical codes are curbs on their brutish instincts that enable them to live together in relative peace (204). Only humans are believed to possess the intellectual power needed to repress natural impulses and so only they can be moral. If the theory of evolution is sound it is believed that morality must result in some part from those processes. Darwin argued that social instinct, that forms man's moral sense, can be found in monkeys and in many other animals. This idea is rejected in modern science and philosophy since nonhurnans are thought to lack the cognitive and linguistic abilities required for such autonomy. It is sometimes thought that human moral behavior is derived from western traditions of monotheism in which morality follows a set of rules that impose restrictions on behavior. If this is true one would think that world behavior would be more ideal.

**Man remains a mammal:** Darwin also indicated that any animal, endowed with social instincts, would inevitably acquire a moral sense or conscience, as soon as its intellectual powers had become as well developed or nearly as well developed as in man. Modern philosophy tends to reject any link between moral behavior of humans and other animals. This is no doubt due to the great difference in cognitive mind structure, rather than a complete discontinuity. Modern philosophers take the position that humans are moral by nature and place no significance on the claims of religious belief.

A discussion of religious behavior must include the earlier mentioned Evangelical groups that have sprung up in America. Leaders are said to have determined that they can easily find out what believers want most and then offer it to them, but at a price. They have been very successful when using good marketing tactics. It is said that their market favors practical convenience rather than serious commitment to deep religious life. They have been accused of becoming corrupt with power and wealth. They build large attractive facilities and make every attempt to see that they are well supported. Their advertising is every bit as professional as one might use for commercial products. Religion is indeed a competitive market and success goes to those with the best marketing skills.

This has been the underlying position of many successful American religious leaders for some time. Their influence and popularity gives them an opportunity to engage in politics for, as fully explained in earlier chapters, they can readily influence voting choices from the pulpit. Although the sincerity of their personal beliefs is not known, it is clear that the process functions very much like business. The sociologist Alan Wolfe suggests that more emphasis might be placed on "parking facilities and provision for baby sitting" than on interpretation of the scripture.

Behavior of man must be constantly monitored and causes and trends analyzed in order to encourage world improvement. Modern science and globalization both play major roles. It is disturbing that for the first time in American history one in every one hundred American adults is behind bars. In three decades of growth, the nation's prison population has nearly tripled. A recent report (205) emphasized that this ratio exists because one in a hundred adults committed a serious criminal offense. It is patently clear that religion is not the answer to human behavior. Factual education and more careful career training may well be. The $30,000-$50,000 it costs to hold a prisoner could be well spent not on warning of religious penalties administered by an angry god but on training in behavior and on useful occupation.

Human hope and imagination with respect to the religious mind appears to have no limit. A decade or more ago a book was published by a disabled author, claiming that God was disabled, referring to Luke 24 in which Jesus invites his disciples to touch his wounds. The book was titled *The Disabled God: Toward a Liberatory Theology of Disability*. It became an important reference with respect to disability studies and contributed much to a recent United Nations Convention on Rights and Dignity of Persons with Disabilities. The author, who earned her doctorate and studied theology and disability in religion, claimed that with all the professed concerns of Christianity, with the poor and oppressed, the disabled were ignored. Although published experiences and their spiritual interpretations often lead to effective recognition of human problems, much more needs to be done to building a more comfortable human coexistence independent of religious considerations.

The plea for improved human behavior and the missionary-like advertisement of a caring personal god is often in the hands of the small church pastor. In Cathedral City, just south of Palm Springs, the pastor of the First Southern Baptist Church crawled atop the church roof and painted the word "eternity" in red (206). He said that he had long thought of his roof as a potential canvas for his doxologies, a "celestial screen saver" of sorts. However, the city officials put a nuisance notice on the doorknob of the church and a battle ensued. He claimed that the government singled him out because they did not like his faith, but the town prohibits signs on its roofs. The strength of his will is summarized in the pastor's words: "The message is worth standing up for, and worth dying for." He finally had to submit to the city government.

**The forgotten world:** Despite the advancing mind and the world movement into more complex technology, there exists an underlying problem that seems to escape the eyes of world leaders. It is human behavior with respect to caring for fellow man. Are they blind to the indigent? Are they spending too much time in their own interest to see the plight of the world? Are they unable to determine how to deal with major problems of society? Are they really equipped to be leaders? A recent news article discussed food shortages (207). It dealt with "Empty Bellies . . . . across the Globe." A special adviser to the United Nations secretary general, Ban Ki-moon, said, "It's the worst crisis of its kind in more than thirty years. It's a big deal and it's obviously threatening a lot of governments. There are a number of governments on the ropes, and I think there's more political fallout to come." It is said that it is not only the poor, but it is also affecting the working middle classes. At a recent World Economic Forum, the president of El Salvador said, "How long can we withstand the situation? We have to feed our people and commodities are becoming scarce." According to an advisor to the Indonesian Ministry of Agriculture, "the biggest concern is food riots." The senior analyst for the World Food Program said, "the human instinct is to survive, and people are going to do no matter what to survive. And if you're hungry you get angry quicker." The article emphasizes the problem of higher cost of food over the globe and gives many examples of the plight of the hungry. It even speaks of the sale of patties made of mud with oil and sugar consumed by the most destitute in Haiti. It speaks of offering children to a stranger just to have them fed. The behavior of society, especially the governments that represent them, has not reached a level of advancement that one might call fully civilized.

Can this be the condition of our wonderful advancing world, the modern world of aviation, space exploration, and wide belief in a benevolent, humanlike, all-powerful god? Can it be the same world in which corporations award bonuses of hundreds of millions of dollars to executives? Can it be the same world in which the Vatican is rich with artistry of precious metals and gems? Is it the same world in which humans are consuming all the oil they can pump out of the Earth, for power, for homes, industry and pleasure? Is it the same world in which religious institutions support leaders in luxury and sometimes with private jets? Is it the same world in which the gambling industry operates in the billions of dollars? Is resolution to be placed in the hands of an imagined god or man?

**Inherent conflict in man:** Many men feel that physical conflict is normal human behavior, that it is a matter of human nature to compete and to fight. They point to the ubiquitous wars among people and nations. Do wars require a justified purpose or do men rationalize their existence? An article recently described cockfighting in Santa Dominica (208). The Dominicans claim that if you put two roosters together, they will fight. If you put a thousand hens together, they will never fight. Although cockfighting is illegal in all fifty states, it is an important pastime in the Dominican Republic. According to the Humane Society, however, tens of thousands of people are involved in cockfighting in the United States, and an estimated forty thousand are involved in dogfighting. They say it is completely integrated in their laws and completely integrated in their traditions. Most Dominicans are also highly religious, but while the women go to church, the men go to cockfights at the galleras. Nearly every neighborhood and village has a gallera. There are fifteen hundred registered with the country's National Commission of Cockfighting.

Dominicans believe that they are "warriors and love competition," and it is something that they carry in their blood. Cockfighting is also popular in Indonesia, Thailand, Mexico, Puerto Rico and in many other countries. This unusual news story verifies the position of anthropologists, psychologists and other scientists familiar with human development. Clearly man emerged from animal ancestors and is continuing the character of pre-emergence even today. Some men are marginally ahead of others in terms of natural male competition, but in the course of long time, they will all tend to advance. Man will increasingly become more civilized, more humane and will more fully recognize reality, as described in earlier chapters.

**The search for Utopia:** One cannot discuss human behavior and the search for freedom and purpose without mentioning an earlier book, Esalen, America, and the *Religion of No Religion* (210) written by Jeffrey Kripal. Esalen was a famous spa located in the Big Sur region of California that was renamed Center for Theory and Research. The author believed that spirituality and science should not be contradictory. He wanted to picture America not as a Christian Nation, or a globally hated superpower, but as a state whose potentiality has not yet been realized. He sought the day when people might say, I am spiritual, but not religious. The author may be surprised to know that this is increasingly the cry of man all over the world, perhaps the path to more fully recognized secularism.

It is said that Esalen, formally known as the Esalen Institute, synonymous with the attractive, the sophisticated, and the rich, was intended to represent an experiment in Utopia, somewhere between the revelations of religion and scientific modernity. The names of many important scholars, politicians as well as the counterculture were connected with the project. Participants apparently included Ronald Reagan, Mikhail Gorbachev, Boris Yelsin, Carl Sagan and Joan Baez. Billy Joel, Timothy Leary and Episcopal Bishop Pike, who was tried for heresy after repudiating the dogmas of the Virgin Birth and the Holy Trinity, were said to be involved as well. It was intended as a spa of good food, intellectual discussion in a free living atmosphere. There were experiments in ESP and other para-psychological concepts. They apparently investigated the occult, hallucinogens and space aliens as well. They planned to explore trends in behavioral sciences, religion and philosophy that emphasized human potentialities and values of existence. The attempt to warm up the cold war relationship through "hot tub diplomacy" was just a secondary part of the deal.

Esalen represented one of many such projects that have appeared throughout the world at various times to explore the expanded mind of man under conditions of maximum relaxation. Many believe that it was intended to further the evolution of religion worldwide toward "no religion" along the lines discussed in these chapters. The atmosphere itself reminds one of the Bohemian Club, a summer romp in its Grove on the Russian River north of San Francisco. It was an "escape to relative freedom" where one might observe a certain level of human behavior, enjoy a full program of music, theater, and art and exchange the deepest of ideas. Companionship always included the top names of America and often of other important nations of the world. However, with all the openness and change of scenery, one must always return to reality to face exactly the same question of existence that one started with, the very questions of choice of belief and inherent human behavior, as openly considered and discussed in this book.

The word Utopia is made up of Greek words that mean "no place." It was the title of a famous book by Sir Thomas More, first published in Latin in the sixteenth century. It was intended to criticize economic and social conditions of the times, representing impractical and visionary reforms. Plato's Republic was an earlier writing, also describing Utopia, some few hundreds years before Christ. In the nineteenth century *Erewhon* (nowhere with a backward spelling) by Samuel Butler, Looking Backward by Edward Bellamy, and H. G. Wells' *A Modern Utopia*, all involved similar concepts of Utopia, either serious or satirical, reflecting idealized government or human behavior. Although this book envisions gradual evolutionary improvement in the structure and function of man, it is still far from ideal with respect to behavior and expectations. Man is many generations away from significant improvements that may be contrasted in future history, but he is expected to gradually learn the difference between practical and impractical direction as well as real and imaginary notions. Reconsideration of the belief in a humanlike god is clearly in the offing. Resolution of more complex aspects of personal and group behavior should gradually follow in further human advance and development. Although we are certain that gradual change is inherent in the evolution of man on his planet, none of the great-grandchildren of readers is expected to live to witness many such improvements. Its real progress will be hidden in the next few centuries. However, it is very important for man to get started in the right direction.

**The inherent human spirit:** Freedom from the bounds of imaginative religions will tend to march the same path as political freedom. Although the course of future globalization and political freedom must be the subject of another study, it has a close bearing on the movement toward secularism as discussed here. With similar structure, there are substantial differences in human minds with respect to learning, memory and general retention and application of knowledge. This book assumes a major movement in world education over the next several generations irrespective of geography, economy and political division. It is the engine that will not only power movement toward global organization and management but also one that will tend to power and perpetuate natural political and religious change.

This chapter particularly concerns the vulnerability of some minds to an imagined fatherly role model in the sky, to promises of security, to overcome the fear of death, and to all the various idealized claims offered by religions. It is a supplement to the early chapters on how and why the human mind responds to religious influences. The mind is easily subject to many temptations well known in society. Some are "weaknesses of the spirit" that cause normal people to behave in a totally regretful manner. Some are "spur of the moment" acts, whereas others may be acts of long contemplation, indeed, working up to the overt exercise of judgment and decision.

On May 6, 2009, CBS Station WFOR-TV reported that a priest, Father Alberto Cutie, was photographed on Miami Beach kissing and fondling a woman. The Archdiocese of Miami relieved Father Cutie of his duties at St. Francis de Sales parish and at the radio station Radio Peace. If true, it is another example of the deep conflict between basic instincts and promises made to God through the church. Whether in belief or in physical behavior, the present human mind remains vulnerable, subject to slow change. Vulnerability to imaginative belief is expected to moderate prior to vulnerability to other inherent emotions.

The human spirit requires occasional free expression and excitement, and vents its feelings in many ways, sometimes privately, sometimes with friends, sometimes uncontrollably, sometimes abnormally interfering with the peace and quiet of other more conservative members of society. A rapidly growing group known as Pentecostal Christians (155) has appeared in America. They seem to congregate in large numbers at megachurches and openly engage in song, sermons, and interaction, that are intended to bring them in close association with their imagined creator. Their devotion is expressed strongly based on the teachings of the Bible (209). This type of service is far in excess of belief and worship. It becomes a form of widespread proselytism, human expression of emotion, of imagination, and perhaps entertainment. Pentecostalism is defined as a Christal body that emphasizes individual experiences of grace, spiritual gifts, as glossolalia and faith healing, expressive worship and evangelism (Pentecost is a feast on the seventh Sunday after Easter commemorating the notion of descent of the Holy Spirit on the apostles, also called Whitsunday).

A recent publicity release by "pure.imagination," an Internet site, appearing under the rubric of "Urban Ministry," announced a new book by Donald E. Miller called *Global Pentecostalism: The New Face of Christian Social Engagement*. It posed the question, "How and why is Christianity's center shifting to the developing world?" Pentecostalism is claimed to be the fastest-growing religion in the world. The book is intended to provide a portrait of this wide-ranging religion, a major new emotional movement. It is said to be based on four years of travel over five major continents and includes interviews with Pentecostal pastors and leaders around the world. If there is a major shift in Christianity to undeveloped areas, it is because of their limited comprehension and therefore high vulnerability to the promises of Christianity, as frequently described in this book, and also because the church is spending a great deal of money, time and effort in dealing with the indigent and their medical treatment. It is unfortunate that service to humanity cannot be separated from the desire to serve an imagined god.

**Muslim edicts:** In the Muslim world, man engages in frequent prayer, openly and aloud from the upper areas of minarets. There were recent public objections to this practice in Switzerland where they claimed disturbance of normal conditions of peace and quiet. This is another example of incompatibility of social tolerance. One must take great care in interpreting the ways of religion for there are obscure reasons for behavior that are hidden to the outside mind. In the Muslim religion, edicts are occasionally issued in an attempt to bridge the gap between the expressed principles of the faith and modern life. This need is considered frequently in the fast moving modern society and has been the subject of many publications and the heart of many disagreements. The Muslim edicts are called fatwas and can be based on pure ideology rather than science, current knowledge and social practice. A recent article (211) spoke of a controversial breast-feeding fatwa that declared the Islamic restriction on unmarried men and women being together could be lifted if the woman breast-fed her male colleagues five times to "establish family ties." This was followed by the urine fatwa, which claimed the drinking of urine of the Prophet Muhammad was deemed a "blessing." These fatwas apparently were a source of embarrassment in Egypt because they were issued by representatives of the highest religious authorities in the land.

Fatwas are intended to be issued only by religious scholars who study the Koran and teachings of Muhammad for appropriate guidance. Political leaders claim that there is a crisis in Islam because too many fatwas are being issued—thousands are issued every month. The conflict in Egypt serves as a reminder of the challenges that face Islamic communities as the true nature of the religion is debated, and as modifications and interpretations are constantly required to meet the needs of the daily modern life of its members. The matter is being taken very seriously. Abdullah Megawer, the former head of the Fatwa Committee, said, "It is a very critical issue for us. You are explaining God's message in ways that really affect people's lives." There apparently is some confusion about what is right and wrong, since a fatwa is non-binding and there is no central doctrinal authority. There has therefore been an "explosion" of sources of fatwas from Web sites that respond to written inquiries, to television shows that take phone calls and to radical terrorist organizations that set up their own committees. As in many religions, the original basis of the Muslim religion is subject to some dispute.

The thousands of sayings of the Prophet Muhammad, known collectively as Hadith, are said to be the proper basis of fatwa issuance. A few typical fatwas are mentioned in the referenced article. One involved advice that all wives must be treated equally and, if not, one could ask for a divorce. In this case the husband was spending more time with a younger, eighteen-year-old wife. Apparently men were permitted to have four wives in Islam. Surely, these issues, without serious consideration and without central authority, could eventually degrade the intended purpose of religion and could contribute to the changes predicted in this book.

**Unified man would do better:** This study is not intended to solve world organization but to speak to the factors that lead to the problems of disunity and to world conflict based on strong misplaced emotional response. The establishment of Utopia is not yet possible by man, but the exercise of logic and reason in the achievement of better world organization is part of expected ongoing human development and advancement. A review of the world condition today will disclose disorganization, conflict, lack of application of some of the simplest principles of proper behavior. Despots and war lords are still commanding areas in their self-interest, they are disturbing the rights and the way of life for an appreciable portion of the population. There are too many borders, too many diverse political ambitions, too many guns, and too many differences in spiritual beliefs. The human mind suffers greatly from vulnerability. It must be freed and more fully educated and offered some semblance of world equality and human rights, both politically and economically. The world requires educated leadership, leadership that understands the real needs of civilization, not the spiritual needs, poorly satisfied by imaginative concepts, but emotional and material needs satisfied by reality and by participation in the many benefits of the natural world.

**If heaven could only help:** The world economic health has suffered greatly in the past year or so, a problem that hits individuals where it hurts most. Unemployment has risen, businesses have closed and the end is not yet in sight. A recent article suggested that "bad times are good for churches" (212). Prayer requests aimed at getting or keeping jobs have doubled. Congregations, including hedge fund managers, have filled some of the churches so completely that closed circuit TV facilities and hundreds of new chairs have been required. Pastors are said to abandon standard sermons on marriage and beatitudes to preach on the theological meaning of the downturn. Reverend Bernard, founder of the Christian Cultural Center in Brooklyn, said "It's a wonderful time, a great evangelical opportunity for us. When people are shaken to the core, it can open doors."

Evidently a Businessmen's Revival was initiated in Manhattan as early as 1857 with a noon prayer meeting among traders and financiers. A recent article (213 ) spoke of ongoing efforts to preach to Wall Street. However, David Beckworth completed a study some time ago called *Praying for Recession: The Business Cycle and Protestant Religiosity in the United States* that disclosed some interesting details. It indicated that during each recession cycle between 1968 and 2004, the rate of growth in evangelical churches increased by fifty percent. By comparison, mainline Protestant churches continued their decline during recessions, though more slowly. This seems to bear out a view expressed in earlier chapters about how highly induced human spiritual emotion and closer connection with everyday social interests are increasingly motivating the average believer.

**Forgotten witchcraft:** Behavior cannot be explained fully without a reference to witchcraft. As recently as the seventeenth century we are told that the State of Connecticut was awash with witches. Today it is not clear just what a witch is. However, it seems to fall in the category of a religious personality and the exercise of witchcraft was then punishable by death. A case recently came to light in connection with lectures on the Connecticut Colony witch trials, such as the Salem trials of 1692

(151). The colonists of Hartford apparently accused thirty-nine-year-old Mary Sanford of witchcraft based on wine drinking and dancing around a bonfire. The amazing consequence was that Sanford was hanged leaving behind five children and a shaken husband. She was eventually acquitted. Existing relatives more recently filed a petition with the State Legislature to exonerate their distant grandmother and ten others who were executed. These seventeenth-century happenings are extremely difficult to understand in the light of modern times. However, the case has taught members of current society about mob mentality. Community leaders had presided over such trials and the accused were usually the least educated and the least powerful. Women of the times often fit that bill. Only two of the many victims of execution were men. The case was discussed openly on recent Internet blogs and the descendants were appalled at the comments, reminiscent of the same mob frenzy that occurred at the time of the trials. As indicated often in this writing, people remain concerned with imaginative visions of demons and powerful spirits and will continue to do so until the whole concept of imaginative religion is clarified by a society of reason and reality.

**Forces of the mind:** This book allows for ample continuity of the many strong influences that religious belief has bestowed on man over the ages, but human discovery is the kernel of change. Man has not only discovered that no humanlike god exists but that life can be enjoyed without theism. The forces of change that propel man are largely forces that are generated in the mind and spread to all that man thinks, feels, and does. Through the processes of cultural evolution, defined in an earlier chapter, man will eventually overcome the need for disguising the fear of death. It is a form of evolution that does not depend on biological or genetic change and can therefore be effective in less than one or two centuries.

**Sanctity of life:** The issue of life itself suggests the value assigned to the process of living. Death lurks over the entire world in major armed conflicts. A recent news release indicated 120,000 homicides take place in a single year in America. Although there is no world war in process at this time, there are myriads of smaller wars, of ethnic cleansing, of ethnic conflict, of competition for land, of competition for power, and the like. Weapons are more plentiful than ever in past history, they are being produced by the millions and their effects are more lethal than ever. The suicide bombing taking place in the Middle East often causes dozens of deaths at once. They are related to the strength of imaginative belief because the bombers are convinced by some agent that life on the other side of death may be more rosy and rewarding. It has grown to be a tragic waste of life in increased numbers. The Kamikaze flyers of World War II were impelled to commit suicide by their own nationalistic beliefs and not so much by religion. In both cases the termination of young life cannot be justified for any reason. Even the imperatives of war have to be extremely serious to begin to justify the loss of young lives. About four or five thousand young lives were lost in Iraq and environs in these past few years, losses that were unnecessary in the opinion of large numbers of American citizens. The planning and execution of individual lives become very important. News reports (214) suggest that there is an alarming increase in the suicide rate of persons in midlife in America. The statistics are surprising and frightening. It demonstrates the critical need for understanding, not God, but the function of the mind and the world around it.

**Spurred onward by a sense of reality:** One of the claims of religion has been that a relationship with a god, in which one believes, brings joy and comfort and eliminates some of the feelings related to depression. This therapy may still be the case but it doesn't seem to be working as well as first thought. It is a baffling public health issue which yet has no useful answer. For women in the mid decade of life the suicide rate is said to have increased thirty-one percent in a five-year analysis. A total of thirty-two thousand people committed suicide in 2004 in America, about half of which were in the midlife region. Authorities speculate on the cause but suspect the increased use of prescription drugs and higher social pressure. In view of the larger membership in Christian churches, religion appears to play some significant role. Either spiritual education is not working to satisfaction or it is sending the wrong message. This book suggests that science and neuroscience be appropriately added to the curriculum in all education processes so that scientific reality and mind function are always included with the studies of other disciplines.

The conflict between religion and other disciplines is rampant and of course would be expected. Religious laws are often in opposition to civil laws as discussed earlier. However, this book suggests that the distant future will see a gradual adjustment toward unity of understanding, a merger of spiritual with reality. Issues such as religious freedom with regard to international law will tend to be self correcting. Concern about the balancing of religious freedom (215) expressed by eminent religious leaders will become less significant as the spiritual mind of mankind finds maturity.

Vulnerability of the mind may well depend on specific mind structures. Although globalization will accelerate the process, world society is not yet well mixed genetically and therefore there may be some major significant differences in human structure and function. Geneticists are now able to make some specific determinations relative to racial and other backgrounds. For example, researchers at Washington University and the University of Maryland (216) have recently discovered that altered versions of genes may exhibit an important medical difference. They found that there are differences in muscles used to control responses to nerve signals. The discovery raises questions about whom to treat with beta blockers. The gene variant may help explain why some healthy people cannot exercise vigorously because they make chemicals that act like beta blockers and their hearts beat less forcefully. They also indicate that it might explain why some people do not respond to drugs such as used to treat depression for those people may already be making versions of antidepressant drugs and adding more would not help.

**The expectations of prayer:** It has been interesting to read about a sense of dependence on human prayer in this advanced modern age. A recent news article (217) describes how the Internet is being used to request prayers. On one site called prayabout.com, a wife asks others to submit prayers that her husband will listen to his psychiatrist to resolve his major issues. Other sites called ourprayer.org and ipraytoday.com, allow anyone to request that strangers pray for them without cost. It suggests that the prevailing belief is that the more people pray for something, the better are the chances of it happening. The Unity Church provides a prayer service, a prayer room filled with operators, with a toll-free number suggesting that anyone can pray for them. It claims to receive two million requests a year. Prayers are requested for a wide range of purposes including tests and examinations, for businesses and jobs. The presidential prayer team is a nonpartisan company that sends out daily messages to about forty-eight thousand users suggesting prayers to help the president and president elect and others. Apparently there are many attacks on the prayer websites by atheists. Instead of answering the site, many of them simply point to the website called godisimaginary.com. There have been important studies on the effects of intercessory prayer claimed to heal people, at least one of which has concluded that prayer by strangers, at a distance, does not make people "better faster." The most frequent request for prayer seems to be for "inner peace." A comprehensive book by Sloan (38), mentioned earlier, is generally negative on the medical power of prayer.

**The benefits of Christian emphasis:** A recent article in the New York Times titled *A Gym Designed to Cater to Christians* described a great effort to "protect their behavior" and "help others with their spiritual journey." Founded by a former drug dealer a gym called Lord's Gym was opened in California. The idea seems to be based on the desire to avoid sexual temptation and to be with people of the same moral values. It specifically mentions a mother/daughter pair who reads and discusses the Bible while walking on the treadmill. A photograph was shown of a similar gym in Florida where the wall behind the kickboxing class exhibits a mural suggesting the coming of Christ. The opening of other Lord's gyms around the country indicates how deeply religion has penetrated daily life and how adherence to the Christian tradition relates to control of behavior. It emphasizes the clash of moral values in a rapidly changing society and leads one to question projected human behavior in a world of accelerated globalization and widespread secularization. As humanity moves into the real world it must be accompanied by the corresponding development of a more rational mind. These events tend to confirm the long-term direction of natural forces acting on planetary human life and the requirement for expanded human education as generally promoted in these chapters.

**Invest in the prosperity gospel:** Another front page article (218) reported on a Southwest Believers Convention in Fort Worth at which "prosperity gospel" preachers suggested to the nine thousand or more who attended that if you have sufficient faith in God he will multiply offerings a hundredfold. It indicated that the Senate Finance Committee is investigating whether evangelists used donations to enrich themselves and abuse their tax-exempt status. One ministry was reported to have an annual income of a hundred million dollars. The stakes are high.

A named professor of religion, researching the practice of prosperity preachers called them "spiritual pickpockets". However, in accord with further human mind development and change as envisioned in this study, such currently accepted practices will go the way of long abandoned human and animal sacrifices.

**When it is too good to be true:** This chapter has dealt with the subject of spiritual behavior and vulnerability, the ready response of the mind and emotions of man to ideal spiritual claims and promises irrespective of their fundamental validity, and an account of some of their various effects. Because of the unlimited basis of creative promises and benefits implied in most every religious belief, no mind is capable of turning them away if it judges them to be available or if it feels a high probability of availability. Religious belief provides a perception of high-level truth because it is also tied in with imaginative belief in the creation of the world and the creation of man himself. Divine help can be solicited by a mere thought of the mind despite evidence that at no time has there been any firm indication that such help has ever been forthcoming. This fact does not seem to phase those with deep religious belief. Psychologically it is very much like successive expectations in casino gambling. Logic, experience, and factual evidence are simply unrecognized and overlooked.

The believing and faithful mind is ever optimistic that its prayerful requests may be immediately fulfilled or filled in the foreseeable future. A mind is often "alone" and desperate and vulnerable to some sort of outcry, supplication, or prayer as a last resort. Making a prayerful request contributes to a sense of satisfaction. However, when proper education provides factual information about observations and measurements regarding the reality of God, the universe, and its "unguarded" human occupants, the basis of judgment changes radically and vulnerability begins to subside. As claimed in these chapters, spiritual sensitivity, and response are now well understood by scientists, intellectuals, philosophers, and by many millions of others. Its understanding worldwide is, however, still very limited, awaiting an improved and effective world system of education. The ideal claims and promises of religion are difficult for man to refuse. The old adage "when it is too good to be true, it is usually not true" does not work well in terms of spirituality, for man has been taught a kind of immunity from rational judgment, a sense of papal-like infallibility, when any serious issue comes before his highly desired, well-advertised, and widely recognized god.

# Chapter Eight

## The Mission of Science

> Despite the efforts to pervert it, the question that every young person asks, "Who am I?" the powerful urge to follow the Delphic command, know thyself, which is born in each of us, means in the first place "What is man?" and in our chronic lack of certainty, this comes down to knowing the alternative answers and thinking about them. Liberal education provides access to these alternatives, many of which go against the grain of our nature or our times.
>
> —Allan Bloom, *The Closing of the American Mind*, Simon & Schuster

**Science is the preferred method of analysis:** Earlier chapters suggested that the still evolving mind of modern man knowingly tends to seek factual as opposed to imaginative knowledge, a high priority, and will for that reason find eventual realism in interpretation of man and his world. Finding realism is the very mission of science, but relatively few people are proficient in its full understanding. As with religion, retained information about science for most people consists primarily of highlights of interest and of what one hears or has been told. Without a background of proper study it is difficult to acquire full confidence in scientific knowledge and meaning. But science should be recognized as the representation of all knowledge. It is now becoming more complex but it increasingly provides new understanding and answers many of the fundamental questions of man some of which relate to human existence and religious belief. This book has therefore, emphasized the importance of education that should includes sufficient science to enable all people to reasonably appraise factual knowledge and understanding, especially that which concerns human function. It has particularly explained how the modern mind views religion and the basis of both real and imaginative learning. It has provided some common examples of behavior in society to demonstrate the power and acceptance of imaginative forms of ideology.

**Man is still in his early stages:** Although man has always had an endless interest in the world about him, there was a time when there simply was no "science." On occasion there were great men of philosophy, thinkers who tried to understand man and his world. But only a few centuries ago, as late as the 1600s, man was still quite ignorant about his relationship to the natural world and much like religion, philosophy had wide variations in interpretation and was subject to serious errors and misjudgment because of limited knowledge. That same period also saw the beginning of a shift from the supernatural to the natural and human reason became the highest authority. But man emerged from animal ancestors more than a million years or ten thousand centuries ago and it is surprising that only in recent centuries has there been any broad rational understanding at all. This book suggests that human rationality, a product of the powerful and relentless evolutionary forces of nature, is still in development but may now be in a state of acceleration. The element of force/time remains a significant consideration with respect to a measure of progress. One must be aware that, in cosmic time, planet Earth has only existed for about half of its predicted terrestrial life and that man, relatively new on his planet, still has far to go.

The more educated readers are already quite aware of the many stages of natural change that have contributed to mankind's reaching the present position. They are therefore cognizant of the invisible forces of change constantly acting on life. They know that living in the distant past was very much different than it is today and they can therefore project a totally different life for those who follow us into future time. This book is not only interested in that projection but also in noting applicable current signs of such change and the views of other visionaries. As a first step in reader appreciation it suggests willingness to replace any speculative information with factual modern information no matter how desirable or undesirable it may appear. Going forward from the present is comparable to beginning a new life starting from an advanced point in the human development process. Clearly, since his beginning, man has accumulated a great deal of understanding as reflected in his huge bank of current knowledge, well beyond the capacity of any one mind. But there is much more information yet to be acquired and human change will continue to closely follow it.

**Closer and closer to understanding:** The acceleration of knowledge and change in spiritual belief are prime elements of the religious theme of this book. Its thesis is particularly directed to progress in increased understanding and greater reasoning ability, both of which are important contributing factors in achieving eventual ideological unity, most likely in some form of worldwide secularization. Timing is considered to be a function of the extent and intensity of human education.

This chapter is intended to emphasize technology and scientific understanding, the prime bases of all knowledge as well as their reliable acquisition. The final chapter of the book that follows will deal with how it all plays out with respect to ongoing globalization and the future condition of the planet. So far, this study has provided a simplified explanation of the mind of belief and has sufficiently reviewed the origin of man and the nature of his world to assure minimum understanding. It has observed that man has finally begun to penetrate the confining processes of his emotional mind and recognize the difference between belief called "faith," based on hope and desire, and "reality," based on confirmed observations and measurements. The distinction is intended to clarify certain early mind function, particularly its tendency to build and rely on simplistic imaginative visions for achieving happiness, contentment and satisfaction. But as science advances further, it is fully expected that the mind will increasingly encompass a sense of reality based on factual determinations equally resulting in human satisfaction and happiness.

History has recorded many theological concepts and religious teachings, particularly developed over the past two or more millennia. But man in different regions of the globe, has interpreted reverence somewhat differently and has had different views of origin and of an all powerful god and indeed, interpretations of how man should best relate to them. History has also recorded the discrete incremental steps of progress made by man over centuries and millennia, each step bringing man closer to understanding, not only of what to believe but also an understanding of belief mechanisms themselves. This stepwise progress now continues in increments of advancement and change based on modern science. For example, society has only recently felt a surge with respect to the serious critique of imaginative belief called the new atheism, a reaction by modern visionary thinkers, scientists, philosophers and intellectuals. Forces of change initially demonstrate their influence on searching leaders and teachers of man.

**Man's wider angle lens:** It is significant that after more than two thousand long years of religious development, the totality of man has only reached a position of spiritual uncertainty, of mixed beliefs, of strong emotional desire, and of inner skepticism about his god. The minds of ignorance, speculation and imagination have brought conflict and confusion to society throughout the growing world. The most surprising aspect of conflict seems to be how much imaginative religious views still dominate findings of factual science. But modern man is rapidly reaching an improved understanding, not only of the complex belief processes but also of the function of the vast universe in which he has been involuntarily thrust. Thanks to science, he is now capable of a far more rational re-explanation of earlier notions. But it is clear that man has only begun to reconcile his present knowledge with the knowledge of history. It is equally surprising that only a small fraction of the population has thus far been able to free itself from the emotional spell to appreciate reality.

At no time in past history could man look to the future with more expectation, expectation for continued positive change and greater human advancement. Globalization, increased population, and man's continual encounter with more complex technology have made his task more difficult. It has substantially extended the scope of his objectives and his educational requirements. Man must now work more diligently to utilize his talents and connect technologies across the globe whereas in the past, he was content to deal with relatively limited information confined to limited regions. It is no wonder that traditional religiosity in various forms has gained a solid foothold throughout the world for man with limited scientific education.

**Understanding oversimplified:** Education has particularly encountered reluctance to engage in the perceived complexity of mathematics and science. However, scientific knowledge is an extension of general knowledge and it cannot any longer be by-passed and avoided. Without factual knowledge man will tend to favor the simplest most emotional and imaginative interpretation of life and revert to visions and concepts of the distant past. Educated and intellectual modern man is increasingly in agreement on new factual technology and on matters that relate to his material world. But that information does not yet sufficiently extend to ordinary man. On matters that concern structure and function of the body and mind his progress is not yet fully shared and even less fully understood. Although the most informed have acquired reasonable understanding in particular fields such as medicine and psychology, the world as a whole is yet awash with misinformation. Many have found refuge in some form of different beliefs but the underlying truths remain far from accepted reality. His attempt to realize a true global community is currently in a state of disorganization because of economic, political and social differences. These differences are reflected not only among governments but also among individual cultures. Although man's desire for worldly goods and happiness is universal, he is still lacking accord on the very basis of life and on accepted standards of behavior. The area of least accord is one that concerns his emotions, particularly the responses of the mind to spirituality discussed in these chapters. It is the critical area of understanding about his origin and expectations. This area of great confusion exists largely because man has been taught differently in different parts of the world and his teachings have all been based, not on invariant factual knowledge, but on what man has desired and imagined and indeed, on the specific prevailing beliefs of an area of his planet. With human migration and better means of communication, information has become increasingly "scattered" with a preference for beliefs that promise the most with the least effort.

**The philosophical jump to conclusions:** Although the current era finds society at the beginning of a steady movement toward reality, we must again consider the forces of change as part of the dominant forces of nature, the same forces that have created life and have advanced man through his emergence, the same forces that have produced desire and imagination and have produced his quest for factual human knowledge on which his continued existence depends. Dependence on imaginative forces alone could not have reliably brought man to modernity. Intellectuals, philosophers, scientists, and other similar minds are the leaders with respect to the future path of understanding for man, as has always been the case throughout past history.

Descartes imagined that the existence of God was "proven" because man could not have the god idea unless it came from God. Spinoza imagined God to be a substance on which all other substances depended, and the intellectual "love of god" the highest good of man. In the centuries that followed philosophy concentrated on the experience of man. Later, Hume said that knowledge is limited to experience, that we can only know objects of sense perception, and can only know probability, not truth. In the nineteenth century, Nietzsche invented the "superman," considering the desire for power to be the basic instinct of man.

Kierkegaard claimed that the goal of human experience is the "knowledge of God." Civilization then became the center of attention. Kant said that man gets impressions of all things through his senses, but that minds shape and organize those impressions. In the twentieth century, philosophy took two directions, one based on service and the other on more specific concern about man. The works of James and other philosophers directed man's concern to himself, the philosophy of "pragmatism" with adjustment and improvement of society as man's chief goal. Philosophy, with great effort, tried to provide a unified view of the universe. James regarded philosophy as "an unusually stubborn attempt to think clearly." Philosophers Jaspers and Sartre followed with their concern with the universe in terms of human emotion and man's place on Earth as a rapidly changing phenomenon. Human thinking has indeed changed in every stage, and this book describes the result of the most modern understanding, indeed current changes taking place in spiritual thought and how it will become a more acceptable way of society as early as the forthcoming century.

Up until the eighteenth century, there was little or no distinction between science and philosophy. They both looked for basic principles and systematic investigations with which to gain knowledge. Science later became directed to specific subjects, whereas philosophy became concerned with general laws and structures of all reality. Philosophy strove to examine the logic behind religion and to test methods of thought that are used by religions. It tried to clarify human meaning and to judge the evidence behind it. Both Aristotle before the renaissance, and Hegel, in a much later century, placed religion high in their thinking. Hegel felt that religion included all aspects of science and that it even went beyond science. Today, philosophy and science are becoming more tightly integrated and science is rapidly developing the tools with which philosophy and religion can finally be mutually understood.

From ancient times, it was thought that a humanlike god existed for man, a god that was all-powerful and knowing, and existed everywhere, that God was intimately concerned with the behavior of man. Consequently, a broad voluntary moral code was developed around it. The moral code was often broken by those who most promoted it. This vision was not only expanded in time, but reached far beyond to the multiple beliefs of present man. Meanwhile, modern man has acquired a vast amount of new knowledge, not only about the function of man and his mind, but also about the nature of the world around him. Scientists, intellectuals and philosophers now tend to have a more rational view of forces that control man and his universe, and a more rational basis for his beliefs and, even more importantly, his behavior.

**Only man can save man:** It is clear that the seminal idea of the existence of a god was simply plucked out of the primitive emotional air and instilled in the imaginative mind as a useful psychological crutch. Belief in a humanlike god, still pervasive, is therefore thought to be a slowly fading ideology of past minds, being replaced by belief in a mechanistic force that has no mind and no relationship with man. The modern theory of origin of the universe is based on many keen scientific observations and is growing as an accepted new reality. Because of factual observations and measurements, this book has taken the liberty of projecting that trend to a future world of unified secularism, one of the most significant influences that has ever moved society. Remaining vestiges of imaginative belief are expected to continue to cause conflict, but the proportions of belief will eventually reverse. Observations and measurements of man must build its bank of factual knowledge to provide man with an indisputable basis for a firm direction.

Today, the mission of science remains very clear. Its mission is certainly not to engage in religion-bashing but to continue to make careful inquiry to determine and verify the true nature of man and his world. The fact that science has difference with certain baseless religious claims is invariant. Science has no preconceived notions nor does it reflect any partiality. It is dedicated to the fulfillment, betterment and extension of human knowledge based on measurement, observation and repeated confirmation. It is pursued for the prime purpose of expanding verified knowledge and understanding. The methods of science have even been used to examine and evaluate the nature of religious belief, as has been discussed in earlier chapters. Science is a very important discipline, the most essential effort of man that only seeks true knowledge, usually as the result of repeated tests made on the same question from different viewpoints. Scientific theories are often proposed, based on such physical experiments, subject to constant appraisal and to future confirmation. New information is always verified by substantial peer review. All the information acquired must fit together to create a complete framework of certain knowledge on which all the movements and advances of man are reliably based. As evidenced by the volumes of papers and literature produced in the scientific realm, everything possible is covered, including the analysis of thought processes and the mechanism of human imagination. Inspection of most any current issue of a technical journal, such as Science or Nature, will provide evidence of the relatively deep penetration of science into the materials that make up man and his world and the complex function of all that it involves. The constant building of data, much of which is not yet fully analyzed and integrated, speaks to further expectation of much important future knowledge. The rapidly increasing number of new students dedicated to scientific research and inquiry represented by all nations of the world is destined to become the eventual basis of human ideology.

**Forces act on mind and body:** This book speaks philosophically about the observed mechanistic origin of the universe and symbolically about the indicated transcendent forces of nature. The December 12, 2008 issue of Science, in an article titled "On Growth and Force," emphasized the influence of physical forces on the character of evolving basic life. One section said, "A remarkable example of how cells respond to their mechanical environment is how stem cells can be directed to differentiate into specific cell types by the stiffness of the extra-cellular matrix in which they are embedded. Stem cells cultured on a very soft synthetic matrix (mimicking brain tissue) will develop into neurons, whereas they develop into muscle cells and bone cells, respectively in matrices of increasing stiffness . . . . the results indicate that these cells are actively probing their environment by a force feed-back mechanism. But cells can also mechanically act on their environments, sometimes with major development consequences." This very significant note, slightly out of context, is one of thousands of notes on reported new research that will continually contribute to advances in knowledge of the make-up and character of life and of changing life, advances that, up to now, were not even imaginable. The emphasis of science is changing in the direction of molecular biology as man desires to know more, not about an imaginative god but about life itself. No one can predict the new technical position to which current science will take us in just a few more years.

Even scientists may be vulnerable: It is apparent that most religious adherents never seem to become sufficiently knowledgeable about the methods and findings of science to understand and to make an effective appraisal of their beliefs. There are many theologians who have read and studied science in substantial depth. But the real meaning of science is sometimes subconsciously tuned out by the effects of strong religious emotions. The influence of indicated dependence on a guiding humanlike spirit is extremely powerful. Those whose lives are primarily dedicated to scientific investigation, however, tend to disregard the emotion of faith in favor of verified facts. As a result, one finds that intellectuals and scientists generally have a different view of belief compared with most men, as frequently indicated in this book. Scientists that claim knowledge of many disciplines generally agree with the findings of this study. There are occasional scientists who effectively pursue dual views as unrelated matters. Scientists who have an early religious background cannot easily shed its lasting influence and sometimes conclude that the scientific explanations merely verify views derived in imaginative belief.

**Rational fact and irrational emotion:** Many ardent believers harbor doubts about science since its findings often clash with strong, deeply set early emotions. If they strongly believe foremost in a humanlike god who monitors their every thought and move, they will not accept or appreciate the factual nature of science. In recognition of difference they must feel that either science or their belief must be incorrect, although many simply tend to remain on the fence without challenge. Most uncertain people are still likely to favor imaginative religion since religion plays a deeper role in daily emotional life and particularly in their desire for continued life after death. Many of their understandings are assumed to be arbitrarily based on emotional feelings and passed along, generation after generation, gradually reaching a more significant collision course with expanding modern science. At present, there are many more religious adherents than there are scientists, and this condition understandably encourages active continuity of religious belief. On the other hand, the rapidly increasing number of secularists is expected to slowly change the eventual balance.

There is really no theoretical clash between the disciplines of science and the disciplines of religion, for they are two totally different things. Religion is the study of the spiritual feelings of human life with respect to his being as imagined, whereas science is independently a study of the factual physical world as observed and measured. There appear to be many claims of the various religions that are different from the findings of science, but they are based not on testing or rigorous examination but on simple faith and belief that has no material basis and has been called imaginative in this writing. Despite the existence of different disciplines, there is a certain unity of knowledge called "consilience" by Wilson (219) and this unity may ultimately bring greater reconciliation. Science will eventually be united with religion through the order of natural law. This book has shown the way through unification of knowledge of factual science with knowledge of the physical state of the spiritual mind. But this is only the beginning. Future minds will refine this view much more clearly and make it more easily available. This will occur as a greater percentage of the world population becomes aware of the function and natural direction of the human mind.

As suggested in earlier chapters, concepts pertaining to religion are closely allied to speculation about the origin of man and the nature of his universe. With respect to religion, science has developed a logical position on the origin and development of man and his world. It is a position that has been consistent with the established factual knowledge of man and with his history and present status. Within the limits of earlier remarks, science acknowledges that only a natural mechanical force exists to influence the function of the universe and all of its parts. At present, the material and dynamic energy of the universe is assumed to follow the natural laws of physics. Science, with all its modern tools, can find no evidence of the existence of any god or a human soul that extends beyond death. If either existed man would know about them. As a practical matter, science assumes that all life returns to the chemistry of the planet and that the dynamic consciousness it once represented merely vanishes with it. Among all the myths and hype about other, more desired interpretations, biological death has been determined to be final death. There are of course many notions that tend to soften this blow to human ego, but no wish or desire can change the methods of nature. They just are what they are, and man must accept them, abide by them, and make the most of them in his finite lifetime. Man should be delighted that he has learned the facts of life and death. He should celebrate his good fortune by any means at his disposal, means that far exceed the celebration of an imaginative notion. It seems very doubtful that science will change much in this view even in the foreseeable future. In the distant future man may well learn how to extend the processes of life in successive steps but will meanwhile remain true to these principles for ages to come.

The need to know: The John Templeton Foundation, whose motto is "Investing in the Big Questions," has always reflected a strong religious bias and has been instrumental in awarding a major annual prize for outstanding work in science and religion. But, up to now, all the winning works seem to have reflected what we have termed imaginative belief. It is my fondest hope that the organization might grow under the influence of a more rational supporting philosophy within the province of this type of study. In a recent two-page spread, its display ad (220) was entitled "Does science make belief in God obsolete?" Thirteen widely recognized persons were shown as contributors to answers, and they are here summarized briefly as follows. Steven Pinker said "yes", if we mean the entire enterprise of secular reason and knowledge. He then proceeded to explain his position with more profound reasoning. Christof Cardinal Schoenborn, Archbishop of Vienna, said "no and yes", no as a matter of reason and truth, that the knowledge we have gained through science makes belief in an "intelligence" behind the cosmos more reasonable than ever, but yes, as a matter of mood, sensibility, and sentiment.

William Phillips, a Nobel laureate in physics, said "absolutely not." Now that we have scientific explanations for the natural phenomena that mystified our ancestors, many believe that we no longer need to appeal to a supernatural god. I am a serious scientist who seriously believes in God. Pervez Amirali Hoodbhoy, Physics Chairman of the University of Islamabad, said "not necessarily" but you must find a science-friendly, science-compatible god. First, try the pantheon of available Creators, inspect thoroughly, and if none fits the bill, invent one. The god of your choice must be a stickler for divine principles.

Mary Midgley, a philosopher, said "of course not," belief or disbelief in God is not a scientific opinion, a judgment about physical facts in the world. It is an element in something larger and more puzzling, our wider world view. If we did start to doubt other people's consciousness and truthfulness or the regularity of nature, we would lose not just science but our sanity.

Robert Sapolsky professor of sciences at Stanford, said "no." Despite the fact that I am an atheist, I recognize that belief offers something that science does not. Christopher Hitchens, author, said, "No, but it should. Belief in a divine presence or inspiration was often merely assumed to be a part of the natural order. It could be argued though, if I were a believer in God I would not attempt it, that a commitment to science by no means contradicts a belief in the supernatural." Keith Ward, a priest of the Church of England at Oxford, said "No, far from making belief in God obsolete, some interpretations of modern science provide positive reinforcement for belief in God." Victor Stenger, professor of physics and astronomy at the University of Hawaii, said "yes," but there was once a strong scientific argument for the existence of God based on complexity of world design. People felt that it must have come about by the action of great power and intelligence.

Jerome Groopman, professor of medicine at Harvard, said "no, not at all," as a physician and researcher I employ science to treat disease. As a person of faith, I look to my religious tradition for the touchstones of a moral life. Michael Shermer, publisher of Skeptic Magazine, said "it depends," on whether one emphasizes belief or god. Science does not make belief in god obsolete, but it may make the reality of god obsolete. Millions of people believe in astrology, ghosts, angels, ESP, and paranormal phenomena, but that does not make them real. Kenneth Miller, professor of biology at Brown University, said "of course not," science itself does not contradict the hypothesis of god.

Rather, it gives us a window on a dynamic and creative universe that expands our appreciation of the Divine in a way that could not have been imagined in ages past. Stuart Kauffman, director of the Institute for Biocomplexity and Information at the University of Calgary, said "no" but only if we continue to develop new notions of God, such as a fully natural God that is the creativity in the cosmos. All these differing views merely confirm the representation of confusion discussed in this book. A more complete coverage of their commentaries can be obtained on the Internet under templeton.org/belief.

Science is the discipline carefully defined above, whereas God is an image of the mind, one that can exist as a humanlike god, or a god in the form of all-powerful nature if preferred as discussed fully in these chapters. In some cases, language meaning seems to be in question, but despite expressed differences, when one examines the full backgrounds, experience, and prior works of these responders, one finds that their responses fall in line with applicable commentary and expectations contained in this book. Man is still in a very early stage of spiritual unity and therefore still subject to widely varying detailed views. The purpose of this major public ad is not clear, it seems exploratory, but without the detailed explanations of spiritual belief, carefully organized and referenced. Most responsive comments of that nature cannot have a great deal of meaning to the reader. In brief, everyone would like there to be a benevolent personal god to look to, but relatively few have yet reasoned that it might only be a wish without evidence of reality. The full study of sciences will indeed, make belief in God obsolete if every aspect of science and every aspect of belief is taken into account and a rational judgment rendered. This book suggests that such belief will continually deteriorate based, not on specific proofs and comparisons but on projections of the effects of observed forces that are changing man and his world.

**Man's reliance on the scientific method:** The scientific method has brought us all of the knowledge of man to date, and will continue to be the most accepted way of understanding man and the world around him. It has given us knowledge of the universe and the position of man within it. It has given us substantial knowledge of the inner workings of man himself. Without such pursuits, man would still be filled with incorrect imagined interpretation of his own structure and function as well as those of nature around him. Unfortunately, many members of society, important and otherwise, have not had an opportunity to learn what constitutes full reality. Modern society does not always appreciate how much the view of man has changed. Most of the significant imagined notions have vanished among modern man, with the exception of the imagined humanlike god. That is largely because those that are ardent believers, in effect, maintain a powerful emotional posture in such a form that little time or opportunity can be devoted to reason. The teaching of factual science is the only major avenue of reason. Scientific and philosophical minds therefore tend to lead advancing understanding and, indeed, the new secular movement.

True science is independent of politics and religious dispute. Dedicated scientists often relate well in the midst of serious argument in other disciplines. The November 28, 2008 issue of Science suggested that science can be effective in defusing global conflicts. Ambassador Pickering, in an address at the AAAS on the subject of science diplomacy, claimed that shared interest in technology can and will build a positive relationship with the strongest competitors and foes, while damping the possibility of volatile confrontation. He spoke of disarmament, nonproliferation, energy, climate, development, health, food, water, and immigration as issues that require diplomacy, but none could be accomplished without good science. All these issues, no doubt, including spiritual considerations, will come to a head within the next several decades, and cooperation among all nations and all religions will be paramount.

**Galactic understanding:** Only in very recent times has man had any real scientific understanding of how the universe is structured. But despite the great volume of astronomical knowledge, there is still much to be learned. This book has stated that all known life is confined to our own planet Earth, that man consists of biological material, extremely sensitive to the character of the environment in which he evolved. It has stated that our relatively small Earth is one of the planets of the huge solar system dependent for survival on the sun's energy, that the solar system is a negligible and insignificant part of our great Milky Way Galaxy, a relatively flat spiral array of billions of bodies, nearly a hundred thousand light years across. It also stated that there are billions of such galaxies in the universe, which appears quite unlimited in size. Based on scientific measurement, we speculated that the universe was derived from some sort of "singularity" in a great explosion about fifteen billion years ago and continues to expand in space (221). All this has been demonstrated by scientific observation and measurement and to date there is no scientific principle that denies such a view. Future photography with the amazing Hubble and similar telescopes combined with a greater depth of study of mind function will result in human understanding that will increasingly point toward the claims of this writing.

**Achieving the possible:** It was as recently as 1957 that the Russians launched Sputnik, the first artificial satellite (222) to circle the Earth. It was the first time a manmade object was propelled into space, climbing out of the return influence of terrestrial gravity into stable orbit above the atmosphere. It altered the nature and scope of the Cold War of that time. Russia clearly intended Sputnik as an indication of its technical prowess with all its military implications. Walter McDougall, a historian at the University of Pennsylvania, wrote that "No event since Pearl Harbor set off such repercussions in public life." After Sputnik, there was no stopping the momentum of space competition. This led to the Apollo moon program, involving a total of a dozen men who walked on the moon's surface, which ended in 1972, and no one has since been there. This was another example of the application of many aspects of scientific knowledge and another milestone of achievement by the underlying forces of progress.

As to our own galactic area of the universe, we are learning infinitely more about the Milky Way, its extra-solar planets, and about the evolution of life on planet Earth. The universe is so vast that most scientists believe that evolution of life might have taken place, or is in the process of taking place, somewhere else in our galaxy if favorable biological conditions similarly exist. They are therefore searching space for four conditions to be met, namely, the presence of a host star like our sun, enough heavy elements to form a planet like ours, sufficient time for biological evolution to take place, and an environment free life-extinguishing supernovas. Such a galactic Habitable Zone (223) has now been identified in an annular region at a specific distance from the Galactic center, about three quarters of its stars being older than the sun. Obviously our own solar system is included in that habitable zone. The eventual finding of some sort of life form, no matter how simple, will forever erase the special creationist character given our own planet with respect to life. Our task is still a very difficult one because of the immense distances involved. However, it is not the purpose of this book to define the universe or to speculate in detail about its structure. Its mention here is to provide a reminder of the magnitude of its size and the nature of its complexity relative to the immediately surrounding world we know and occupy. Although religious belief can exist without a great deal of material information, such knowledge does play a role in understanding as a basis for confident interpretation of the world and the nature of human existence. It is important to understand that further knowledge of the laws of physics and the discovery of more particles and more details of material properties does not change the view of the god problem expressed in this book.

**Each day brings new understanding:** The science of astronomy that gave us much of discovery and confirmation of knowledge of our outer world is relatively new and much of it was acquired well into the twentieth century. The Eddington confirmation of Einstein's prediction that massive bodies can warp spacetime and bend light, and the Hubble discovery of the first galaxies outside of our Milky Way, were disclosed. The Tombaugh discovery of Pluto and the Jansky detection of radio waves from the center of our galaxy took place only a few decades ago. Similarly, cosmic ray research by James Van Allen led to the discovery of radiation belts surrounding the Earth. Schmidt later identified the first pulsar, thought to involve emission from matter falling into black holes, and Bell detected the first pulsar energy emitted from rapidly spinning cores of burned-out stars. The big bang microwave afterglow and gamma ray bursts from the explosion of massive stars followed soon thereafter.

This much abbreviated history marks only the beginning of many successive scientific discoveries in recent times, and our own spacecrafts were sent to confirm observations with various types of planetary surface images sent back to ground stations by telemetry. Extra-solar planets were discovered, and distant supernovas showed that expansion of the universe is accelerating, driven by a mysterious "dark" energy. A supernova is a star explosion having reached its maximum intensity. The Wilkinson probe, measuring cosmic microwave background, gave high resolution for measuring age and composition of the universe. The

Sloan Great Wall was recently discovered as an enormous collection of galaxies over a billion light years in size and about a billion light years distant. The Spitzer telescope, promoted by Lyman Spitzer at Princeton, discovered faint infrared glow remaining from the first stars of the universe. It was observed that powerful "black holes" exist at the centers of galaxies and emissions can be detected from hot gases streaming into them.

It is clear that the science of astronomy, supported by new space vehicles, is advancing very rapidly and our view of the universe and its parts is not only known in reasonable detail, but has advanced rapidly over time. Telescopes are now getting very large and far more sophisticated, and are used in various combinations for increased clarity. Here, again, the continuing forces of progress have enabled man to learn more and more, year after year, about where he is in the amazing structure of the universe and how he might fare in its reactions and predictable future changes. It is clear to most highly recognized scientists such as the late Carl Sagan, a well-known nonbeliever, who has contributed so much to man's understanding, that humans live in an extremely small and insignificant part of a huge mechanistic world. Even the great minds of forward vision indicate a preference for the imagined ideal concept of a benevolent supreme being of many religions, if only it were possible.

**Evolution and creationism:** In the ongoing arguments, purportedly between science and religion, perhaps the most heated area has been the science of evolution and creationism. Evolution has already been mentioned in prior chapters. There have been many books and pamphlets published on the subject and only an explanation based on scientific discovery and research is acceptable. One of the best pamphlets on the subject is one prepared by the National Academy of Sciences, Institute of Medicine. Evolution is one of the most important discoveries of man. It successfully explains the observation and experiment in a broad spectrum of scientific disciplines. It represents the very foundation for modern biology and has opened the doors to entirely new types of medical, agricultural and environmental research. It has led to the development of technologies that can help prevent and combat diseases. The attempt to introduce non-scientific concepts into the processes of education by faith-based groups just muddies the educational waters and undermines efforts to continue factual scientific education.

Every adult is patently aware that biological traits pass from parents to offspring. Within the natural forces already discussed, this is the very basis of evolution. Non-scientists do not have the depth of knowledge to understand evolution with conviction. There are even occasional scientists whose backgrounds are so deeply religious that they will find ways of twisting science to argue in favor of some version of creationism. The reason behind this is explained in the first chapters. However, the processes are so clear-cut and proven that evolution in some form must become universally accepted by all men.

In ordinary conversation, a theory is considered a hunch or speculation. In science, an established theory is based on a comprehensive explanation of an important feature of nature and is supported by many established facts over a long period of time. Theories make it possible to make prediction about yet unobserved phenomena. Scientists therefore no longer question the underlying facts of evolution as a process. The concept has already withstood extensive testing by many thousands of specialists, not only in biology but in medicine, anthropology, geology, chemistry and other disciplines. Research and discoveries in various related fields have reinforced each other without fail over a century or two and continue to show evidence that evolution is a firm reality.

**Creation is like magic:** I emphasize the theory of evolution particularly, because imaginative religions still tend to include the creation and development of life in their visions of belief. Creationism has no basis in reason or logic and it has been criticized in many solid arguments, including the notion that a creator must also have a creator. It is not my mission here to set forth all the arguments on both sides of the issue, but only to say that creationism is merely another phase of imaginative belief. Once the human mind is subject to the imaginative framework of a religion, there is no limit to its imaginative extension. One element of the imagination builds on another until the entire framework has been established to suit, with the same form of imaginative thought. This has happened in religion over many centuries. Every phase of major religious beliefs has been carefully placed and repeatedly reviewed and studied. However, the breakdown of one part of its structure tends to destroy the entire framework. In this writing, the concept of a humanlike god, the basic central backbone of most religions, is in itself subject to question. Once that imaginative vision is replaced with concepts based on reality, all the surrounding imaginative components also vanish,

Creationists claim that many features of living beings are far too complex to have evolved through a natural process. They say that irreducibly complex systems must have been created by an all-knowing, intelligent designer. Scientists not only ask who designed the designer but see most products of nature as being complex only with respect to the capability of the present human mind. They also suggest that most of the design concepts, studied and learned by humans, are related to observations of nature. The notion of creation is so foreign to the mind of a competent scientist that he finds it difficult to respond. Creationism is certainly not comparable to a science, but based on beliefs outside the known natural world. Science can only deal with naturally occurring phenomena and most often admits little or no knowledge about details of the imaginative worlds as described in this book. The many questions raised in defense of creationism have been satisfactorily answered in terms of the physical sciences. Many scientists believe it is useless to argue with a nonscientist about evolution because he does not have enough training in scientific method to understand the strength of the argument. This point was also made earlier with respect to the underlying basis of imaginative belief. Where creationists work their way into government or positions of management, capable scientists and teachers increasingly find their jobs at risk. As mentioned elsewhere in this writing, religion and science are two different disciplines, two different processes of thought. They cannot be compared on a one to one basis. It would be like comparing a material thing with an idea or the law of gravitation with a fairy-godmother.

**Man is also a mammal:** Science involves the investigation of everything in order that man develops a good understanding of his world and its function. For example, in recent years, science has studied the growth of animal meat in the laboratory with a view to providing expanding food needs. The organization called People for the Ethical Treatment of Animals (PETA) is offering a million-dollar prize for the first person to come up with a method to produce commercially viable quantities of in vitro meat at competitive prices by the year 2012. The meat thus produced is expected to be equivalent to animal meat that has never been part of an autonomous animal. The objective is to make the production of meat more ethical and humane. Could it mark the end of domesticated farm animals for our food supply? If done in this manner it would be equivalent to normal growth but absent a brain and sensing means. If man were absent a brain and sensing means, would there still be a god?

With this method of meat production in the offing, people seem to prefer the more measured approach. They desire to ensure the least possible cruelty to animals, raising them in ways that are both ethical and environmentally sound. Historically, the cultural and historical bond between humans and domesticated animals exists only because of the uses we have found for them. It is suggested that the world would indeed become barren if the herds and flocks disappear in favor of meat grown in a laboratory tank.

A question often arises as to whether man is playing God again. The new process verifies the basic chemical nature of living substances and the position of animals in the well established complex food chain of the world. Most religions have arbitrarily claimed that animals do not have souls and are therefore classified as being at a lower level. But what happens as one moves up in the chain toward man in evolution? The ape and the chimpanzee are equivalent to man in most respects, as confirmed by their almost identical genetic structure. One is only time-removed from the other. Why do they not have souls and an appropriate relationship with man's assumed god? With the advancing technology of molecular biology, man is learning a great deal about living chemistry. The genetic manipulation of life has already become routine and the cloning of humans in some manner may not be far behind (224).

**Vanishing minds of progress:** The mission of science was never made clearer than by the words of the late Princeton professor, John A. Wheeler, whose obituary appeared within only a few months of this writing. Working with Einstein and Bohr, both world-famous names in theoretical physics, he was considered the successor of the new relativity theory (225). It is said that he was the first to assign the name "black hole" to a certain defined cosmic condition at the centers of galaxies. He contributed directly to the design and construction of the first nuclear weapons and to methods of achieving nuclear power. At the blackboard, Wheeler started at the upper left hand corner and soon filled the entire board with mathematics. He had chalk in both hands, giving the impression that he just could not write fast enough, as fast as his mind dictated. Knowledge about physical principles poured from his mind. Einstein, Pauli, and Feynman were often present, all since becoming Nobel scientists. When I mistakenly thought one of them had fallen asleep, he would abruptly respond to the slightest mathematical slip on the blackboard with brilliantly aroused enthusiasm.

**Science and religion will merge:** I often thought about the personal beliefs of these top physicists, and only occasionally would they offer signs of their inner ideologies. Their religions were certainly part of their deep understanding of the material world and they seemed to think little beyond physics. Wheeler once remarked, "I have to admit that I never stop thinking about physics. I have never been able to let go of questions like, How come existence? How come the quantum? What is my relation to the universe and its laws? Can space-time be all that there is? Is there an end to time?" He said that questions of this sort have nothing to do with his religious convictions, which "center on guides to living, guides to civilized intercourse among humans. The deep questions that I wrestle with belong to science as I define it, not religion." Although I never heard him say so, I think his personal beliefs were quite similar to his fellow scientists, Einstein and Sagan.

In looking back at his life, Wheeler indicated that it was divided into three parts. First, that "Everything is particles" looking to build everything with neutrons and protons and the like, out of the most fundamental of particles. The second part was "Everything is fields," when he viewed the world as one made of fields in which particles were mere manifestations of electrical, magnetic and gravitational fields and space-time. He lastly talked about "Everything is information," as he focused on the idea that logic and information is the bedrock of all theory. It is this type of thought that has brought us to our new modern view of religion. Human viewpoints change radically with learning and with time, and modern man of reason can see answers that have never previously been possible.

Wheeler felt that the greatest part of being a theoretical physicist was to develop concepts and equations that tie observations and experience together. Yet, he felt that was not the whole thing, there was a place for invention and for theory, guided by aesthetics. He, like Einstein, saw the beauty in nature and the beauty in mathematical equations that defined its laws. But he said there was still more for him, for he had not been able to stop puzzling about the riddle of existence. I think often of these great minds who pieced together the mechanisms of the material world, but Wheeler had little time to theorize on biology or on neuroscience. I think of these great minds whose functions were as those I described in earlier chapters, minds that are now returned to the Earth, molecule by molecule and atom by atom, never to again assemble to provide thought and theory. Hopefully, those atoms will ultimately migrate through many phases and symbolically become part of future minds that must begin anew to contribute to human knowledge.

**Meaning and what you make of it:** As we consider the questions posed by Professor Wheeler, science has brought us many of the practical answers but not all that those complex minds would desire. There are many more details of interest, such as the makeup of the very smallest of particles, and current scientists are in hot pursuit. Science has, however, given us sufficient information to answer many questions about practical life. It has given us information that allows man to rationally look at religious beliefs as I have done in this book. It has shown that the meaning of life rests with man on Earth and with the relationships among man, and not with any humanlike god. Although scientists theorize about the origin of our universe and much of its subsequent history, there is still the matter of earlier time and earlier history that is not fully understood. There is also the matter of subsequent time with which we continually speculate. Meanwhile, man has all it can do to make his existence as meaningful as possible for himself and for fellow man, while passing through his limited period of life.

**Indelible engrams:** A prestigious Templeton Prize, relating to Science and Religion, was earlier awarded to Father Michael Heller, a Polish Catholic priest and cosmologist, who is said to have developed original concepts in the study of the cause and origin of the universe. His work was mentioned in earlier chapters. The March 13, 2008 Global Spiral printed a portion of his statement made upon accepting the award: "I always wanted to do the most important things, and what can be more important than science and religion? Science gives us knowledge and religion gives us meaning. Both are prerequisites of the decent existence. The paradox is that these two great values seem often to be in conflict. I am frequently asked how I could reconcile them with each other. When such a question is posed by a scientist or a philosopher, I invariably wonder how educated people could be so blind not to see that science does nothing else but exploit God's creation."

This quotation speaks for itself and it must be interpreted as presented. Father Heller has evidently complied with his Catholic training and education. Despite his acknowledgement of a paradox in the full sense, he simply concluded that the universe was created by his humanlike god and that the methods of science conveniently fit it, at least to the extent that they do not violate the assumed premise. That view is not the interpretation of most recognized cosmologists. His vision is precisely that which was discussed at length in the first two chapters of this book as imaginative. Once the mind is charged with a deep belief, there is a very slim chance of conversion, irrespective of the level of education. The imagined existence of a humanlike god and his imagined creation of the entire universe and all of its contents, is certainly incompatible with the reason and logic of the modern mind. Indeed, the world and its people all have meaning. It is the meaning seen by the rational mind. As to the purpose of life, the word is not applicable for purpose is acquired and defined separately by each individual. It is the same with or without a god.

**Studying to believe:** Father Heller studied theology at the Tarnow Seminary in Poland. He was ordained as a priest and obtained a doctorate in Philosophy from Catholic University of Lublin. He conducted research at the Institute of Astrophysics at Oxford and became a full professor at the Pontifical Academy of Theology in Cracow. He was later elected to the Pontifical Academy of Sciences in Rome. It appears that his entire education was acquired under the direct influence of the tenets of the Catholic Church. It would indeed be surprising if a vision of a Roman Catholic god, so deeply ingrained by dint of such an education, could ever escape the memory and emotion of any human mind. This would be predicted by the explanations of the initial chapters of this book.

**Intuitive conclusions:** Great scientists, such as Einstein, Sagan, Hawking, and many others, looked at the "god problem" with much care and concern. Great theologians too numerous to begin to name, also studied the god problem. However, it is clear that scientists looked at the matter from the "outside "whereas theologians do so from inside. By that I mean that the scientists, no matter how they began their spiritual experiences and beliefs, have scientifically reached the conclusion that there is no demonstrated humanlike god in the known world structure and go on with their science, looking for an explanation of the mechanistic universe and the laws of nature considered factual. Theologians, on the other hand, become familiar with, latch onto, and retain the idea of a personal god and imagine its existence along the lines of traditional belief. This book does not claim that all scientists are atheists and that all theologians are believers, for there are always a few, but relatively few, who do not completely fit that rule.

**Miracles deter scientific progress:** Science is a catch-all title for most everything man does. Collectively, it includes the continued investigation and confirmation of knowledge which has been referenced repeatedly in this writing. However lacking in its definition at the outset, it has always been the basis of progress and will continue to be forever in the future. As stated earlier, the clash between science and religion is only apparent. Each is a field of its own and one does not negate the other as a discipline. The clash arose because some of the claims of religious belief violate natural laws on which science depends. They are therefore lacking in factual truths. These "miracles" might include popular beliefs such as creation, virgin birth, rising from the dead, ascension into heaven, the miracles of wine, loaves, and fishes, and the parting of the Red Sea. There are many others so unsound that they cause some to interpret those stories as mere legends with symbolism. Science tells us that these events just cannot occur and have never occurred. Although religion tends to maintain them as beliefs, the position of science is gradually winning out. The argument will reach ultimate settlement in the forthcoming era of secularism when man will be forced to concentrate on human behavior on a global scale.

A casual scan of the several hundred references in attached will provide the reader with a broad review of current thought and writing relative to religious belief and contrary thought. It includes the works of deep believers, of believers on the fence, and the works of respected people who are certain about the imaginative aspects of a humanlike god. These chapters have demonstrated that, although imaginative belief in a humanlike god still dominates society, largely because of its emotional appeal, the rational minds of reality are moving steadily in a secular direction. There is sufficient movement and evidence to predict the eventual conversion of all of society to some form of secularism as defined in earlier chapters.

The role of science is indeed an endless human role and is only at its very beginning. It is a role of constant discovery and new understanding based on a process of constant confirmation. In a sense, it is unidirectional for it allows those that follow to more specifically determine the meaning and basis of life. Its direction alone adds to the soundness of the theme of this book. Only a quarter century ago, my nephew, at the University of Pennsylvania, successfully modified the "sacred" germ line of life. Ralph Brinster injected cloned rat genes directly into a fertilized mouse egg to produce a rat-like mouse. A television documentary called it "the Brinster technique." Ordinary human cells, called soma cells, are often exploited, but the "immortal" germ cells of fundamental human reproduction are treated with special care. Although man may expect to experience natural mutations, this work led to a declaration that "engineering of the human germ line should not be attempted without the consent of all members of society." A resolution to that effect was signed by religious leaders and scientists and entered into the *United States Congressional Record* of June 10, 1983.

**The way things are:** Christopher Reddy, director of the Coastal Ocean Institute, reminded us in the March 13, 2009 issue of Science that Thomas Jefferson wrote "an enlightened citizenry is indispensable for the proper functioning of a republic." He added that if we believe that science has a rightful role in our society, then it is the scientific community's responsibility to enlighten the public as to why and how. All nations of the world must have rational and intelligent leadership in order to move mankind through the next stage of human awareness, to reach unity and peaceful coexistence. Scientific understanding is the most essential base and should not be diverted by imaginative notions.

From the viewpoint of science and logic man will eventually be aware that he lives in a totally mechanistic universe in which cosmic time has no importance to him. It is not what he would choose based on how his mind has developed. He simply cannot easily visualize how the mechanical universe can exist without a humanlike or some other form of god. He thinks in god terms because he only relatively recently came into existence on the planet and his evolving mind has been influenced by the history that became part of his primitive development. His mind will continue to develop and become far more complex as described in earlier chapters and his future science will eventually provide more rational conviction of his position. His future science will merely further confirm the mechanistic character of the universe. It will find no evidence of any god or any other humanlike force with which man can communicate. It will find no applicable intelligent design or any other capability that is exclusively related to the development of man. With chemical origin and biological evolution, it is truly amazing that man has been able to learn as much as he has about his own life and it is most fortunate that his evolved mind structure and function makes it possible to experience such emotions as love, beauty, and desire in addition to feelings that allow him to achieve survival and propagation. He will find no heaven, no hell, and no soul but that basic human biology results in birth, growth, death, and redistribution of its fundamental elements. He will find that constant reaction and change is the character of the universe and that all life is influenced by its forces. He will lastly find that human life will adapt well to this reality and progress within the limits already discussed, for he will continue to seek reality and survival until he undergoes significant new morphology. These are not the prevailing traditional views of man but will in time become the accepted interpretation of all man. I do not fear that the expression of these views will result in "burning at the stake" but perhaps I should look forward to a reconciliatory statue in the Vatican garden or in the town square as described for Galileo and Bruno in chapter five.

**Man should enjoy his real world more:** This book has reviewed appropriate scientific aspects of imaginative belief and has claimed that the anticipated new world of secularism, although a godless mechanistic world, will provide beneficial ideological uniformity and eventual human satisfaction. Society, raised and nurtured on the existence of an imagined benevolent humanlike god, will adjust but rather slowly. The world is already beginning to experience a slow transformation from a clash of religions to a clash of cultures. By the time man experiences a uniform society of secularism the cultural differences should be reasonably dispersed through processes of education, globalization, mind development, and, indeed, through the constant mixing of genes in the process of diverse human reproduction. These are processes in which all branches of advancing science are destined to play a major role.

# Chapter Nine

## Forces of Globalization

*The Singularity is Near,* by Ray Kurzweil. Excerpt from a conversation between Bill Gates and the author

Ray:. . . . knowledge goes beyond information. It's information that has meaning for conscious entities: music, art, literature, science, technology. These are the qualities that will expand from the trends I'm talking about.

Bill: We need to get away from the ornate and strange stories in contemporary religions and concentrate on some simple messages. We need a charismatic leader for this new religion.

Ray: A charismatic leader is part of the old model. That's something we want to get away from.

Bill: Okay, a charismatic computer, then.

Ray: How about a charismatic operating system?

Bill: Ha, we've already got that. So is there a god in this religion.

The first wave of continental philosophy of religion assumes a certain self-distancing from religion, and as such, is

> symptomatic of the spirit of secularism that pervades the thought and culture of late modernism.
>
> There are a number of contemporary indications that this secularist mindset has undergone a decided shift. Indeed, one of the most surprising aspects of our postmodern culture is the global resurgence of religion though, as noted by Gianni Vattimo in introduction to the book that first brought this to the attention of philosophers, this return of religion is taking place more in the parliaments, terrorism and the media than in the churches, which continue to empty.
>
> —The Hermeneutics of the Kingdom of God: John Caputo and the Deconstruction of Christianity, Jeffrey Robbins,
> The Global Spiral, February 2008

**Humanity will continue to learn and progress:** The final chapter of this study offers a review of its principal concepts together with appropriate comments on expectations and effects of continued planetary globalization. It discusses the position and progress of migrating human life in our finite world, not under the influence of an imagined humanlike god, but under the powerful mechanistic forces of nature. It emphasizes certain aspects of the condition of man and suggests important steps be taken for his most favorable onward march. As in earlier chapters, it also includes some examples of behavior that support the theme at hand, behavior that demands review and reconsideration with respect to the optimum path for society.

The introduction of this book began with a discussion of the work of William James and his observation of responses of the spiritual mind. At the end of his "varieties" book referenced in that discussion, he indicated that his own beliefs did not correspond to the traditional pattern of the broader world about which he wrote. Understanding of the new psychology and neuroscience was not yet sufficiently advanced to allow him to offer adequate technical analysis. This book has therefore added much of the underlying causality, explaining how improved understanding of the mechanisms of the mind not only accounts for the past behavior that he described but also for much current behavior and for the anticipated behavior of man looking forward. With considerable humility, this writing was suggested as a fitting sequel.

**Belief problems solved:** This study has also outlined difficult problems inherent in imaginative spirituality and consequential social stigma. This is largely because most belief systems are fundamentally incompatible with human reason and with reality. In analyzing causality, it noted that such belief systems are still extensively taught and still variously fixed in the majority of human minds. It suggested that belief in, and reliance on, an imaginative humanlike god have represented significant factors in human behavior as well as the direction and pace of human development. It further suggested that society is slowly moving toward a new "age of human awareness," an age of greater understanding and appreciation of reality. It discussed the early creation of a variety of belief systems to satisfy emotions and desires, and claimed that adherence to traditional beliefs has not only occupied substantial time and effort without material reward, but that religious difference and conflict have also led to destruction, uncertainty, and serious interference with peaceful and harmonious human coexistence.

**Changes with globalization:** Despite continued human interest and strong desire to engage in imaginative belief, this study has generally shown that spiritual views are gradually changing subject to natural evolutionary forces and that long-term advancement and further globalization will result in a more unified world. Traditional belief systems are expected to suffer from increased recognition of reality, improved human awareness, more factual education, decreased human fear, and more restrictive teaching of underlying imaginative notions. The direction of movement was claimed to be consistent with the natural forces acting on evolving man and suggested that its effects should be recognized and encouraged in his best interest. Meanwhile, man is expected to continue migration over his relatively small spherical planet, requiring frequent adjustment for differences in language, beliefs, knowledge, cultures, and economies, but gradually "settling in" to a condition of greater equilibrium and unity, more fully acknowledging his prime responsibility to fellow man.

Approaching the second decade of the twenty-first century, man continues to undergo changes that were not readily predictable. The world is currently experiencing an economic downturn after many years of relatively healthy growth. Its condition seems to be the result of a "domino effect," involving temporary loss of economic confidence, a cycle of depressed evaluations, curtailed trading and spending, and a period of abrupt economic fear following a period of gross over-optimism. Society is well aware of such cycles of varying magnitude, and this one is expected to begin to right itself within a year or more, when progress, however muted, is expected to resume. In terms of the underlying religious theme, it should be noted that earlier chapters included mention of how such an economic and psychological downturn results in substantially increased church attendance as though a guiding spirit might be available to helpfully intercede. However, the flight to spiritual support is expected to vanish as quickly as it appeared when economic conditions return to normal.

**The struggle for order:** Most nations that constitute the continually growing and changing global pattern are also struggling with their individual economies, and some governments are additionally experiencing problems of stable leadership. Heads of state, once established, are often reluctant to give up power despite replacement in democratic elections by capable new leaders. The natural processes of human behavior, organization, and peaceful democratic occupation of the globe still have far to go to expect a voluntary system. With further globalization there should be a tendency toward unifying cultures, not only with respect to belief, political ideology, and economics, but also increased expression of demand for equality, human rights and more effective leadership. These are indeed the compatible objectives of the forces acting on human processes in addition to the material universe.

High density populations will tend to lead movement. People will continue to seek opportunity as well as more favorable geography and living conditions. Boundaries will tend to crumble under the forces that fuel human migration. There will be a continuous process of cultural mixing and readjustment in which the need for a more central world government will be strongly indicated. It is already evident that peaceful world governance requires more specific cooperation of a multiplicity of "sovereign" nations in order to influence and control individual national misbehavior without critical threats or armed conflicts by any single nation. Meanwhile, the world is approaching a condition in which weapons of mass destruction will become more available and the race for still higher destructive power will reach unmanageable proportions. It is likely that man will eventually be successful in reaching that position. However predictable, it is yet difficult for governments to guarantee security or prepare adequately in advance for enhanced destructive power.

**The need for peaceful coexistence:** Diplomacy and compromise are major factors in the process of peaceful coexistence more powerful and more lasting than any other weapon. The American use of special envoys should continue on a full-time basis in order to deal in proper and timely fashion with rising problems in unstable regions of the world. Other countries should follow the same course, continually engaging in diplomatic resolution and coordination of the multiple conflicts of the many intertwined competing nations that constitute "jig-saw orientation" on a relatively small globe. World problems require increased effort to establish "conflict resolution" by processes of sharing and compromise hopefully backed by the will of citizenry as the principle of democracy slowly matures worldwide. It is increasingly important that diplomacy be conducted by mature, well-trained officials and not by lesser political appointees.

Increased occupation of the global sphere, greater international competition, and more advanced communication are important factors in dealing with enhanced human activity and interaction but will also add to national conflict and misunderstanding. The underlying responsibility for world improvement is a direct function of authoritative leadership of high respect and proven talent. National lack of trust and the fear of terrorism have increasingly entered the globalization process, in recent decades but causal justification is not entirely clear. It seems to violate the very religious principles that seem to be embedded in it. As a practical matter, competing nations often present an air of secrecy and opposition, jockeying for the best position both in the world view and the view of their citizens. These "straight talk" limitations must be reasonably settled or overcome before major continued globalization can successfully proceed. Globalization without such consideration will tend to proceed haphazardly without responsible conflict resolution, understanding, or planning. It could conceivably result in major international disorganization from which a "disturbed state of nations," once established, might not easily recover.

**War leads to war:** America continues its committed engagement in two formal wars, in Iraq and Afghanistan, carried over from its military response to the 9/11 terrorist attacks on its home soil. Terrorism then suddenly became a more significant factor in the peace and security of all nations. It is not a formal war to settle an issue between two nations but involves wanton destruction without public explanation or declared purpose, a cowardly approach with a sense of impunity. Man has indeed, lost direction with respect to evolving morality. He might even have lost some evolved animal morality. Because globalization processes have not yet sufficiently mixed world cultures, conflicts carry not only an unexplained level of cultural incompatibility but also underlying religious friction. It has recently been difficult to avoid a sense of Muslim/Christian and other more parochial religious conflicts. World society is also concerned with the ongoing development of weapons of mass destruction, the technology for which has filtered down to the hands of suspected terrorist nations. This is only the beginning of widespread nuclear and thermo-nuclear capability, the present leaders in methods of mass destruction. Because of the seriousness of the issue with respect to security and peace of mind, it must be considered an issue that requires urgent worldwide resolution in a highly effective and lasting manner. If it does not reach a satisfactory level of understanding those nations with sufficient power will attempt to destroy threatening developments before they can reach deployment. At the same time, worldwide poverty and disease are being given inadequate attention and require a much more substantial world plan and commitment by all nations to achieve acceptable results. Since religious belief and ideology often are indirect and sometimes hidden factors in the pursuit of all these issues, the theme of this book is a most important consideration with respect to achieving effective world serenity. The perceived purpose and eventuality of man is always an underlying question but it is increasingly clear that world issues must be resolved without consideration of a humanlike creator. The plight of man must be considered the sole responsibility of man.

**Survival of the wise and righteous:** The entire world is indeed, concerned with the rise of rebellious factions, insurgencies that destroy the idealism of peaceful globalization. Much of the disturbance is found within nations themselves. The disturbance generated in the mind of individual terrorist is quickly joined by other like minds everywhere. The degree of damage and destruction heaped on innocent members of society seems to be without remorse. The basis for rebellion and resultant terrorism is not always clear however, and current society is actively in search of answers that may find the most effective direction of resolution or repression without the strain of continued stand-offs and periods of bloodshed. Military intervention now seems to be the only effective means of protection.

Fathli Moghaddam's book, *How Globalization Spurs Terrorism*, may be germane to the present condition. However, the fear of disturbing elements of civilization by greater occupation of the geographical world is a somewhat baseless fear. The friction between touching nations, touching religions, and touching cultures is of greatest concern. But modern minds should be capable of remedial improvement. The process of globalization is powered by the subtle prime forces, which this book has frequently emphasized and they will continue despite any areas of human concern. There may be some "deep wounds" inflicted by lopsided globalization, but the processes will proceed with or without care and understanding. The time for "drifting migration" of large masses is, however, expected to be at an end, but more national planning should be exercised to support the best interests of moving society.

Indeed, man requires an effective earthly form of "god authority" for peaceful guidance and further orderly occupation of the planet.

**Global damage control:** The "irrational" reactions that might arise from threatened identities must be considered. Society must eventually rise above destructive means to seek and maintain recognition among world elements. It is the primitive function of animal instincts, of greed and territorial ambition, ambition that continues to interfere with processes of peace and understanding normally available through effective compromise and negotiation. It may be naive to expect ideal behavioral improvement, but the Earth is replete with graves of militaristic leaders of history who have exercised their animal instincts, who have sought to fulfill personal ambitions. These are not always traits of strong voices involved in reasonable reorganization of the globe but reflections of misguided personal objectives with blind support of followers based on desire to share power. History has shown that nations can be "fired up" by a single leader or by a relatively small group of leaders who boldly enter conflicts to their eventual disadvantage or destruction. Whatever the nature of victory or defeat, it often costs many lives of man and untold suffering for very little long-term achievement. The time has come when the state of globalization and cooperative influences, financial and otherwise, must be brought to bear with strength and meaning by unification of a sufficient fraction of the "righteous" nations of the world. Citizens of a misbehaving nation may be held "captive" in mind and body by its military dictatorial government. Righteous nations with their "infinite" combined strength and influence in numbers must learn how to effect damage control on a timely basis with minimum harm to innocent civilians. All occupants of the planet must be taught responsibility not only with respect to behavior but also with respect to positive contribution to the world as whole. Human development does

not yet seem to have advanced quite far enough to engage in acceptable comprehensive world order.

**Diplomacy, compromise, and warfare:** The behavior of nations can be as variable as human emotion. The ambitions and policies are sometimes underhanded and in serious question. The character of a sovereign nation may follow the traits and ambitions of its leadership, which may be contrary to much of the rest of society. Its vulnerable citizens may be influenced by despotic mandates or by effective governmental propaganda. As in the case of Nazi Germany, it is difficult for citizens to sense the true intentions of governments in early stages and just as difficult to escape their influences at a later time. If nations feel seriously threatened by a course of action, they must timely face the options of compromise or force. If negotiation is unsuccessful, an act of war is then left as the only option. Warfare among modern elements of society will be increasingly severe and should be avoided through the exercise of new, more advanced minds of diplomacy, compromise, and human understanding.

Warfare in the twenty-first century can vary with methods of destruction ranging from conventional explosives to nuclear blasts, including explosive forces at the devastating thermonuclear level. Destruction can no longer be confined to the military as in the past. Although weapons of mass destruction have been under discussion and placed in various arsenals for decades, none has actually been employed since World War II. Using advanced forms of weapons, such as in missiles with multiple warheads, can create unthinkable chaos and loss of life, human and otherwise. They are capable of wiping out cities and towns in a single stroke. New destructive power should be made highly visible to the entire world. A reasonable and moral society would be reluctant to use such weapons against its worst enemy except as a defensive measure of last resort. These prospects should be given new

emphasis to encourage diplomatic and negotiated means of acceptable coexistence and assurance of order among earthly inhabitants.

**The supreme power of united authority:** Human differences have involved a wide range of issues and a wide range of dangers and risks. Threats can be imminent or distant creeping shadows, but in today's real world must quickly be brought to resolution and understanding. They should not allow long-term "brooding and festering" in which substantial portions of life remain difficult and marginal. They can be based on many degrees of human trustworthiness. Deep suspicion without transparent negotiation can lead to a first strike whether or not justified by its underlying circumstances. It is therefore the responsibility of all parties to act openly and without serious question if tragedy is to be avoided. In the modern world of education, reason, and understanding there should be no purpose in destructive warfare.

This book has therefore suggested the establishment of a powerful and reasonable new form of United Nations, one that can timely and effectively create and enforce international rules of behavior and encourage international development with increased authority and protection. The principle of sovereignty of nations may even require occasional breaching under some circumstances. This approach has not been used for many reasons, but it is now an essential objective for future man and can be created in a gradual extension of the existing United Nations organization with appropriate changes and additions. Man is now patently aware that he must see a new day of international cooperation and compromise. A new level of trust must be established among man with extraordinary understanding and leadership, a new day of world integration of human cooperation in the common interest. Man may pray with great sincerity to his imaginary god for world improvement but he must eventually rely on his own talents for his proper behavior and governance. He must work out systems and methods of understanding relating to human function and they must be made uniformly applicable to the entire world. An ideal United Nations must be able to serve the interests of its members with honesty and selflessness, avoiding the introduction of personal interests that may interfere with or impair the purpose of serving its members reliably and effectively. This book believes that man is gradually approaching the capability of such idealistic eventuality.

**Man is responsible for his world:** The days of "hordes at the gate" and the fear of "ambition of kings" are gone forever. The international ballgame is constantly changing. Today one modern weapon, directed with great precision, can destroy an entire small country in a single blast. Future technology may even allow such a weapon to be easily planted surreptitiously in the midst of a peaceful region. The power of diplomacy and negotiation must be considered in more idealistic organization and governmental function. Negotiation and compromise are the most powerful tools for coexistence of man, tools that, if wisely employed, are capable of achieving peace, satisfaction and freedom from fear. It benefits both international and intra-national confidence. A common transcendent UN authority can also eliminate pockets of selfish rule of warlord societies. It can eliminate concerns with apparent concessions or unbalanced nuclear power. A change may be required in the established models of behavior and government. A system of corrective human forces can be initiated, backed by unity of righteous nations and relentless pressure to penetrate all of humanity. Occupants of the planet do not yet appreciate the reality of their position. The education system of the world must provide for substantial training and preparation for world organization and peaceful rule of law despite the stark differences in cultures, practices, traditions, religions and languages on the planet. The concept of an effective singular world is rapidly upon us. It might take the form of planetary governance, the extension of the effective sphere, range, and degree of all government activity to global proportions. It would be based on coordination of the most advanced human understanding of global issues such as climate change, weapons of mass destruction, medical and scientific technology, pandemic protection, energy sources, monetary systems, travel, communication technology, space exploration, natural disaster reaction, and the like. Its usefulness would be equally applied in times of conflict and times of peace, in times of serenity and in times of natural disaster. I feel sure that the United States does not want to control China, India, or any other country. However, it desires all citizens to have the opportunity for a reasonable life of personal freedom under governmental guidance and assurance.

**The changed role of women:** The Moghaddam book mentioned above suggests a vital role for women in developing a better future in society, especially in the Moslem world. It is considered part of the unstoppable power of change. The successful role of women in America and in other countries has indeed expanded greatly in recent decades. Many women have achieved the highest level of national, corporate, and university function. Women are not only a strong contributory part of advancing government and business but are increasingly key participants in new research and discovery. One should not forget the intimate interchanges between mother and offspring, not only in umbilical chemical communication, but the effects of cultural evolution following birth. It seems that religious laws of the Moslem world have not yet allowed women the necessary freedom to fully follow suit. This book has constantly applauded the infinite strength of the basic forces of change that act upon man and predict slow positive changes in the direction of greater freedom, participation, and equality, especially for the female gender in areas where women have lived under restrictive social conditions or without deserving opportunity. These forces are inevitable, forces that will continue to act despite their occasional opposition and resistance.

**Planetary leadership:** The world is much too often in turmoil with respect to social, economic, or political condition. There never seems to be a Goldilocks era in which it is all just right. Even ordinary travel now involves costly elevated security with precautions and restrictions that have not been seen in history. Freedom and human rights have indeed been compromised. Borders are being increasingly violated and immigration control and monitoring have been substantially weakened. The rule of law is not fully operational. Pirates have even appeared in the shipping lanes off east Africa. Drug lords have become stronger and menacing in several areas with respect to its supply and distribution. The drug trade is extremely profitable and difficult to control because of its wide usage in most countries. Neighboring countries in North America are experiencing overwhelming armed strength and bold aggression among powerful drug cartels. Poppies are supporting the insurgents of the Middle Eastern and Asian countries. The problem side of inherent desires such as for drugs, greed, prostitution, and the like, requires continued appraisal, education, monitoring, and practical enforceable rules. Religion has not been the answer. Religious fundamentalism might even be included in the list of problems. Full correction is not an immediate expectation, but factual education and rule of law must continue with unprecedented intensity. A pattern of improvement should follow over time as it does with all training processes applied to animal life. The ratio of responsible and knowledgeable teachers to students must be increased in a rapidly growing world population. The quality of teachers and responsible leaders is in serious question and must first be given urgent world attention.

The routine government of sovereign entities, especially those with shaky borders and fragile provision for self dependence, is increasingly difficult and problematical. Some of them cannot even serve their own citizens in a proper manner. Much of the social turmoil takes place within nations that tend to refuse adequate outside assistance based on the claim of sovereign rights. These are the concerns of healthy and peaceful fellow nations and must accordingly be addressed with appropriate power and authority in order to defuse global turmoil. Energy and food options are constantly of concern and not yet in satisfactory focus. The inequitable distribution of natural resources has always raised social issues. Organized leadership must rise worldwide from the most able minds of man. Problems must be more orderly addressed in terms of national and international democratic function. Routine activities such as banking, investment, manufacturing, farming, trade, environmental control, military and security often seem disorganized and out of administrative control in a richly developed world that has enormous ability and long experience. Evil is indeed still inherent in animal-based man, who sometimes uses his developed talents to even exceed the dangerous threats of his uncivilized animal ancestors. World organization based on the best rational minds must be directed to moderate self-interests. It is all within the capabilities of modern man. Waiting for heavenly leadership to intervene, or for the fear of god to make appropriate correction, is certainly a losing proposition for all mankind.

**The miraculous advancement of knowledge:** As can be seen by scanning the scientific and technical journals, the human mind is probing deeper and deeper into understanding with respect to both material and biological processes. It is not only advancing the depth of investigations, but greater understanding is also taking place. Progress has been far greater than is appreciated by the average mind, and a worldwide plan of better distribution of such knowledge is sorely needed. Successful programs of basic research, discovery and development, in earlier years were represented by just a few advanced research laboratories. IBM, Sarnoff, and Bell Telephone Laboratories single-handedly advanced cutting-edge world technology now largely assumed by laboratories scattered around the world both in industry and at universities. Competition and distribution of knowledge are most essential for the progress of man. Expansion of molecular biology, neuroscience, and behavioral disciplines, is especially important for enhancing human relationships and requires more attention and appropriate funding. Research on improving the life of man should be given particularly high propriety and be more vigorously pursued. World progress cannot be allowed to stagnate even in economic downturns. Appropriate international coordination of human needs and objectives should not be competitive but cooperative. At the same time, belief systems must be separated from consideration of basic human needs, for human progress cannot any longer be based on imagined divinity.

The problems of the complex, newly "interconnected" world have become too difficult and too numerous to be accommodated in a single individual vision. An organized cooperative assembly of capable minds and nations now seems most essential. A new and different approach to world order must be considered for successful resolution and advancement to the next major step in planetary occupation. Problems must be divided and given lasting solutions with effective persuasion and established authority. The occasional conferences of representative nations are insufficient and increasingly ineffective. Responsibility is not properly evaluated, monitored and executed. The "do it my way" attitude of individual leaders and controlling groups has grown to permeate the policies of nations at great risk. Government representatives are increasingly looking to their own financial interest and continuity of office. The approach must be better planned and pursued with a spirit of national selflessness and an unequivocal common desire for continued human advancement and improvement. As globalization advances, action without adequate diplomacy becomes increasingly risky. There is a question, however, of whether individual and group understanding and behavior will become sufficiently advanced in time to avoid serious glitches and unresolvable world conflict in achieving desired progress. Society tends to allow excesses that create bubbles vulnerable to sudden periods of deflation.

**Going with omnipotent nature:** The forces of change are continuously pressing on society in various ways irrespective of how they are individually constituted and, as always, man will naturally respond and eventually emerge quite different from his past. Forward-looking problems will not alone be resolved by improved global economics and diplomacy. Other powerful forces that move man's mind are intrinsically tied to education, behavior, unity of belief, and the relationship with man's cosmic world. Man is learning how to adjust to changing climates, to changing weather, to changing crop yields, to terrestrial and cosmic events, to new diseases, to expanding food requirements, and, indeed, to changing human function. These chapters have shown that, while continuing his normal advance in knowledge and technology, he is also slowly learning. Man will be facing a "maturity" that he has never known in the past, a maturity that he must even now begin to contemplate more seriously.

In all of its chapters, this book has emphasized the forces that constantly and relentlessly act upon man. They are largely unidirectional and in accord with observed human progress. They are akin to the forces of development such as those that allow man to continually break athletic records. The recognition of reality and changing spirituality has been discussed at length in these chapters. However, there is another implied change in the development of man which is perhaps yet too far off to include in these discussions. It is the most important change of all, the recognition of human meaning and purpose in a mechanistic universe and the continued natural evolution of improved behavior. Just as man will act to require factual reality, he will also act to encourage more peaceful coexistence. It is the direction of human development based not on a benevolent humanlike god but on the relentless natural forces that underlie successful life itself. Although man has no control or influence with respect to modifying the universe, he has considerable influence on the direction and timing of human change.

**Greater investment in man:** Only when man is properly informed can he go forward with a sense of unity, satisfaction, and confidence. Beyond the emotions that produce his imagined god, man must also become more aware of the role of other important emotions in reaching a position of satisfactory behavior, of peaceful organization and coexistence. The entire process of moving into a new age of awareness places different demands on all of society. Internationally, all governments must concentrate on diplomacy, on developing trust, belief in a common world program, on productive rather than destructive competition, on education about the natural forces of human change, on resolution of differences inherent in conflicting cultures. Effort must be sufficient to make a measurable difference in the rate of human progress. It requires an intense force of world education that has never before been applied. It represents a giant step in a "powered" advancement of man in the occupation and globalization of his inhabitable earthly sphere. Where can man find the knowledge, understanding and capability for negotiating the simultaneous advance of all nations?

**Comprehensive education:** A recent book on the future of education. (226) confirms the tasks of neuroscience and its importance in formulating new educational practice. It understands how the "historical forces" shape a constantly changing environment of cognitive tools. One reviewer said it is difficult enough to keep the tension between brain and external symbolic material at the center of a line of intellectual inquiry, but it is much harder to build a public campaign for change around it. The author apparently feels that past methods of education do not suit present society, a claim also implied in this writing. He too believes that the futures of young people will be shaped by technologies that we cannot yet imagine. He draws on evolutionary psychology and cognitive science as well as history to review the purpose and objectives of education and presents a series of vignettes for imagined education in the next half century. He feels that the outdated cycles of the past must be reconsidered in terms of modern human capability, that a new curriculum must be formulated based on mastering the cognitive tools provided by advancing society. He feels that the reductionist tendencies of academic disciplines tend to narrow educational focus. This book has already suggested that education for future students, a massive transformation of knowledge, must first be focused on the natural dynamics between brain and "realism."

**The future is for the young:** For those who were born just recently there will be major observable effects of the forces of change within their projected lifetimes. Effects on man over the past century have already been discussed above. But they affect both man and his physical world. Oceans may rise to flood major populated regions such as Bangladesh forcing millions to crowd into the already overpopulated highlands. Glaciers may disappear from the Andes Mountains the highest range on the planet. Irrigation may be adversely affected and arable regions may undergo major moves and readjustment. These would be geographical eventualities completely beyond the control of humanity. But there are many events spread over time that also remain invisible, changes that are caught by history but not entirely subject to human analysis. Because of increased population and globalization the future rate of change may seem greater than those of the past and adjustment may be correspondingly difficult.

**Time is not endless for man:** If humanity reasonably survives on its planet long term, the natural forces which this book describes will completely transform it. Speculating on conditions of far future time tends to provide better realization of the prospects for human life. A recent book entitled *Surviving 1,000 Centuries* (227) attempts to make some projections. Although such a time span now seems quite infinite, it is only about ten percent of the age of emerged man of the Earth. The authors of this publication suggest that the planet must eventually stabilize at a population of about eleven billion and will have installed global governance as earlier suggested in these chapters. If nothing else, it is appropriately thought-provoking with respect to current views of the future. Needless to say, the human mind will not only be far different by then, but religious belief will have become ancient history.

**Recognizing opportunity:** A study of inequality in America by two economics professors at Harvard University (228) claimed that retrogressive differences have arisen in the last one-third century, not because of the fast pace of technology but because of the surprising stagnation in the level of education. It suggests that young Americans no longer have the educational advantage enjoyed by their parents and grandparents over the rest of the world. Technological change has made increasing demands on workers and the supply of qualified workers has not increased nearly fast enough. Their book is aptly called *The Race Between Education and Technology*, similar to the race discussed in these chapters relative to education and belief. This book suggested a "marathon" between factual education and imaginative belief systems, but it applies as well to all disciplines that would raise the level of world knowledge and understanding.

Thomas Friedman (229) also tells it like it is. Despite a worldwide economic downturn, he suggests that opportunity is once more knocking at the door, that the country not only requires a new push in areas that will recover its economic condition but also investment in science, research and technology leading to new medical breakthroughs, new discoveries, and eventually to entire new industries. Quoting inventor, Dean Kamen, in the face of recovery measures, he said "You can bail out a bank, but you can't bail out a generation." He wisely advises granting full scholarships to needy students who want to attend public universities or community colleges. This book has provided assurance that, through the forces of nature, man increasingly seeks understanding and that such understanding will result in a more suitable world. It is therefore imperative that society makes it possible to respond to his advice not only in America but on a global scale.

It is the responsibility of all leaders of all nations to see that the minds for which they assume responsibility are not just treading water, but are rising well above the minimum level of survival. What was once a simple biological assembly of animal neurons is now a powerful human brain, a thinking machine capable of continually advancing the opportunities and experience of man. It is the principal machine on which man must depend for his successful future existence and growth. The great missionaries of the past are no longer required to travel the globe to flood eager spiritual minds with imaginative promises. Time has proven that society now needs broad factual education and not imaginative images. Well-trained teachers must teach the truths of man, the reality of the universe, solid information that will eliminate spiritual conflict and allow the world to decide its own ideology valid information and reason. Those whose minds are still affected by the spell of imaginative ideology will find confusion not guidance in its further promotion. All members of society must share in nature's knowledge just as they share in its other resources. They must understand the transcendent power of nature and its relationship with man, how this relationship can be made more compatible in man's best interest and comfort. Mechanistic nature and its human product require compatibility little different than that of a caring mother and her offspring, and a relationship of understanding is essential to propel man forward. Every savant is patently aware that "knowledge is power" in the practical world and that knowledge was derived from nature. Factual knowledge is a form of information storage battery, a controlled source of power that can be used at will. The power of every man is proportional to his level and extent of learning. This proportion is verified at every great center of learning and is evidenced by following the careers of accomplished men. But the average level of world knowledge is still very limited. The twenty-first century must therefore become a century of intense positive catch-up.

**Cooperative advancement:** Globalization, said to have its origin on the continent of Africa, has seen a continuous movement since human emergence from animal form. Perpetually changing and expanding world population means not only a changing mix of cultures, but also changing responsibility with respect to new caretakers and new managers of the planet. There are already indications of major movements in population, of increased interaction of races and cultures, of major technological advances and indeed, changes in basic human capability and understanding. University enrollments in recent years have included a large percentage of foreign students, some of whom are expected to return to their native lands to educate and mix with others. This form of globalization should be encouraged.

Areas of the continent of Africa, the birthplace of humanity, appear to be embroiled in a state of turbulence but can hopefully achieve a position consistent with the pace of other continents. We see India advancing in a mixture of new wealth and intellectual power in a background of poverty and ignorance. We see China gearing up to become a world supplier of goods and a financial power backed by military buildup. We see a recovery of Russia as a competitive power with energy independence. We see South America and Mexico growing in recognition, taking full advantage of their proximity to regions of greater development. Despite declared incompatibility, we see Middle East countries with the potential of accelerated development as never before possible. With all the positive changes, we see simultaneous conflict in many separate parts of the globe, conflicts that suggest limited central leadership and resolution. For the most part, society still has too much reliance on the guidance of an imaginative god but requires far more reliance on educated fellow man, irrespective of color, creed, or nationality. The target for world unity must be the entire planetary population. The needs and interests of citizens of all nations are essentially the same if expressed freely, out of reach of their narrow cultures and governments. In the age before us it seems more possible than ever to unite minds of the world with respect to non-government issues in order to lay the groundwork for more ideal living conditions and ongoing globalization. Nations that misbehave, that do not recognize required change and do not act in concert, cannot be a contributory part of world progress and may be left behind, or may even be absorbed by the combined power of other more successful nations. The best and most knowledgeable of our current futurists see a major different world by the end of the century, one that requires all of the preparation implied in these chapters to provide a better and more fulfilling life for those of the next century and beyond.

Despite the current downturn, transportation will soon be revolutionized in terms of more efficient air traffic and electronic navigation. Man and machines will uniquely merge to revolutionize technology from the place of his employment to his home, and from outer space to beneath the ocean and land. Space technology will be introduced for intercontinental transportation of people and freight and will permit extensive zero-gravity experiments, never before possible. Freight may be carried unmanned at super altitudes. Electronic communication technology of all forms will become interspersed and the world will suddenly become "one." Biology will blossom in all areas from human makeup to medicine and from food products to microscopic life. As in all new ages of man, the forces of nature are constantly pressing man, and man will gradually respond and be renewed, becoming far more mature as a global society. Much of the future technology is already known and merely has to be suitably transferred from the laboratories to the population.

**The real status of man:** It is sometimes rather shocking to uninformed man to be told about his true position in the universe, about the chemical origin of life and about his dependence on the mechanistic transcendent force of nature. But it is what it is, and it is what the most able and respected searching minds appear to have determined, it is what man must accept. The nature of the universe and the appearance of its life have become ever clearer. The imaginative spiritual explanations of early uninformed man should accordingly be played down and reconsidered. Real life should be glorified and celebrated and the beauty and harmony of the universe suitably revered. The forces of nature must be respected and shown appreciation for providing all the benefits and opportunities of life. Man has roamed the Earth for over a thousand millennia but has only now factually learned about the distant past that led to the knowledge of where he now stands. He has only recently learned about details of his own structure and function, and is only now learning about the complex processes of his emotional mind and of new ideological interpretations made possible through his "awareness of awareness." Man has now learned about his very being. He has learned about the mechanistic behavior of the cosmos and how it relates to the existence of man and his relationship with the motion and influences of planets and stars, their chemical composition, processes and lifetimes. He has learned about the forces of nature and his expectation of change, change in his own chemistry and in his surrounding cosmic world. He has learned about monitoring and tracking all the forces that can determine his most beneficial future. Indeed, the knowledge that man has acquired now allows him to create a more realistic satisfactory and rewarding life with few cosmic or theological concerns.

**Survival of the best informed:** This book has made reference to many learned individuals of advanced vision, persons of superior intelligence and learning, persons who have long considered the basis of human existence and agree with the premise of this writing. It has also made reference to respected religious minds of contrary view that have variously become "wired differently" in their learning processes, minds whose sincere imaginative views will remain intractable even in the face of factual scientific explanation for the duration of their lives. As the result of its ubiquitous teachings and cultural development of the past, spiritual understanding of most members of the world population is still very much overwhelmed by the prevailing emotional desire for a protective and guiding god figure and for the widely popularized benefits thought to be derived. However, like the task of missionaries of old, modern education is expected to teach minds that will follow in the footsteps of existing informed men of rational vision. The current process of change is similar to many significant transitions that have formed the long history of developing man. New change is always dependent on past change, a continuous, never-ending learning process. New, temporary gods may occasionally rise again, but all gods will eventually disappear as they always have in past history. Ideology will change as it always has. Deep spirituality, however imaginary, presents a somewhat greater challenge than many other forms of past changes. It is because it bears heavily on the strongest and most fundamental instinct of man, that of perceived immortal survival.

**Again man's responsibility for man:** This study promises that man will safely pass through the current age of lethargy, ignorance, conflict, and spiritual interpretation and will ultimately rely only on his own determined knowledge, the basis of realism, and not on spiritual hearsay. Globalizing man will therefore continue to approach, and eventually reach, a unifying state of spirituality in the form of world secularism as frequently predicted in these pages. At the same time, he will engage in the building of a more organized and stable earthly world, one that involves much of the power now attributed to his imagined god, a world that is benevolent and secure, one that monitors and punishes deviations that cause suffering and torment, one that encourages spiritual unity and peaceful coexistence, all in his finite planetary home of reality and its environs. He is rapidly learning that his future is in his own hands and not in the hands of any humanlike spiritual being, he is subject to only one force, the force of nature, under which he was born, lives and will die, and he is fundamentally capable of meeting the challenge. Society will realize that it must organize much improved planetary leadership for its protection, growth, guidance, and must monitor behavior for the benefit of all man.

**Like insects on a large ball:** Even in the immediate years ahead, human communication will more fully cover every part of the finite globe as society slowly adjusts to new advances in its technology. Much of the world will involve direct, person to person contact, either by cell phone or by satellite. Meanwhile, migration and travel will also expand reaching proportions beyond anything yet seen in the history of man. Like busy ants on a large ball, modern humans are already in constant motion over the planet for many different reasons, some because of displacement and refuge, others because of business needs, and still others in connection with education, military, government, research, sightseeing, or recreation. Areas of the globe that have been barely penetrated by migrating civilization in the past are now being included as part of the frequent travel scene. Antarctica, for example, has never seen more civilian visitors than in recent years. Language and understanding are becoming increasingly more intelligent and compatible over the entire planet. There is still major movement of population for greater political and religious freedom and for economic opportunity. The globalization process is indeed continuous, and never more effective than now in radically changing the living character of the planet since the very first migration of man. Continually increasing population may even tend to occupy most all of the desirable and livable Earth's surface.

**Mind development through communication:** Communication may play a more significant interacting role than travel. Worldwide conferences are now frequently arranged using video technology of the most advanced nature. Personal communication devices not only carry voice, but also data, video, music, news, TV, memory, control functions, warning signals, geographic position, and have potentially unlimited connection to every corner of the planet. Service groups are now providing advice from opposite sides of the globe. Undeveloped areas are thereby meshed with the most developed areas. A review of the rapidly increasing technology of recent decades confirms the pace of man's progress, a reliable measure of expectations of new understanding that man will experience in his further development. Technology already exists for many further unprecedented advances scheduled for the years immediately ahead. Indeed, human progress continues to evolve as a natural consequence of the unstoppable, though sometimes indefinable forces acting on human life as earlier described.

Intra-cultural communication will bow to greater inter-cultural exchanges and processes of merger contributing to a more effective new form of cultural evolution. The ingenuity of the human mind has not declined but will become even more pronounced and better utilized in an expanded society. Indeed, the potential advances of future man cannot be accurately estimated by current minds.

This final chapter is not intended as a complex analysis, but as a broad reminder of what is already known and what may be expected in realistic extrapolation, the changing path of society. Information and understanding of the reality of factual knowledge will tend to infiltrate society as with a leak in a huge dike. As may be expected with new space telescopes the world will gradually be flooded with greater understanding and appreciation of the universe and the planet. The world will also be flooded with understanding of human origin, brain function, and human emotions with all their strengths and limitations, and indeed their consequences. This understanding will eventually trickle down to ordinary man where it will become effective on a global basis.

Despite the complex human make-up of the world with all its current conflicts and its changing inequality, man will increasingly act toward the organization of the interests of the common planet and to formulate appropriate rules and laws of society that will restrict behavior uniformly throughout the world. Nations will increasingly become unified, for man will gradually realize its importance and value in providing satisfaction, happiness, peace and security for all of society. Guns and explosives will become select and more advanced, but common weapons will become the bane of human existence. The world condition today consists of a wide mixture of new knowledge, a mixture that will eventually be integrated to a more unified state much like a new vat of different colored paints becomes a single hue. This book emphasized the powerful effects of rapidly acting cultural evolution in combination with more slowly acting genetic evolution in the constant intermixing and development of man. In the current age of man, an age of religious confusion, of latent ignorance, of easy imagination and misinterpretation, of animal-like interaction, of failure to envision stable organization and management of human needs, spirituality would be expected to be out of hand and imaginative beliefs expected to rule man's still vulnerable mind.

**The lessons of history:** In its educational processes, society must be very careful to place science and the humanities in their correct relative importance. Mark Lilla, a distinguished professor at Columbia University, in his review of *Motherland* (230) in the July 29, 2007 New York Review of Books, warns that Russian philosophy did not even begin until it left behind the superstitious, pre-scientific world of the nineteenth century and the romanticism that replaced it in educated minds. It had to separate "values from facts" and "personalities from truths," to be considered as more than poetry, as people remained attached to pre-scientific worldviews and romantic dreams, even while living in the midst of modernity. This was a good example of a period in which an entire nation rejected the Enlightenment for idealism. Russia had gone through centuries of religious orthodoxy and political repression. The author considered it an enormous mistake to abandon the subtle "equipoise between reason and skepticism" that characterized the French and English Enlightenment at their best. Lilla said that, like Napoleon's troops, the modern ideas of Bacon, Descartes, Locke and Hume "were turned back at the gates of Moscow and beat a slow retreat through the snow."

**Keep looking forward:** The Russian experience applies to all civilizations. It shows the influence of excessive spiritual influence and the critical need to "sweep out the cobwebs and to rationalize society." While Russia was in the throes of a type of reform, America appeared to be bogged down with a spirit of theism that could have produced a similar condition (231). However, the country was protected by the diversity of American interests despite its retreat to biblical evangelicalism. History began to move faster at once. Following the World Trade Center attack, the military might had to be tested. The country was suddenly thrust into a strange mixture of religion, politics, economics and aggression. The Neocons who quickly encouraged a policy of war footing were determined to defend America against terrorism. At the same time, the plight of Israel ironically became an active concern of the American Religious Right. History now needs to right itself in reality.

Advanced nations are now entering more sophisticated efforts such as space exploration, molecular biology, genetics, advanced military weapons, and nuclear power. Nevertheless, civilization is drifting. In *The World Is Flat* (232), Thomas Friedman also characterized the changing world. He spoke of China and India and other countries becoming a significant part of the service and manufacturing providers of the world. The new growth and expansion of these "undeveloped" parts of the globe would tend to flatten the globe, causing human beings to make considerable adjustment relative to the past. He claimed that these developments are desirable and unstoppable as this book has claimed with respect to the march toward realism. The huge influence of globalization on the individual is unprecedented in the history of man and by the end of the century will produce a far less spiritual man, but one who is much more integrated with all the beneficial functions of man.

With expansive world modernization and globalization there are major human problems that require attention by the developed and advanced portions of civilization as recognized in earlier chapters. A recent editorial (233) emphasized the increased poverty in the world, indicating that the World Bank reported that in recent years there were almost one and a half billion people living below the poverty line of less than a dollar and a quarter a day. The poverty is said to be so abject that it is difficult for the industrial world to even comprehend it. The concept of new economic growth such as in India and China does not take the real poverty of the world into account. The leaders of the world appear to be blind to the real problems facing the expansion of human life on the planet. The emphasis on religion and spiritual satisfaction appears to bypass human responsibility to manage and care for all of society.

The group of eight industrialized nations are said to continue supporting roles for rich countries. The concerns of the religious that center on their visions of salvation and satisfaction do not extend to practical life on the planet. This book has again advanced the need for some form of more powerful and knowledgeable central world government, the United Nations of the World, a forward-looking power without which the planet will slowly approach disaster. Most of the religions of the world preach goodness and righteousness and the benefits derived from their relationship with their benevolent god. How long will it take man to realize that he is collectively managing the status of his world and has long spun the wheels of responsibility behind the shrouds of imaginative belief?

**Biology now bigger than physics:** Futurist Freeman Dyson (234) recently explored the projected emphasis of the coming century. He claimed that biology is now bigger than physics and is likely to remain the principal science through the twenty-first century. He claimed that biology is now more important than physics as measured by its economic consequences, by its ethical implications, and by its effects on human welfare. He further claimed that the domestication of biotechnology will dominate all of our lives for the next fifty years.

Dyson sees the epoch of Darwinian evolution based on competition between species having ended about ten thousand years ago, when a single species, Homo sapiens, began to dominate and reorganize the biosphere. He too speaks of "cultural evolution" replacing biological evolution as the main driving force of change as I have claimed in earlier chapters and in prior books, suggesting that cultural evolution is running about a thousand times faster than biological evolution. Guided by precise understanding of genes and genomes instead of by trial and error, modern man will be able to modify plants to increase yields and improve nutritive value and resistance to pests. Within a few more decades he will be able to design new species of microbes and plants according to our needs. He calls this "green technology" as opposed to the gray technology of the past, a technology that will halt migration from villages to cities and narrow the gap between rich and poor countries. The working components will be the sun to provide energy where it is needed, the genome to provide plants to convert sunlight to fuels as needed, and the Internet to end the intellectual and economic isolation of rural populations.

The space program has also made it possible for spy satellites to scan the entire globe, to inspect people, structures and movements in remarkable detail. Satellites are regularly circling the globe in something like ninety minutes. Other satellites remain stationary for service in specific areas of the globe. They have made it possible to monitor the geographical structure of plants, trees, mountains, rivers and oceans to provide scientific data on the condition and changes on the surface of the Earth. Much of human life can be monitored. It is something like realization of the belief of early civilizations in an omnipresent god that sees all from above. Monitors are equipped not only with optical instruments but also with radar type instruments of various wavelengths that often provide more information than man can easily handle. But it is only the beginning of modern technology that brings man closer together and in tune with his planet.

**Planetary development:** Much has been written about world development, about globalization and the spread of democracy. All eyes are on the largest developing countries, countries that will clearly expand internally and in their association with other world powers. Among the larger Asian nations already mentioned are China and India. There is a great deal of speculation as to whether democratic India or communistic China will produce the most successful global partner both in terms of economics and in terms of freedom and human rights. It may be of special interest to readers of this book to review the incredibly strong early missionary efforts to convert China. In the June 7, 2007 issue of the New York Review of Books, Jonathan Spence wrote a review called the *Dream of Catholic China*. He spoke of the Jesuit "soldiers from heaven" and "conquistadors of souls." Assigned to the Society of Jesus, they underwent rigorous training in Europe. They had several years of Latin and Greek, three years in art, philosophy, mathematics, and astronomy followed by four years of advanced theological studies. They were sent to all parts of the world to meet the most dissolute sinners, the most pertinacious heretics, preach to heathens, hear confessions, work with barbarians and even dispute with Lutherans. Many young men petitioned to go to Japan or China. It was said that some were in search of martyrdom because the shoguns made death for Catholic faithful a regular occurrence. The Jesuits chose the most brilliant of missionaries who quickly learned Chinese and became friendly with the Chinese intellectual elite by the depth of their knowledge. Throughout the sixteenth and seventeenth centuries, Jesuit missionaries even made their way into the emperor's inner circle as advisors and teachers (234).

The China project is not only a good example of early globalization but also an example of the great effort to teach the world about God. It was a major human desire to spread the "word of God" globally because it was considered God's will to do so. This was the time period in which America was beginning to be colonized by people who escaped the religious oppression of Europe in preparation for its war of independence. It was an age of discovery and occupation of new lands, not only because of trade in new and different commodities but movement of people in the interest of spreading religious beliefs. Despite many pockets of imaginative beliefs, China today is an atheistic country. It is experiencing rapid growth and development and enjoys favorable trade balance with the rest of the world. It will no doubt become more and more democratic in structure in the course of time based on the desire and needs of expanding free society. In the aftermath of the Cultural Revolution, a period of turmoil of the 1970s, new leaders ushered in a period of modernization of industry, education, science, and military defense that produced a favorable socialist-capitalist economy. Economic liberalization was not accompanied by political liberalization, for the country is an economic powerhouse under repressive communist rule. Education and the inherent will of man to be free are expected to bring further change.

The other giant Asian country has an equally interesting and progressive history. Following World War II, the British, under pressure, gave India independence. In turn, India was divided religiously. Pakistan became the home of Muslims and India the home of Hindus, East Pakistan became Bangladesh. Burma separated from India and Sri Lanka gained independence. Sarawak joined the federation of Malaysia but Singapore became independent and the Republic of Indonesia won international recognition. In the widespread unrest, ethnic cleansing, and fighting over the greater region, about a million lives were lost.

India itself has now become the most populous democracy in the world. Its low-cost labor has become a major factor in world service through the use of the Internet. This history is typical of movement and change around the globe with respect to freedom, culture, and religion. Change is expected to continue until secularism is finally realized and all mankind can concentrate on human improvement and freedom with uniform understanding and proper government. Social movement will only slow when improved economic equality and more unified culture are reasonably achieved.

**Changing America:** No country is more representative of world changes as the result of continued globalization than the United States where, in one more generation, existing minorities may become the majority (235). The principal reason will be the relatively high birthrates among immigrants and the increased influx of foreigners. Hordes of immigrants flock to America daily, primarily across illegal southern borders and to a much lesser extent across northern borders. The 300 million population of a decade ago is expected to increase to 400—500 million at mid-century. America is still a major melting pot of nationalities and races. Proximity to the Hispanic countries of North America and the history of opportunity in America will tend to accelerate change. The sixty-forty ratio of non-Hispanic whites is expected to reverse to forty-sixty. A fifty percent increase in the number of African-Americans is expected, while the Asian population is expected to double. The forces of change by these world-shaking movements of people and geography will result in a "different" country.

Although earlier commentary suggested that increasing Hispanic population tended to make up for the decline in the Catholic population, it is expected that the overall Christian population will continue to decline in America as it has in Europe. The Hispanic population is estimated to triple, to account for about a third of Americans, but religion will largely depend on their depth of education. Continued genetic mixing and cultural normalization may actually contribute to stabilization. However, the religious character of the United States should eventually follow the secular trend of the much more mature European continent, generation by generation.

The problems of imaginative belief in globalizing society are principally of two forms as viewed today. First, the wide differences in the character of beliefs stemming from old ideas and traditions is beginning to clash with modern society, creating and emphasizing a sense of stigma despite ecumenical efforts. Secondly, the requirement of pure imagination consciously or subconsciously opposes the function of the mind of logic and reason, not only creating a sense of neural confusion but deterring the long-term process of natural improvement of control of emotions. Stigma is a very important factor in achieving more ideal coexistence in society and was therefore emphasized in a separate chapter. As explained in early chapters, emotion and reason are of equal importance in terms of the development of man. Man is realistically always living in the present as his past fades away and his future arrives at equal speed to become his present and then quickly moves into his past. Society often claims that life isn't good enough, that man has to have some being to look to. Some say they know belief is imaginative but still desire to feel connected to a superior being. They look with proven futility to someone or something to improve life and protect them from the ills of the world.

**Planetary upkeep:** The forces of globalization suggest major changes both in people and in the Earth itself. The planet and its occupants are constantly interacting and more so every decade. Only now are the real details of these affects being determined and published. Even the great oceans of the world are showing the effects of civilization (237). Scientists are building the first worldwide portrait of dispersed human impacts on the ocean, revealing a planet-spanning mix of depleted resources, degraded ecosystems and disruptive biological blending as species are moved around the globe. Scientists tell us that more has been learned about the planet's oceans in the last century than has been known in all of the world's history. These accelerated changes are mentioned here to emphasize the human interest and concern with their physical world. It is only the beginning. Many more occupants of the planet must become involved and participate. This book claims that concern with his Earth will contribute to man's sense of reality.

**Globalization and the Vatican:** A recent news article (236) indicated that every Pope since 1945 has supported "robust global governance" by the United Nations. With recognized spiritual differences over sex, abortion, and birth control, papal concerns with poverty, disarmament and environment merit laudable comment. No one on the global scene is said to make an argument for freedom of the press, religion, and dissent with more clarity than Pope Benedict XVI. He repeatedly warned against the "bitter fruits of relativistic logic" and refusal to admit the truth about man and his dignity, his "supremacy over the animal kingdom." Together with putting a human face on international politics and economics, which John Paul II called globalization of solidarity, the Vatican still exercises an important function in organization and control of the planet's residents. However, these are all universal objectives that should be separated from belief in a humanlike god.

He claimed that "dictatorship of relativism" not only menaces the Catholic Church and institutional religion, but all of the most vulnerable society as well. Relativism is considered a theory that knowledge is relative to the limited nature of the human mind and the condition of knowing and that ethical truths depend on the minds that hold them. Within the theme of this book, overtly expressed at the very outset, it is unfortunate that the Vatican Christian belief is based on miraculous occurrences and does not recognize modern reality. Its imaginative ideology is incompatible with the advancing mind. It is equally unfortunate that powerful Vatican level voices are not available in support of realism, the eventual global interpretation of human existence.

**Authority to protect man:** A recent news story (238) pointed out that change is inevitable, that it is in fact taking place in a quiet manner. It mentioned that independence is being declared by many states in violation of the principles of the Treaties of Westphalia dating back to 1648. The unfettered power of a state within its own borders was the inviolable basis of international law. NATO circumvented the Security Council, which was considered too divided to act when it waged war for the first time to prevent Milosevic from ravaging Kosovo. Its legitimacy stemmed from the consensus that acts of genocide can never be a purely internal matter. When a government abused the rights of its citizens through slaughter or ethnic cleansing, sovereignty could in effect be suspended and transcended.

The new European state, Kosovo, has now been born. It is recognized by Western powers and is the first major fruit of the idea behind R2P, the "responsibility to protect." It is said that the rights of human beings are at last catching up with the rights of states. The global Center for the Responsibility to Protect is the spread of R2P principles. It is said that an R2P generation is coming, it is slow but it continues. The reorganization of the United Nations with full representation and participation should include a rapid strike force of supreme power, far greater than any single nation. It should intercede on behalf of humanity when governments fail to recognize their minimum responsibilities to its citizens. It is their only protection against corrupt and failed governments.

**Reality is for the young:** Approximately a quarter century ago, a survey by the National Opinion Research Center (22) indicated five to eight percent of the American population described itself as unaffiliated with any religion. The most recent poll indicated that about sixteen percent of American adults are not part of any organized faith. This general movement in the population that prides itself as being Christian is very significant. This book takes the position that it is in line with the anticipated beginning of a decided secular movement, a movement that has already shown progress in other parts of the world. As implied elsewhere in this writing, the first step in such a change is that of breaking away from the structure of organized religion based on the inability to accept all of its claims and promises. Once that position of freedom is established, other aspects become doubtful and eventually beliefs fall in line with established factual human knowledge. It is indeed difficult for the advancing human mind to tolerate conflicting ideas for very long in the modern world, and the movement to spiritual freedom is a very significant step in the direction of realism.

The "unaffiliated" appear to be largely under fifty years of age. The one in five males that have no affiliation do not necessarily regard themselves as agnostics or atheists but often as nothing in particular. This book predicts an increase to one in three by mid-century and perhaps one in two before the century ends. The difficulty in accuracy is that people cling to the idea of spiritual belief for secondary reasons enumerated in other chapters. They are reluctant to claim their freedom because of uncertainty and because of the stigma discussed earlier.

**Reconciling countercultures:** In Jordan, the Brotherhood is legal and many students believe that Islam "is the solution" (239). They feel that their government is not Islamic enough. The young people seem to spend a disproportionate amount of time discussing religion and politics of the state. It would seem more productive for the thirty thousand students at Jordan University to become more engaged in the sciences and in other more productive and perhaps international interests leading to cooperation and peace. They have developed mixed opinions about the West that can only be corrected and advanced through closer contact and broader education. The West is regarded as a counterculture, one to be challenged by the power and strength of the more intense rules of Islam. It is time for cultures to be "reconciled" all around the globe.

**Personal contributions:** The group called Doctors without Borders is a wonderful example of human cooperation that should be emulated and expanded. It certainly does not need to be a faith-based service. The contributions of faith-based ministries can only be admired and appreciated for their selfless penetration and sacrifices. However, realism is a religion keyed to the reality and progress of man in contrast to the quality of imaginative belief and should keep proper pace with the natural advancement of world intelligence. This book suggests the introduction of a worldwide PHD program or some equivalent, a suggested short-cut name for reference to global "Poverty, Homelessness and Disease." It may also have an add-on PMS program, similarly referring to the optimization of "Physical, Mental, and Social well-being" for all mankind. Undeveloped civilizations must be given full and effective attention by developed civilization, not just casual and occasional emphasis. Full world participation and responsibility has indeed been missing and cannot be made dependent on ineffective appeals in the name of a non-existent god for its success. Contrary to popular views, man must be responsible for man, and "waiting for God" will only result in eventual abandonment and failure.

A program of PRE, meaning "Promotion of expanded Research and Education" would also assure continued advancement of knowledge and its application. Factual education throughout the world is not only essential for true inter-human understanding but also necessary for eventual "unity of belief" as frequently explained in this book. Education with respect to both knowledge and ethics is the key to human success. These educational programs are intended to allow all members of humanity to gain true understanding of life processes independent of beliefs and to become involved in continued global development in the interest of human advancement. These programs are also intended to encourage the growth and successful application of all existing world programs that serve and support mankind under the principles of equality and freedom. Programs such as suggested here, combined with a new UN structure, would keep the world condition in good order and would tend to encourage peace and satisfaction on which the developing natural world must increasingly depend. Utopia is not an expectation, but world rationality and reasonable function are within sight for all nations.

The world is still haphazardly organized and managed. Leadership is often uncertain and unreliable. It is frequently managed selfishly and without effective recognition of the physical and mental needs of its charges. Much leadership time is devoted to continuity of office from the first day of election. Successful politics and good management for smaller nations do not seem to enjoy simultaneity. Why is such a condition tolerated by the powerful intelligent residents of the planet for even one minute? The building of a strong central United Nations organization is sorely needed with power to negotiate, monitor and forcibly control human rights and peace throughout the world. The condition of the world is the result of misjudgment of origin and purpose and abject human carelessness. Why is the Israeli/Palestine conflict allowed to continue for half a century or more under the tools of modern knowledge? Where are the talented world leaders to more beneficially steer man's life on the great planet? Does mankind not yet know what is right and just? Has he not yet occupied the planet long enough?

Past experience has taught us that progress and change will eventually bring the science and technology of weapons of mass destruction to all peoples and to all nations. This thought was clearly in the mind of the late Robert Oppenheimer, head of the original atomic bomb development who later became director of the Princeton Institute for Advanced Study. Even before the hydrogen bomb was completed, he suffered the embarrassment of a congressional hearing and the loss of secret clearance as the result of his public expression of concern about its very introduction to man. The hydrogen bomb represents the greatest of all known man-made power of explosion, and man may even find means to advance its explosive power substantially further. Needless to say, the gradual reduction and elimination of highly destructive weapons, supervised by an agreed upon strong central authority, would begin to settle sensitive international tensions (240).

**No faith in divine intercession:** Any benevolent humanlike god, if one existed, would be the first to accomplish that mission. But clearly, that possibility will never occur. World organization has never been successfully structured in any central fashion and it would become a "first" in human recognition and development. The power to monitor, control, and provide security should rest in the hands of a far more powerful revised United Nations, one more powerful than any of its separate components. Such a central institution would contribute to the responsibility that man had incorrectly assigned to an imagined god in the past, for the management of more peaceful and fulfilling human life on Earth. These human programs are all in keeping with the central principle of realism. A world governed humanely and democratically under prevailing secularism is not out of range of the capability of present society. The resolution of the god problem to stimulate and provide a serious planetary move in the interest of man himself is perhaps still the main prerequisite and, hence, further justification of this book.

Among the defining programs represented by PHD, PMS, and PRE, the letter designations used above, are activities intended to assure that all members of civilization who require it, are benefited in some manner, directly or indirectly, and all are faced in a positive direction. The time for asking a "helpless" imaginative god for assistance for nations and its citizens must come to an end in the interest of assuring real health and happiness administered by capable man. It is time to explain the factual existence of man and his universe to all mankind in realistic terms, a matter long overdue. Man cannot continue the spiritual charade generated by his natural tendencies and past ignorance. Even if there were an all-powerful humanlike god, it certainly is not effective, and mankind must assume the responsibility to organize and manage its affairs in accord with observed and appraised requirements. The power of man, supported by the transcendent forces that created him and his vast universe, far exceeds the power of any imagined god. It is partly hidden within his powerful mind but must eventually be interpreted and applied with reality.

**Solving old problems:** As we look at the globe, like a traditional god might do from afar, we see many problems, some of which rest in Africa, the very birthplace of humanity. Much of the continent has been developed in past years through the efforts of courageous missionaries looking to save human souls. Man has now entered a new era of spiritual interpretation and must concentrate on human improvement. A recent editorial (241) was directed to such improvement. It spoke of the AIDS epidemic that should be controlled by medical means and proper behavior and not by reliance on prayer. It spoke of the responsibility of the United Nations and the International Monetary Fund to influence its development in the modern world. Africa is still very poor, and millions of children die of diseases such as malaria. More than two million children are said to die before they are one month old. Africa, considered the cradle of humanity as confirmed by every measure of history, requires the assistance of the rest of the world. Now, over the long term, it must move up the chain of commodities and into the world of manufacturing in order to become an effective participant in the world economy. Western aid is crucial in development of transportation and power generation as well as bringing new technology to agriculture. Africa, as all other continents, must find its proper place in the globalization of the planet. The conversion of religious interests and serious development by other nations would constitute an excellent beginning. The competition among ethnic and religious groups must be resolved by rational powers, primarily by a more modern and more responsible form of the United Nations and their various service committees. Education must be high on the priority list for it will induce much self-help. Africa south of the Sahara and north of the developed southern tip has many natural assets and much potential but requires active planning and guidance to become part of the modern world. It is time that larger and richer nations invest in its potential as partners and not in greed.

In parts of Angola, Congo, and the Congo Republic, a surprising number of children are still accused of being witches, and then beaten, abused or abandoned (242). Thousands of children lie in the streets of Kinshasa, the Congo capital, having been cast out by families. Officials in one town identified 432 street children who were called witches and abandoned, often as the rationale for not having to feed or care for them. A common belief in the Bantu culture is that witches can communicate with the world of the dead and eat the life forces of others, resulting in sickness and death. Stories of the treatment of children are gross and horrendous. The church in one town serves as a sanctuary for children, but it is not enough to rely on religious organizations alone. Basic rationality is at stake to a larger degree than that described in the earlier chapters for more developed society. All levels of mind development must be given attention in the globalization process. It is the responsibility of man for fellow man that is at stake and modern society must provide the tools for resolution.

It was refreshing to read an optimistic report by Roger Cohen (243) about globalization in Africa. Free elections will be held in Ghana, and its capital is buzzing with construction. New communication technology enables its people to text money to poor relatives still in the bush. Cohen doesn't buy the proliferation of new authoritarianism. Although four million people in Ghana still defecate in public, he feels that African ascendancy is about to flourish. With it factual knowledge and belief systems will now surely follow, as filtered and redeveloped by the more advanced regions of society. No greater acceleration of human advancement can be observed than one so far behind in the midst of modern world society. It is destined to become the final continental link between the "cradle of humanity" and its latest achievements. Globalization is a natural consequence of man's occupation of the planet. It is part of the superior force that propels humanity on the planet. No force will interfere with that process. It is limited only by reason of man's immobility, a limitation that is gradually declining over time. Geographical mixture of the population in turn means mixture of race, creed and culture as well. This book has already mentioned expectations of world mix in gene structure resulting in more uniform human characteristics throughout the world over an extended time period.

**Man desires peace and comfort:** Recent war, violence and political dissatisfaction in Islamic states have led to a great deal of displacement of families that practice Islamic religion. These movements are in line with the principles of globalization already discussed. People move for many different reasons, often for oppression and freedom from mandated practices. They often move to a more "comfortable" location. They move for education reasons and sometimes have relatives who have settled in distant lands much earlier. There have been new pockets of immigrants reported in America, England, Germany, and in many other countries where immigration is reasonably possible. Although a few have expressed radicalism in sympathy with former countries, they have generally shown moderation and integration with local citizens. This process is expected to result in mixtures that will fill the prediction of this book that civilization will gradually tend toward greater genetic and cultural equality and toward higher average levels of education and understanding This process has been in effect for centuries but is now accelerating for the reasons enumerated in this and in other chapters.

Germany's vibrant postwar economy has been attracting foreign workers for the past several decades. Typically the city of Cologne must integrate as many as 120,000 Muslims who desire to have a suitable mosque built for their adherents (244). A city famous for its most spectacular Gothic cathedral denied a building permit and has provoked an ugly backlash of ethnic and religious bigotry. Debates are going on about striking the right balance between assimilation and cultural diversity, not only in Europe but in many other world areas. Civic leaders in Cologne are not so much concerned about social balance or even about security, but about demonizing a religious minority and keeping it invisible. Cologne has about thirty mosques, most of which are in back courtyards and factory buildings hidden away, and Islam now looks to its minarets, domes, and glass walls to affirm identity.

**Troublemakers feed on dissatisfaction:** The recent attack on Mumbai, the old Bombay (245), in which scores of citizens were killed and wounded, suggests continued need for world reorganization, not under imaginative religious division but under the unity of secularization. In 1993, Hindu mobs burned people alive in the streets for "committing" the crime of being Muslim in Mumbai. The latest attack was said to take place by young Muslims intent on slaughtering Hindus. They attacked the plush hotels and bars, the principal railroad station and a Jewish establishment. They want Pakistan and India to go to war over Kashmir. They want Indian Muslims expelled. They want mosques torn down and temples bombed. The incident is typical of the current terrorist philosophy, a militaristic expression of dissatisfaction without rational negation, or even discussion. It is an expression of the hate of stigma rooted largely in difference of spiritual belief but often in how cultural clashes are developed. Widespread world education and unity of belief in peaceful secularism is an excellent starting point.

This book has emphasized the process of change in human development not only in a material sense but in viewpoint and attitude. A recent article by Mark Mellman, a recognized American pollster (246), confirms how quickly change can occur in a civilization. He claimed that in an earlier poll taken a decade or more ago Americans thought that the federal government would try to do too much and not do it well. Americans today believe that the federal government will not do enough to help ordinary people. Previously wary of government, the same people are now calling for it.

A similar poll reflects a shift in the public view of morality. A few years ago a survey indicated that absolute standards of right and wrong should apply to most everyone in almost every situation. Today they believe that everyone has to decide for himself as to what is right and wrong in given situations. In the case of foreign policy, Americans supported the concept of peace through military strength by a wide margin. Today the margin has virtually disappeared. The article suggests that people themselves have changed and that their shifting values are likely to alter the future course of policy. At the moment the country seems to feel that the government should play a larger role, that moral relativism rather than absolutism is desired and that our foreign policy should be a multilateral approach grounded in diplomacy.

**It is a very old story:** In considering future globalization of the human world with respect to religion, one might review *The Marketplace of Christianity* (247). It is a surprising account of the evolution of Christianity, claiming that the economics of religion has little to do with the collection basket and much to do with understanding the development of competitive religious organizations and politics. It applies the techniques and rules of economic analysis to explain the emergence of Protestantism and examines contemporary religious issues including evolution and gay marriage. It demonstrates how Christianity developed to satisfy the changing demands of "consumers," that is, worshippers. Any organization that includes major world membership is a strong and healthy business and can afford to suffer occasional loss of membership as described in this work. But we have seen the strongest and wealthiest of corporations die natural deaths as society changes with times. And this book has spoken strongly of major spiritual change and reconsideration in every chapter.

An article (248) recently commented on *The Post-American World* by Fareed Zakaria. It speaks of another tectonic power shift following the rise of America in the last century. It suggests that the competitive rise of the balance of the world will be largely based on American efforts and ideas, ideas and accomplishments in science and technology, in commerce and capitalism and in agriculture and industry. It speaks of the future advances of China, India and Russia, discussed earlier, becoming bigger and more aggressive players in the world. American dominance has not been solely military in the past, but more creative and flexible culturally than any other part of the world.

This book emphasizes the fact that the globalizing world is changing at the "speed of light" and that religious belief is slowly changing with it. We will see a totally different world following the recovery of the present economic slump. In a sense, the world, as a whole, is learning very much like the singular human mind in a manner described in the first chapters. With more rapid future spread of human knowledge and the increasing number of minds, one can begin to think of populations assuming the same or similar understanding as that of a single mind and the world must become far more concerned with unfavorable group behavior. There will be a conflict in world adjustment with respect to government function. Democratic government will tend to continue to be the will of the people and must adjust to common principles. Leaders and representatives of the people will have to grow wiser and be held increasingly more responsible. The characteristics of individual power and personal greed that remains part of every current government must eventually be more successfully challenged through the very principles of democratic rule.

**Moving ahead of the curve:** The unipolar status of America will very gradually fade into the unified globe in which borders and immigration concern will assume quite different meaning. Man does not normally think and act that far ahead and yet he must be far more aware of future trends because of increased velocity and it must prepare for the applicable essential principles in order to set the optimum path. Man is the path of human progress and no other factor can influence or contribute to positive direction. The subject of the above referenced article will be discussed again and again over the forthcoming years of change, and governments will characteristically always lag well behind their needs. There will be a few intellectual members of society who can envision direction and contribute advice, but they might not be in sync with those who govern. Leaders of the people will increasingly have to rise from the very best of human minds to pursue world government in the interest of the entire human world with an acceptable degree of selflessness. All current governments must function more ideally with their long-term futures in mind.

In a News Brief (249) contained in the Internet issue of Ekklesia of September 2007, Theo Hobson, a freelance writer and theologian, included an article entitled, *We Need a More Intelligent Religion Debate*. His article began with the statement that he had wished for years that the intelligent media would show a bit more interest in religion, and now he has more than he wished for in the form of books by Hitchens, Dawkins and Grayling, books referenced in the introduction that he calls a new wave of "religion bashers." He says, "All three are in the grip of an ideology that is pretentious and muddled. Atheism of the kind they espouse is pretentious in the sense of claiming to know more than it does. It claims to know what belief in God irreducibly entails, and what religion, in all its infinite variety, essentially is." He adds that atheism of this sort is the belief that the demise of religion and the rise of rationality will make the world a better place. He says that anyone who claims that he knows what religion is must be underestimating its complexity, that it is far wiser than a belief in a supernatural power. He implies that these authors are intellectual cowards for they choose simplicity over complexity and difficulty.

**Imagination needs to be examined:** This book has represented a contrary view, that religion has been on the forefront of the media for many years but the recent increase in concern stems from the fact that visionary intellectuals have felt that imagination has been carried too far, it needs to be reexamined and that an appropriate critique must be made public. All the goodness and benefits of the occupied planet can be preserved and even advanced under realism. Society can be far happier without uncertainty. Imaginism tends to avoid exposing reality and fullness of knowledge of the world and society has now forged beyond that stage. The complexity about which Hobson speaks seems more like "confusion" and "uncertainty," emotion that has reached beyond the limit of judgment and control of reason. It is religion that has become muddled, and the once complex "pure" traditional belief has grown to be more simplistic because of idealistic additions and arbitrary modifications of its original concepts without any fundamental basis. This book answers many of the questions posed by Hobson in its ongoing clarification process. More average minds now recognize the difference between reality and imagined god, suggesting confirmation of the predicted secular trend.

**Less religion bashing and more motherhood:** The series of new books to which he referred, earlier called the new atheism, may be somewhat more "bashing" than is this writing, but their attempt to be properly analytical and explanatory is quite sincere. Should man continue to be afraid of understanding reality? This and several of my earlier books have expressed concern with the long-range future of a civilization of imaginative belief, one that seems to continually overlook reality in the process. The concern is largely one of deterioration rather than an anticipated expansion of the underlying reasoning process of the changing human mind. Religions are not only guilty of much of the criticisms included among those publications but make no effort to offer a disclaimer with their dogmatic claims and teachings. In many instances they even claim that human discussions have taken place directly with their gods without a sign of any doubt or fact. Such a claim looms as grossly pretentious in the modern mind. Exaggerations or baseless claims, that are free to echo many similar early writings, only play havoc with the logical mind and add to the confusion of the spiritual world without any attempt toward resolution.

Hobson questions the rise of secularism with respect to making the world a better place. He may be among those caught in the neural feedback circuitry described in earlier chapters. He must first acknowledge that the natural forces acting on the universe and on man are indeed the unstoppable forces of progress as explained in chapter one. This movement cannot be diverted by man. What first comes to the advancing mind is effective neutralization of the conflicts of many different beliefs, a consolidation to common understanding. Secondly, the primitive fear of god and his alleged monitoring of human behavior subject to heavenly judgment will continue to wear extremely thin and be given new consideration in terms of reality. The flood of secularism across the world in the course of time is very real but may require some supplementary effort by man as suggested in these chapters. The widespread disunity of the mind of man on his finite planet Earth must first be overcome before human advancement will enjoy the fruits of his intelligence and develop more mature social behavior.

**Waiting for global leadership:** Man is only very gradually beginning to understand the imaginative nature of the concept of god and that process is part of his development, which will continue even in the face of religious promotion. Man is already caught in the stream of natural forces. Influential elements of society know that the world can be made a much better place with unity of spiritual understanding followed by suitable agreement in standards of world behavior and effective means for monitoring and punishment on Earth. This in turn must be followed by modern organization and standards of government that offer real freedom, equality and recognition of human rights. This book has recommended a suitably strong and powerful form of the

United Nations, one that can exercise authority and corrective measures without delay until further globalization renders such government obsolete. Continuity of existing methods of global planning will allow only the slowest of progress. It is evident that change is required and waiting for appropriate leadership.

These chapters have indicated that our progeny, a smarter, more intelligent, and better educated generation, will eventually see the disappearance of imaginative beliefs and the adjustment of man to spiritual reality. But that is only the start of major long-term natural change. That is only the beginning of social change required for human happiness and peaceful coexistence. Society must train and utilize the very best minds for its social architecture, its organization and its executive leadership. Recent competition for the American presidency has shown how difficult it is to find the quality and ability for successful leadership of the occupants of only a single nation. Global leadership for all men of the planet requires far greater talent, the créme de la créme, the very best in the world, and its successful efforts should command the maximum in compensation and praise.

**Man's view of man:** When one stands away from an illuminated globe of Earth, sitting in a typical study, one cannot visualize the billions of humans or the effects of their variant minds. One cannot see the clashing personalities, the competitive greed or the conflicting interpretation of spiritual emotion. It is a peaceful globe warmed by the sun that enables life to grow, multiply, and return to its chemistry with reasonable predictability and consistency. Everyone on its surface has less than about a century to mature, to learn about himself and his world before vanishing from the scene. Since full human emergence there were only about ten thousand such successive life slots available. Progress and change has been no less than remarkable. Man has had ample time to contribute to the world or to just bask in its assets. But he has wasted much of his time on promoting and recycling imaginative ideas. He has time to employ his slowly expanding power of reason to add to the central bank of human knowledge or to choose to live in an imaginative emotional world of little contribution and final emptiness. His disposition is a function of the quality and reality of his education. Education is the key to understanding and underlies the potential for satisfaction and happiness. It is largely voluntary and can vary continuously from animal ignorance to intellectual genius, and the world will fare in those terms.

In recent years, there has been increased recognition of the changing state of our rapidly globalizing planet among members of the international scientific community. Professor O. E. Wilson, an eminent evolutionary biologist, recently produced a book on the importance of caring for future Earth and its occupants (250). From a Southern Baptist background, he taught us much about the ascendancy of man from nature and the diversity of products from a long existing cold Earth. The focus of his book was on the proper treatment and survival of life on the globalizing planet. He has been particularly concerned with the lack of biological education which he feels is the root of natural understanding, an argument made frequently in these pages. Wilson seems to have reached the conclusion that attention to the Earth on which we must live, its life and its environment, is more important at this time than concern about religion. In his joint meeting with Evangelical leaders to gain cooperation on this subject, Reverend Richard Cizik, vice president for government affairs of the National Association of Evangelicals, acknowledged that great scientists are people of imagination and so are people of great faith. In the process of continued globalization and in the long future of man, the resolution of his origin and spiritual eventuality will become crystal clear to all as education becomes more comprehensive and widespread.

**Who will tend to the planet?** In Wilson's view, whatever the spiritual beliefs of man, he must understand the full content of human knowledge and recognize that man is responsible for man, and together, must take good care of his only planetary home. Human development involves a continuous chain of individual lives, and optimism with respect to the evolution of the mind is justified by virtue of past history and modern understanding. The ultimate success of the objectives of mankind awaits the contribution, not only by each human link, but by the sum of many generational links that will gradually move toward more ideal principles as recognized by the reality expressed in these chapters.

The advent of science in recent centuries began to supplant religion in a gradual process, but that process is now rather advanced. The ignorance of man has taken centuries to work through his observations, inquiries and inquisitions. A newborn child has a great natural desire to learn and suffers terribly if not encouraged. The desire for understanding and particularly factual knowledge has always been the basis of life, and the processes of human development are long, uncertain but continuous. Religion gained its authority through the consistent testimony of respected men and was used as a base on which to build moral strength. Looking to the future world, the mechanisms of tradition have now given way to modification, to a new society of greater understanding and reality.

These mechanisms will tend to more deeply penetrate the lives of planetary occupants for the very first time. They will not only affect the economic and political association of man but will also modify the spiritual base of all society as described in earlier chapters. Heilbroner (251) once quoted the *Communist Manifesto* published in the middle of the nineteenth century, saying, "the bourgeoisie, during its rule of scarce one hundred years, has created more massive and more colossal productive forces than have all the preceding generations together . . . . what earlier century had even a presentiment that such productive forces slumbered in the lap of social labour?" Since that time society has progressed with the vision of two additional centuries of even more pronounced development and understanding.

**Adaptation to reality:** This book has explained the underlying neural processes, their information acquisition and their influence on belief and behavior. It has described the magnitude of changes that are constantly occurring with time within the mixed state of society and it has projected a human condition of eventual uniformity of spiritual understanding. It has emphasized the changes in the form of successive minds for each mind of a generation is irreversibly triggered anew and each is only slowed by strong expression of reversing logic. The factual position of man, his origin and that of the universe will increasingly be defined and eventually recognized by world society. Change will not be dependent on building the will of the people but on reformation of the mind of greater reason. Reliance on the factual knowledge of man will eventually supersede that determined to have no factual basis. Robert Wright in his most informative book, *The Evolution of God* (252), questioned how religions will eventually adapt to science and to one another in an age of advanced science and rapid globalization and suggested that their history points to "affirmative answers." I think these chapters have adequately described the manner of their ultimate adaptation.

The introduction began with a discussion of the works of William James, and his observations of the behavior of the spiritual mind. At the end of his famous book referenced in that discussion, he gave an indication that his own beliefs did not include theism and that he might return to the subject in a future writing. Since an explanatory writing never appeared, this book has added the underlying missing causality explaining how modern understanding of the mechanisms of the mind not only accounts for the behavior he described but also for the anticipated behavior of man looking forward. With a great deal of humility this writing was suggested as a fitting sequel.

**Irreversibility:** The oozing toothpaste of reality, of cosmic and natural belief, of agnosticism, humanism and atheism, indeed, eventual world secularism can no longer be successfully forced back into the religious tube. The analysis and judgment with respect to belief will increasingly be based on fact, free of outright imagination, driven not by the ideals of emotion but by palpable human development and more courageously by expanding human logic and reason. The underlying force that moves men will be increasingly directed to deeper, more realistic understanding, and more widespread unity of mind than at any time in the past. It must henceforth be directed firmly, with confidence, courage, optimism and purpose. Long overdue, its direction must be aimed, not at bashing imaginative religion, but at achieving sufficient unity among minds to eliminate spiritual conflict based only on the imaginary and on all its multiple derivatives. Despite continuing deficiencies, man can and will achieve a higher goal of meaning and peaceful coexistence on Earth. He must redirect his spiritual celebration to the beauty and harmony of nature and apply appropriate rites and services with a sense of truth, reality and satisfaction as emotions may demand. Without achieving this goal, spiritual interpretation will not only drift to become meaningless to man, but intellectual man will be forced to admit abject failure and defeat with respect to his hard-fought centuries of advancing knowledge and understanding.

**The end of the beginning:** I have chosen to end this book with thoughts of the "beginning" and the "end" of our universe with everything and everyone in it. By doing so it brings intelligent man to a level of understanding of the nature of his world. It is unlikely that the average man thinks much about these matters for its end is of very distant concern and not yet clearly understood. But he must know that even the universe and the solar system have beginnings and endings. Speculation does have a bearing on how one might best look at human existence and spiritual belief. This book has assumed the existence of a transcendent mechanistic force, rather than an imagined traditional humanlike god, as being responsible for our universe and all of its functions. Appropriate concepts of a "beginning" and an "end" that might fit scientific understanding will continue to be the subject of discussion for a long time to come. These concepts are based, not on my own ideas, but on the careful analysis of the latest observations and measurements of all of science, by those who are most competent to study and report on such matters. Although the end of the universe is a scary prospect as viewed by the human mind, it is really no scarier than the natural ending of every human life.

There were very few effective scientific observations and measurements of consequence in the field of astronomy until very recently. With respect to the universe, the most modern paradigm allows man to extract a reasonable interpretation of his observations and measurement in concert with his mathematical and physical laws of understanding. In contrast, the uninformed vulnerable and sensitive human mind has, from its very beginning, created simplistic interpretations that would best suit man's emotional existence. He conceived of imaginative gods that would not only care and guide human life, but also offer it an ideal heaven following his death on Earth. However desirable, such temporary expectations are now regarded as too good to be true, and future more developed minds will certainly have a different, more realistic interpretation as suggested in this study.

This book spoke of the beginning of the universe from a "singularity" and its subsequent expansion in an event that scientists call the "big bang", approximately fourteen billion years ago. However dramatic it all may sound, it is confirmed by human observations, senses and logic. The galaxies and planets were naturally formed and scientists and astronomers have made every possible attempt, within the limits of their technology, to understand the circumstances of origin, its present form and condition and its eventuality. The time sequences are frequently estimated, consistent with understanding of the known behavior of matter. We know that our solar system was formed about four or five billion years ago in our Milky Way galaxy and that life, as we know it, formed and evolved from its chemistry. Man has subsequently learned a great deal about its formation and growth and about its detailed structure. He further knows a great deal about the formation of the brain and about its function as a guide to human thought and behavior. He now also knows about the functions of human emotion and imagination that underlie spiritual interpretations. Many humans think there is some kind of god that controls all this and is cognizant of man as an occupant of planet Earth, a veritable speck of the universe. But select man thinks otherwise, has increasingly studied the behavior of its contents and has speculated on its probable future fate.

**Man might luck out:** Although many warnings have been advanced for an early "end of the world" by uninformed emotional fundamentalists in connection with their spiritual beliefs, we now know that the planet is doomed to eventual destruction only in far distant time. Of course many other things might happen but, in another five billion years, it is thought that our sun will run out of hydrogen fuel and expand to a gigantic size. Its expansion is expected to vaporize the Earth and everything on it. We can wildly speculate on the distant future of life, but there are too many factors for more reasonable consideration. In trillions of years our larger galaxy will be unable to create new stars. There will be slow cooling and decay, leaving only scattered particle remnants in space. Some scientists are alternatively considering a "cyclic process" in which the expanding universe reverses over trillions of years and contracts to experience a fiery "rebirth" similar to the big bang. Although all life would be destroyed in the process, it is possible that life could again appear in the new expansion. These views are not those of science fiction but based on the known behavior of the universe under the laws of nature.

Whatever the projected future of the universe, we can assume it to be a purely mechanistic process even without any further depth of understanding. There is certainly no indication of design or purpose or even the involvement of any humanlike spirit. Mechanistic forces have been shown to be the primary forces of the universe and life is merely a secondary chemical process. The forces that produce beginnings and ends, and are responsible for the incidental appearance of life and its evolution, must be recognized for what they are, and not given an arbitrary anthropomorphic interpretation solely for emotional satisfaction. Human development has now passed all the uncertain periods and ages of earlier existence, now reasonably well known and understood. Intellectuals have now revealed as much as man requires in order to build his ongoing journey into his future. Should man then give up on the prospect of an ideal future? Certainly not. Man has learned that he indeed lives in an amazing world and that his past has merely served as a sound basis with which to envision a far more ideal and satisfying future. It will be a future of increased understanding and more peaceful coexistence. Man must learn that his temporary position on Earth is directed, not to an imagined heaven but to residence on Earth that provides opportunity for a continuously cycling civilization to live in a definable world, an opportunity to serve, not an imaginary god, but his real fellow man.

**How should we leave it?** Finally, this study has carefully considered the "god problem" of Turgenev and Belinsky and has demonstrated the causal factors underlying the writings of William James. Indeed, it has provided final resolution of the spiritual uncertainties of man. It has suggested that the natural mechanistic forces that created and continue to propel the universe and its contents will act to fully replace the uncertain and disorganized imaginative spirituality with uniform factual understanding in ongoing evolutionary development. These are the only conclusions that might be reached by an informed and unbiased mind in the highest court of rational human judgment.

I would like to close with another reference to the late John Updike, who ended his final book with a comment typical of his great literary style: "We are part of nature and natural necessity compels, and in the end dissolves us; yet to renounce all and any supernature, any appeal or judgment beyond the claims of matter . . . leaves in the dust too much of humanity . . . ." Indeed, although this book explains how the forces of the universe will eventually affect the entire cosmic world and how the view of man will change producing everlasting world secularism, yet, inherent emotions and feelings still make it difficult for rational minds to completely and abruptly close the door of reality. Therefore, in consideration of such widespread spiritual dependence with unassailable sincerity, I might, for the moment, leave it just slightly ajar.

# REFERENCES INCLUDED IN THE TEXT OF THIS STUDY ARE SHOWN IN PARENTHESIS THROUGHOUT


(1) THE DECLINE AND FALL OF CHRISTIAN AMERICA, by Jon Meacham, Cover and first page, Newsweek. April 13, 2009

(2) THE END OF CHRISTIAN AMERICA, by Jon Meacham, Newsweek, April 13, 2009

(3) SECULAR EUROPE'S MERITS by Roger Cohen, the New York Times, December 13, 2007

(4) SOCIETY WITHOUT GOD: WHAT THE LEAST RELIGIOUS NATIONS CAN TELL US ABOUT CONTENTMENT, by Phil Zuckerman, NYU Press
SYRIA'S SOLIDARITY WITH ISLAMISTS ENDS AT HOME, by Kareem Fahim, the New York Times, September 4, 2010

(5) AT EXPENSE OF ALL OTHERS, PUTIN PICKS A CHURCH. by Clifford J Levy, the New York Times April 24, 2008

(6) GROUPS PLAN NEW BRANCH TO REPRESENT ANGLICANISM, by Neela Banerjee, the New York Times September 30, 2007.

(7) THE POSEN FOUNDATION, Center for Cultural Judaism, Lucerne Switzerland, www.posenfoundation.com.

(8) IN JAPAN, BUDDHISM MAY BE DYING OUT, by Norimitsu Onishi, the New York Times, July 14, 2008

(9) CAMERAS ROLL, AND FAITH HASN'T A PRAYER, by John Leland, the New York Times, September 28, 2008

(10) HUMANIST CHAPLAINCY AT HARVARD. September, 2008 Newsletter, Edited by Greg M Epstein, Harvard University Chaplain

(11) PRINCETON CENTER FOR THE STUDY OF RELIGION, Summer, 2008 Bulletin, Princeton University
(12) THE SECULAR MIND, by Robert Coles, Princeton University Press
(13) THE WAY THINGS ARE: THE CHANGING PERSPECTIVE OF HUMAN EXISTENCE, by John F Brain (Brinster), Xlibris
(14) THE NATURAL BIBLE FOR MODERN AND FUTURE MAN: THE ULTIMATE THEOLOGY OF THE STILL EVOLVING MIND, by John F Brain (Brinster), Hamilton Books, University Press of America
(15) THE MAN WHO CREATED GOD, by John F Brain (Brinster), Xlibris
(16) THE ABDUCTION: THE SACRED LEGEND OF THE GREAT WALL, by John F Brinster, AuthorHouse
(17) TABLET IGNITES DEBATE ON MESSIAH AND RESURRECTION, by Ethan Bronner, the New York Times, July 6, 2008
(18) ALBERT EINSTEIN'S COSMIC REVERENCE, by John F Brinster, Op/Ed, Philadelphia Inquirer, January 23, 2006
(19) BEGINNINGS OF CELLULAR LIFE: METABOLISM RECAPITULATES BIOGENESIS, by Harold J Morowitz, Yale University Press
(20) KISSING THE EARTH GOODBYE IN ABOUT 7.59 BILLION YEARS, by Dennis Overbye, the New York Times, March 11, 2008
(21) THE THIRD CHIMPANZEE THE EVOLUTION AND THE FUTURE OF THE HUMAN ANIMAL, by Jared Diamond, Harper Perennial
(22) POLL FINDS A FLUID RELIGIOUS LIFE IN U.S., by Neela Banerjee, the New York Times, February 26, 2008

(23) THE EDGE OF EVOLUTION: THE SEARCH FOR THE LIMITS OF DARWINISM, by Michael Behe, Free Press
(24) BREAKING THE SPELL: RELIGION AS A NATURAL PHENOMENON, by Daniel C Dennett, Viking Penguin
(25) THE GOD DELUSION, by Richard Dawkins, Houghton Mifflin
(26) GOD IS NOT GREAT: RELIGION POISONS EVERYTHING, by Christopher Hitchens, Hachette Book Group
(27) FAITH'S LAST GASP, by A C Grayling. Prospect Magazine, November, 2006
(28) THE END OF FAITH: RELIGION, TERROR AND THE FUTURE OF REASON, by Sam Harris, W W Norton and Company
(29) UNCERTAIN CHURCH AWAITS POPE IN U.S., by Laurie Goodstein, the New York Times, April 14, 2008
(30) HUMANIST MANEFESTO 2000: A CALL FOR A NEW PLANETARY HUMANISM, by Paul Kurtz, Prometheus Books
(31) GOD THE FAILED HYPOTHESIS: HOW SCIENCE SHOWS THAT GOD DOES NOT EXIST, by Victor J Stenger, Prometheus Books
(32) CHALLENGING NATURE: THE CLASH OF SCIENCE AND SPIRITUALITY AT THE NEW FRONTIERS OF LIFE, by Lee M Silver, Harper Collins
(33) AFTER GOD, by Mark C Taylor, the University of Chicago Press
(34) AFFECTIVE NEUROSCIENCE; THE FOUNDATION OF HUMAN AND ANIMAL EMOTIONS, by Jaak Panksepp. Oxford University Press
(35) ACROSS THE SECULAR ABYSS: FROM FAITH TO WISDOM, by William Sims Bainbridge, Lexington Books

(36) ATHEIST URGING ITALY TO GET RELIGION, by Rachel Donadio, the New York Times, April 6, 2008
(37) KLUGE, THE HAPHAZARD CONSTRUCTION OF THE HUMAN MIND, by Gary Marcus, Houghton Mifflin Company
(38) BLIND FAITH: THE UNHOLY ALLIANCE OF RELIGION AND MEDICINE, by Richard P Sloan, St Martin's Press
(39) THE FUTURE IS NOW? PRETTY SOON, AT LEAST, by John Tierney, the New York Times, June 3, 2008
(40) JACKING INTO THE BRAIN, by Gary Stix, Scientific American, November, 2008
(41) TECHNOLOGY THAT OUTTHINKS US: A PARTNER OR A MASTER? By John Tierney, the New York Times, August 26, 2008
(42) ON DEEP HISTORY AND THE BRAIN, by Daniel Lord Smail, University of California Press
(43) THE NEURAL BUDDHISTS, by David Brooks, the New York Times, May 13, 2008
(44) BRAIN AND BELIEF: AN EXPLORATION OF THE HUMAN SOUL, by John J McGraw, Aegis Press
(45) SHARING THEIR DEMONS ON THE WEB, by Sarah Kershaw, the New York Times, November 13, 2008
(46) A UNIVERSE OF CONSCIOUSNESS: HOW MATTER BECOMES IMAGINATION, by Gerald Edelman and Giulio Tononi, Basic Books
(47) ORIGIN OF SEX: THREE BILLION YEARS OF GENETIC RECOMBINATION, by Lynn Margulis and Dorion Sagan, Yale University Press
(48) THOUSANDS EXPECT APOCALYPSE IN 2012, AOL News, July 6, 200.

(49) ADAPTING MINDS: EVOLUTIONARY PSYCHOLOGY AND THE PERSISTENT QUEST FOR HUMAN NATURE, David J Buller, MIT Press

(50) THE SINGULARITY IS NEAR: WHEN HUMANS TRANSCEND BIOLOGY, by Ray Kurzweil, Penguin Books

(51) THE JEWISH MIND, by Raphael Patai, Charles Scribners Sons

(52) INFORMATION IN THE BRAIN: A MOLECULAR PERSPECTIVE, by Ira B Black, MIT Press

(53) THE STUFF OF THOUGHT: LANGUAGE AS A WINDOW INTO HUMAN NATURE, by Steven Pinker, Viking

(54) PREFRONTAL REGIONS ORCHESTRATE SUPPRESSION OF EMOTIONAL MEMORIES VIA A TWO-PHASE PROCESS, by Brendan Depue, Tim Curran and Marie Banich, Science, July 13, 2007

(55) BRAIN CELLS DOING THEIR JOB WITH SOME NEIGHBORLY HELP, by Benedict Carey, the New York Times, December 25, 2007

(56) MEMORY: THE KEY TO CONSCIOUSNESS, by Richard F Thompson and Stephen Madigan, Princeton University Press

(57) IN SEARCH OF MEMORY: THE EMERGENCE OF A NEW SCIENCE OF MIND, by Eric R Kandel, W W Norton and Company

FEEDBACK LOOPS SHAPE CELLULAR SIGNALS IN SPACE AND TIME, by Onn Brandman and Tobias Meyer, Science, October 17, 2008

(58) FROM STEM CELLS TO GRANDMOTHER CELLS: HOW NEUROGENESIS RELATES TO LEARNING AND MEMORY, by Tracey J Shors, Cell Press, September 11, 2008

(59) THE POWER OF PLACE: HOW OUR SURROUNDINGS SHAPE OUR THOUGHTS, EMOTIONS AND ACTIONS, by Winifred Gallagher, Harper Perennial

(60) ANTERIOR PREFRONTAL FUNCTION AND THE LIMITS OF HUMAN DECISION-MAKING, by E Koechlin and A Hyafil, Science, October 26, 2007

(61) WHY GOD WON'T GO AWAY: BRAIN SCIENCE AND THE BIOLOGY OF BELIEF, by Andrew Newberg, Eugene D'Aquilli and Vince Rause, Ballantine Books

(62) THE GOD HELMET, Michael Persinger leading researcher, Wikipedia

(63) THE GREAT TRANSFORMATION THE BEGINNING OF OUR RELIGIOUS TRADITIONS, by Karen Armstrong, Anchor/Vintage

(64) DREAMS STIFLED, EGYPT'S YOUNG TURN TO ISLAMIC FERVOR, by Michael Slackman, the New York Times, February 17, 2008

(65) FROM ANTS TO PEOPLE, AN INSTINCT TO SWARM, by Carl Zimmer, the New York Times, November 13, 2007

(66) THE VATICAN AND GLOBALIZATION: TINKERING WITH SIN, by Eduardo Porter, the New York Times, April 7, 2008

(67) HOW HAS CHRISTIANITY CHANGED OVER 2,000 YEARS? Course lecture by Bart Ehrman, The Teaching Company

(68) WHAT PAUL MEANT, by Gary Wills, Viking/Penguin

(69) SIN AND EVIL MORAL VALUES IN LITERATURE, by Ronald Paulson, Yale University Press

(70) THE HIPPOCAMPUS BOOK, by Per Anderson et al, Oxford University Press.

(71) SIGNIFICANT LIFE EVENTS AND THE SHAPE OF MEMORIES TO COME, by Tracey J Shors, Neurobiology of Learning and Memory, September, 2005

(72) SELF—ANCIENT AND MODERN INSIGHTS ABOUT INDIVIDUALITY, LIFE, AND DEATH, by Richard Sorabji, The University of Chicago Press

(73) A PHILOSOPHY OF FEAR, by Lars Svendsen, The University of Chicago Press

(74) GETTING REAL, Private draft communication on nuclear technology, by Theodore Rockwell.

(75) SCIENTISTS STUDY THE FEAR FACTOR, Associated Press, The (Trenton) Times, October 31, 2007

(76) SELECTIVE ERASURE OF A FEAR MEMORY, by Jin-Hee Han et al, Science, March 13, 2009

(77) WHEN GOD BECOMES A DRUG: BOOK 1, UNDERSTANDING RELIGIOUS ADDICTION & RELIGIOUS ABUSE, by Leo Booth

(78) CHIMPANZEES ARE RATIONAL MAXIMIZERS IN AN ULTIMATUM GAME, by Keith Jensen, Josep Call, Michael Tomasetto, Science, October 5, 2007

(79) ORIGINS OF LIFE, by Freeman Dyson, Cambridge University Press

(80) PURSUING SYNTHETIC LIFE, DAZZLED BY REALITY, by Natalie Angier, the New York Times, February 5, 2008

(81) WITHOUT GOD, by Steven Weinberg, The New York Review of Books, September 25, 2008

(82) SYNAPTIC TRANSMISSION: A BIDIRECTIONAL AND SELF-MODIFIABLE FORM OF CELL-CELL COMMUNICATION, T M Jesse' and E R Kandel, Howard Hughes Medical Institute.

(83) NOTHING TO BE FRIGHTENED OF, by Julian Barnes, Alfred a Knof

(84) FINDING DESIGN IN NATURE, by Cardinal Christoph Schönbotn, The Catholic Education Resource Center, www.catholiceducation.org

(85) THE GRAND DESIGN, by Stephen Hawking and Leonard Mlodinow, Bantam Books

(86) GAUGING A COLLIDER'S ODDS OF CREATING A BLACK HOLE, by Dennis Overbye, the New York Times, April 15, 2008

(87) APOPTOSIS: THE LIFE AND DEATH OF CELLS, by Potten, Christopher, and Wilson, Cambridge University Press

(88) ON A RING AND A PRAYER, by Seth Freeman, the New York Times, November 25, 2008

(89) PHILOSOPHERS WITHOUT GODS, MEDITATIONS ON ATHEISM AND SECULAR LIFE, by Louise M Antony, Oxford University Press

(90) ATHEISTS, AGNOSTICS, AND DEISTS IN AMERICA, A BRIEF HISTORY, by Peter M Rinaldo, DorPete Press

(91) THE FUTURE OF THEOLOGICAL INQUIRY IN THE ACADAMY, CHURCH AND MEDIA, Three Day Ecumenical Conference, April 2007, Center of Theological Inquiry, Princeton NJ

(92) CENTER FOR THE STUDY OF RELIGION, Princeton University, 5 Ivy Lane, Princeton NJ

(93) BUILDING A FRAMEWORK OF FACTUAL KNOWLEDGE IN A FAITH-BASED SOCIETY, A Scientific Panel, Organized by John F Brinster, Princeton University, May 30, 2008

(94) THE HUMANIST CHAPLAINCY AT HARVARD, The Memorial Church, One Harvard Yard, Cambridge, MA 02138

(95) KEEPING THE FAITH: IN A SEAT OF SECULAR LEARNING, RELIGIOUS OBSERVANCE THRIVES, by Merrell Noden, The Princeton Alumni Weekly, December 17, 2008

(96) CREATIONISM'S TROJAN HORSE: THE WEDGE OF INTELLIGENT DESIGN, by Barbara Forest and Paul R Gross, Oxford University Press

(97) A TEACHER ON THE FRONT LINE AS FAITH AND SCIENCE CLASH, by Amy Harmon, the New York Times, August 24, 2008

(98) IN TEXAS, A LINE IN THE CURRICULUM REVIVES EVOLUTION DEBATE, by James McKinley Jr., the New York Times, January 22, 2009

(99) EVOLUTION: THE FIRST FOUR BILLION YEARS, Edited by Michel Ruse and Joseph Travis, Harvard University Press, Belknap

(100) THE PRESIDENT OF GOOD AND EVIL: QUESTIONING THE ETHICS OF GEORGE W BUSH, by Peter Singer, A Plume Book

(101) LIBERTY OF CONSCIENCE: IN DEFENSE OF AMERICA'S TRADITION OF RELIGIOUS EQUALITY, by Martha C Nussbaum, Basic Books

TEN TORTURED WORDS: HOW THE FOUNDING FATHERS TRIED TO PROTECT RELIGION IN AMERICA . . . . AND WHAT'S HAPPENED SINCE, by Stephen Mansfield, Thomas Nelson

(102) FOUND: AN ANCIENT MONUMENT TO THE SOUL, by John Wilford, the New York Times. November 18, 2008

(103) POPE, IN U.S., IS 'ASHAMED' OF PEDOPHILE PRIESTS, by Ian Fisher and Laurie Goodstein, the New York Times. April 16, 2008

(104) FINDING JESUS ON FACEBOOK, AND CHECKING PODCASTS FOR A PEW THAT FITS, by April Dembosky, the New York Times, October 26, 2008
(105) SACRED PLACES, Special Issue, U S News and World Report, November 26, 2007
(106) PILGRIMAGE TO ROOTS OF FAITH AND STRIFE, by Isabel Kershner, the New York Times, October 24, 2008
(107) THE SCANDAL OF THE EVANGELICAL MIND, by Mark E Noll. Eerdmans
(108) IN THE NAME OF GOD The Afghan Connection and the US War Against Terrorism, by Lars Ersleve Anderson and Jan Aagaard, University Press of Southern Denmark
GOD'S WAR: A NEW HISTORY OF THE CRUSADES, by Christopher Tyerman, Harvard University Press
(109) LEADER OF PAKISTAN MOSQUE VOWS A FIGHT TO THE DEATH, by Carlotta Gall, the New York Times, July 9, 2007
(110) WHY I AM NOT A CHRISTIAN, by Bertrand Russell, Simon & Schuster
(111) TRAIN YOUR MIND, CHANGE YOUR BRAIN: HOW A NEW SCIENCE REVEALS OUR EXTRAORDINARY POTENTIAL TO TRANSFORM OURSELVES, by Sharon Begley, Ballantine Books
(112) SERVING SHORT-HANDED U S PARISHES, FATHERS WITHOUT BORDERS, by Laurie Goodstein, the New York Times, December 28, 2008
(113) FOR CATHOLICS, AN ON-AIR MIX OF SACRED AND SILLY, by Paul Vitello, the New York Times, July 13, 2008
(114) EPISCOPAL SPLIT AS CONSERVATIVES FORM NEW GROUP, by Laurie Goodstein, the New York Times, December 4, 2008

(115) SCIENCE IS IN ITS DETAILS, by Sam Harris, the New York Times, July 27,2009

(116) SOLDIER SUES ARMY, SAYING HIS ATHEISM LED TO THREATS, by Neela Banerjee, the New York Times April 26, 2008

(117) RELIGION AND ITS ROLE ARE IN DISPUTE AT THE SERVICE ACADEMIES, by Neela Banerjee, the New York Times, June 25, 2008

(118) AS TENSIONS RISE FOR EGYPT'S CHRISTIANS, OFFICIALS CALL CLASHES SECULAR, by Michael Slackman, the New York Times, August 2, 2008

(119) THE FAITH TO OUTLAST POLITICS, by David Kuo and John Dilulio Jr, The New York Times, January 29, 2008

(120) A MATTER OF FAITH—RELIGION IN THE 2004 PRESIDENTIAL ELECTION, David E Campbell, Editor, Brookings Institution Press

(121) ARE SCIENTISTS PLAYING GOD? IT DEPENDS ON YOUR RELIGION, by John Tierney, the New York Times, November 20, 2007

THE CRISIS OF FAITH, Editorial, the New York Times, December 7, 2007

(122) FAITH vs. THE FAITHLESS, by David Brooks, the New York. Times, December 7, 2007

(123) LEVELING THE PRAYING FIELD, by Nancy Gibbs and Michael Duffy, Time Magazine, July 23, 2007

(124) CAMPAIGNS LIKE THESE MAKE IT HARD TO FIND A REASON TO BELIEVE, by Eduardo Porter, the New York Times, December 14, 2007

(125) CONVERT OR DIE? PAKISTANI CHRISTIANS SEEK HELP, by Munir Alunacl, Associate Press, May 17, 2007

(126) DON'T KNOW MUCH ABOUT THE BIBLE: EVERYTHING YOU NEED TO KNOW ABOUT THE GOOD BOOK BUT NEVER LEARNED, by Kenneth C Davis, William Morrow and Company
(127) SO HELP ME GOD: THE FOUNDING FATHERS AND THE FIRST GREAT BATTLE OVER CHURCH AND STATE, by Forrest Church, Harcourt
(128) PROFESSORS SUE ORAL ROBERTS PRESIDENT, by Ralph Blumenthal, the New York Times, October 11, 2007
(129) THE FAITH OF GEORGE W BUSH, by Stephen Mansfield, Tarcher
(130) BELIEVING SCRIPTURE BUT PLAYING BY SCIENCE'S RULES, by Cornelia Dean, the New York Times, February 12, 2007
(131) REAL CHRISTIAN MATCHING, for Christian singles, New York, NY
(132) RELIGIOUS RIGHTS GROW IN TRENTON, by Robert Schwaneberg, The (Trenton) Times, June 12, 2007
(133) MORE ATHEISTS SHOUT IT FROM THE ROOFTOPS, by Laurie Goodstein, the New York Times, April 27, 2009
(134) COOL RECEPTION FOR BIBLE PARK IN BIBLE BELT, by Theo Emery, the New York Times, June 10, 2007
(135) MY SAUDI VALENTINE, by Kajaa Alsanea, the New York Times, February 13, 2008
(136) FROM TINY SECT, WEIGHTY ISSUE FOR JUSTICES, by Adam Liptak, the New York Times, November 11, 2008
(137) HOW DEMOCRACY PRODUCED A MONSTER, by Ian Kershaw, the New York Times, February 5, 2008
(138) ISLAMIC REVIVAL TESTS BOSNIA'S SECULAR CAST, by Dan Bilefsky, the New York Times, December 27, 2008

(139) CATHOLICS AND MUSLIMS PLEDGE TO IMPROVE LINKS, by Rachel Donadio, the New York Times, November 7, 2008

(140) BRAIN RESEARCHERS OPEN DOOR TO EDITING MEMORY, by Benedict Carey, the New York Times, April 6, 2009

(141) A BRIEF HISTORY OF TIME, by Stephen Hawking, Bantam

(142) HOW LIFE BEGAN: EVOLTUTION'S THREE GENESES, by Alexandre Meinesz, University of Chicago Press

(143) A NEW PICTURE OF THE EARLY EARTH, by Kenneth Chang, the New York Times, December 2, 2008

(144) THE PROTEIN PYRAMID, New York Times Editorial, November 10, 2008

(145) THE SCIENCE OF EVOLUTION AND THE MYTH OF CREATIONISM: KNOWING WHAT'S REAL AND WHY IT MATTERS, by Ardea Skybreak, Insight Press

(146) THE PHENOMENON OF RELIGION: PAGAN AND BIBLICAL RELIGION, by Manfred H Vogel, University Press of America

(147) THE MIX TAPE OF THE GODS, by Timothy Ferris, Op/Ed The New York times September 5, 2007.

(148) NASA SNAPS IMAGES OF 'DEATH STAR GALAXY', by Seth Borenstein, Associated Press, Science News, December 18, 2007

(149) THERE GOES THE SUN, New York Times Editorial, September 17, 2007

(150) TAKING SCIENCE ON FAITH, by Paul Davies, The New York Times, November 24, 2007

(151) OF WITCHES AND THE WAIT FOR JUSTICE, by Maura J Casey, the New York Times, April 13, 2008

DO YOU BELIEVE IN GHOSTS? SURVEY SAYS ONE-THIRD DO, Associated Press, The (Trenton) Times, October 26, 2007

(152) FLY ME TO THE DEITY, by Tunku Varadarajan, the New York times, October 29, 2008

(153) THE ORIGIN AND EVOLUTION OF RELIGIOUS PROSOCIALITY, by Ara Norenzayan and Azim Sharrif, Science, October 3, 2008

(154) HEAVEN FOR THE GODLESS?, by Charles Blow, the New York Times, December 27, 2008

(155) FIRE FROM HEAVEN: THE RISE OF PENTECOSTAL SPIRITUALITY AND THE RESHAPING OF RELIGION IN THE TWENTY-FIRST CENTURY, by Harvey Cox, Addison-Wesley Publishing Company

(156) LOTIS THERAPY, by Benedict Carey, The New York Times, May 217, 2008

(157) THE A TO Z OF TAOISM, by Julian F Pas, The Scarecrow Press/Rowman and Littlefield

(158) PHANTOMS IN THE BRAIN: PROBING THE MYSTERIES OF THE HUMAN MIND, by V S Ramachandran and Sandra Blakeslee, Quill

(159) TERROR IN THE MIND OF GOD: THE GLOBAL RISE OF RELIGIOUS VIOLENCE, by Mark Juergensmeyer, University of California Press

(160) WAYS OF KNOWING, Kathryn Johnson, Adam Cohen, Mariam Cohen and Bary Leshowitz, The Global Spiral, July 13, 2007

(161) A SECULAR AGE, by Charles Taylor, Harvard University Press

(162) GOD BEYOND LANGUAGE AND TEXT: ICONOGRAPHY AS SPIRITUAL PRACTICE, by Colleen Burlingham, Inspire,

2007-2008 A publication of the Princeton Theological Seminary, Princeton NJ

(163) GOD AND MAN AT NOTRE DAME, by Kenneth Woodward, the New York Times, April 16, 2008

(164) TEXAS PILGRIMS LOSE THEIR WAY, BUT NOT THEIR FAITH, by Ralph Blumenthal, the New York Times, April 16, 2008

(165) TRIALS OF THE SAINTS, by James Martin, the New York Times, March 3, 2008

(166) TRIALS FOR PARENTS WHO CHOSE FAITH OVER MEDICINE, by Dirk Johnson, the New York Times, January 20, 2009

(167) CLUELESS IN AMERICA, by Bob Herbert, the New York Times, April 22, 2008

(168) A NATION AT A LOSS, by Edward B Fiske, the New York Times. April 25, 2008

(169) EVIL INCARNATE: RUMORS OF DEMONIC CONSPIRACY AND SATANIC ABUSE IN HISTORY, by David Frankfurter, Princeton University Press

(170) A FREE-FOR-ALL ON SCIENCE AND RELIGION, by George Johnson, the New York Times, November 21, 2007

(171) IS 'DO UNTO OTHERS' WRITTEN INTO OUR GENES? by Nicholas Wade, the New York Times, September 18, 2007

(172) AMERICAN GOSPEL: GOD, THE FOUNDING FATHERS AND THE MAKING OF A NATION, by Jon Meacham, Random House

(173) EINSTEIN AND GOD, by Thomas Torrance, Reflections, Spring 1998, a publication by the Center of Theological Inquiry, Princeton NJ

(174) EINSTEIN: HIS LIFE AND UNIVERSE, by Walter Isaacson, Simon and Schuster

(175) EINSTEIN AND RELIGION, by Max Jammer, Princeton University Press

(176) OUR BIOTECH FUTURE, by Freeman Dyson, New York Review of Books, July 19, 2007

(177) FOUR VIEWS ON FREE WILL, by John Fischer, et al, Blackwell Publishing

(178) WHY THERE ALMOST CERTAINLY IS A GOD: DOUBTING DAWKINS, by Keith Ward, Lion Publishing

(179) THE WAR ON DRUGS STARTS HERE, Editorial, the New York Times, February 13, 2008

(180) COURT RULES AGAINST 'BONG HITS FOR JESUS' CASE, by Mark Sherman. AOL Associated Press News, June 25, 2007

(181) WAR ON DRUGS LOSING WORLDWIDE BATTLE, by Misha Glermy, The (Trenton) Times, September 2, 2007

(182) ANIMAL LEARNING AND COGNITION, by Charles F Flaherty, McGraw Hill Inc. THE JOYS AND PAINS OF BEING AN ANIMAL. by Dwight Garner, the New York Times, January 21, 2009

(183) WHO WOULD JESUS SMACK DOWN? by Molly Worthen, the New York Times, January 11, 2009

(184) WHY GOOD THINGS HAPPEN TO GOOD PEOPLE, by Stephen Post and Jill Neiman, Broadway Books

(185) PALESTINIANS' RIFF PREVENTS GAZANS FROM TRAVELING TO MECCA, by Taghreed el-Khodary and Ethan Bronner, the New York Times, December 4, 2008

(186) INDEPENDENT, TAJIKS REVEL IN THEIR FAITH, by Sabrina Tavernise, the New York Times, January 4, 2009

(187) SACRED BARRIERS TO CONFLICT RESOLUTION, by Scott Atran, Robert Axelrod, Richard Davis, Science, August 24, 2007

(188) MEGACHURCHES ADD LOCAL ECONOMY TO THEIR MISSION IN GOD'S NAME, by Diana B Henriques and Andrew W Lehren, the New York Times, November 23, 2007

(189) PRESIDENT OF EVANGELICAL UNIVERSITY RESIGNS, Associated Press, the New York Times, November 24, 2007

(190) CLEVELAND DIOCESE ACCUSED OF IMPROPRIETY AS EMBEZZLEMENT TRIAL NEARS, by Christospher Maag, the New York Times, August 20, 2007

(191) FEDERAL GRANT FOR A MEDICAL MISSION GOES AWRY, by Diana Henriques and Andrew Lehren, The New York Times, June 13, 2007

(192) EPISCOPAL CHURCH FACES DEADLINE ON GAY ISSUES, by Neela Banerjee, the New York Times, September 16, 2007

(193) ABSTINENCE EDUCATION FACES AN UNCERTAIN FUTURE, by Laura Beil, the New York Times, July 18, 2007

(194) TALKING WITH CHILDREN ABOUT SEX AND AIDS: AT WHAT AGE TO START? by Donald McNeil Jr., the New York Times, February 26, 2008

(195) VATICAN ISSUES INSTRUCTION ON BIOETHICS, by Laurie Goodstein and Elisabetta Povoledo, the New York Times, December 13, 2008

(196) THE ABSTINENCE-ONLY DELUSION, Editorial, the New York Times, April 28, 2007

(197) TELLING THE STORIES BEHIND THE ABORTIONS, by Cornelia Dean, the New York Times, November 6, 2007
(198) THOU SHALT HAVE SEX EVERY DAY, by Caroline Howard, AOL News, February 20, 2008
(199) CATHOLIC LAY GROUP TESTS A STRATEGY CHANGE, by Pam Belluck, the New York Times, June 24, 2007
(200) SNAKES ON A PLAIN: LIVING WITH COBRAS OUTSIDE CALCUTTA, by Yaroslav Trofimov, The Wall Street Journal, November 26, 2007
(201) BENDING THE RULES: MORALITY IN THE MODERN WORLD FROM RELATIONSHIPS TO POLITICS AND WAR, by Robert A Hinde, Oxford University Press
(202) A PASTOR BEGS TO DIFFER WITH FLOCK ON MIRACLES, by Erik Eckholm, the New York Times, February 20, 2008
(203) ROCKIES PLACE THEIR FAITH IN GOD, AND ONE ANOTHER, by Ben Shipigel, the New York Times, October 23, 2007
(204) PRIMATES AND PHILOSOPHERS: HOW MORALITY EVOLVED, by Frans de Waal, Princeton University Press
(205) US IMPRISONS ONE IN 100 ADULTS, REPORT FINDS, by Adam Liptak, the New York Times, February 29, 2008
(206) A TOWN AND A BAPTIST PASTOR VIE OVER "ETERNITY", by Jennifer Steinhauer, the New York Times March 1, 2008
(207) ACROSS GLOBE, EMPTY BELLIES BRING RISING ANGER, by Marc Lacey, the New York Times, April 18, 2008
(208) DOMINICANS SAY COCKFIGHTING IS IN THEIR BLOOD, by Katie Thomas, the New York Times, February 13, 2008

(209) THE NEW FACES OF CHRISTIANITY: BELIEVING THE BIBLE IN THE GLOBAL SOUTH, by Philip Jenkins, Oxford University Press

(210) ESALEN: AMERICA AND THE RELIGION OF NO RELIGION, by Jeffrey J Kripal, University of Chicago Press

(211) A COMPASS THAT CAN CLASH WITH MODERN LIFE, by Michael Slackman, the New York Times, June 12, 2007

(212) BAD TIMES DRAW BIGGER CROWDS TO CHURCHES, by Paul Vitello, the New York Times, December 14, 2008

(213) PREACHING TO WALL STREET, The New York Times, December 17, 2006

(214) MIDLIFE SUICIDE RISES, PUZZLING RESEARCHERS, by Patricia Cohen, the New York Times, February 19, 2008

(215) CAN GOD AND CAESAR COEXIST? BALANCING RELIGIOUS FREEDOM AND INTERNATIONAL LAW, by Robert F Drinan, Yale University Press

(216) GENES EXPLAIN RACE DISPARITY IN RESPONSE TO A HEART DRUG, by Gina Kolata, the New York Times, April 29, 2008

(217) IF YOU POST IT, THEY WILL PRAY, by Allen Salkin, the New York Times, November 30, 2008

(218) EVEN IN RECESSION, BELIEVERS INVEST IN THE GOSPEL OF GETTING RICH, by Laurie Goodstein, the New York Times, August 16, 2009.

(219) CONSILIENCE :THE UNITY OF KNOWLEDGE, by E 0 Wilson, Alfred A Knoff

(220) DOES SCIENCE MAKE BELIEF IN GOD OBSOLETE? A two-page ad, the New York Times, May 18, 2008, by The John Templeton Foundation

(221) ORIGINS: FOURTEEN BILLION YEARS OF COSMIC EVOLUTION, by Neil deGrasse Tyson and Donald Goldsmith, WWNorton

(222) WITH FEAR AND WONDER IN ITS WAKE, SPUTNIK LIFTED US INTO THE FUTURE, by John Wilford, the New York Times, September 25, 2007

(223) THE GALACTIC HABITABLE ZONE AND THE AGE DISTRIBUTION OF COMPLEX LIFE IN THE MILKY WAY, by Charles Linweaver, Yeshe Fenner, Brad Gibson, Science, January 2, 2004

(224) HUMAN CLONING CALLED INEVITABLE, BAN URGED, The (Trenton) Times, November 11, 2007

(225) GEONS, BLACK HOLES & QUANTUM FOAM, by John A Wheeler, W W Norton and Company

(226) THE FUTURE OF EDUCATION: Reimagining Our Schools from the Ground Up, by Kieran Egan, Yale University press

(227) SURVIVING 1,000 CENTURIES: CAN WE DO IT? By Roger-Maurice Bonnet and Lodewijk Woltjer, Springer, Berlin Press

(228) THE RACE BETWEEN EDUCATION AND TECHNOLOGY, by Claudia Goldin and Lawrence Katz, Belknap/Harvard University Press

(229) TAX CUTS FOR TEACHERS, by Thomas L Friedman, the New York Times, January 11, 2009

(230) MOTHERLAND: A PHILOSOPHICAL HISTORY OF RUSSIA, by Lesley Chamberlain, The Rookery Press/Overlook Press

(231) GODLY REPUBLIC: A CENTRIST BLUEPRINT FOR AMERICA'S FAITH-BASED FUTURE, by John J Dilulio Jr, University of California Press

(232) THE WORLD IS FLAT, A BRIEF HISTORY OF THE TWENTY-FIRST CENTURY, by Thomas L Friedman, Farrar, Straus & Giroux

(233) IN DEFENSE OF DEATH, by David Brooks, the New York Times, January 13, 2009

(234) A BRIEF HISTORY OF CHRISTIANITY IN CHINA, Frontline, PBS, July 17, 2008

(235) IN A GENERATION, MINORITIES MAY BE THE U. S. MAJORITY, by Sam Roberts, the New York Times, August 13, 2008

(236) THE VATICAN'S RELATIVE TRUTH, by John Allen Jr., the New York Times, December 19, 2007

(237) HUMAN SHADOWS ON THE SEAS, by Andrew Revkin, the New York Times, February 26, 2008

(238) A CHANGE TO BELIEVE IN, by Roger Cohn, the New York Times, February 21, 2008

(239) JORDANIAN STUDENTS REBEL, EMBRACING CONSERVATIVE ISLAM, by Michael Slackman, the New York Times December 24, 2008

(240) THINKING THE UNTHINKABLE—A WORLD WITHOUT NUCLEAR WEAPONS, by Carla Anne Robbins, the New York Times, June 30, 2008

(241) AFRICA'S CHANCE, Editorial, the New York Times, November 2, 2007

(242) AFRICAN CRUCIBLE: CAST AS WITCHES, THEN CAST OUT, by Sharon LaFrantere, the New York Times, November 15, 2007

(243) NEWS GOOD ENOUGH TO BURY, by Roger Cohen, the New York Times, August 20, 2008

(244) CELEBRATING, NOT HIDING, Editorial, the New York Times, July 6, 2007

(245) WHAT THEY HATE ABOUT MUMBAI, by Suketu Mehta, the New York Times, November 29, 2008

(246) ANOTHER COUNTRY, by Mark Mellman, the New York Times, September 17, 2008

(247) THE MARKETPLACE OF CHRISTIANITY, by Robert Ekelund Jr, Robert Hebert, and Robert Tollison, MIT Press

(248) A CHALLENGE FOR THE U.S.: SUN RISING ON THE EAST, by Michiko Kakutani, the New York Times, May 6, 2008

(249) WE NEED A MORE INTELLIGENT RELIGION DEBATE, by Theo Hobson, Ekklesia News Brief, September 4, 2007

(250) THE CREATION: AN APPEAL TO SAVE LIFE ON EARTH, by E 0 Wilson, W W Norton

(251) VISIONS OF THE FUTURE, THE DISTANT PAST, YESTERDAY, TODAY, TOMORROW, by Robert Heilbroner, (New York Public Library lecture) Oxford University Press

(252) THE EVOLUTION OF GOD, by Robert Wright, Little Brown and Company

# PARTIAL LIST OF ADDITIONAL REFERENCES USED FOR THIS STUDY NUMBERED WITHOUT PARENTHESIS


1 U S RELIGIOUS LANDSCAPE SURVEY (2007), Pew Forum on Religion & Public Life, Washington DC

2 ENCYCLOPEDIA OF CATHOLIC SOCIAL THOUGHT, SOCIAL SCIENCE AND SOCIAL POLICY, by Michael Coulter, Stephen Krason, Richard Myers and Joseph Varacalli, The Scarecrow Press/Rowman and Littlefield

3 MEN, MINES AND MOSQUES: GENDER AND ISLAMIC REVIVALISM ON THE EDGE OF EUROPE, by Kristen Ghodsee, Occasional Paper of the Princeton Institute of Advanced Study, Princeton NJ

4 A NATION OF RELIGIONS: THE POLITICS OF PLURALISM IN MULTI-RELIGIOUS AMERICA,Stephen Prothero, Editor, The University of North Carolina Press

5 GOD'S WAR: A NEW HISTORY OF THE CRUSADES, by Christopher Tyennan, Harvard University Press

6 THE COURTIER AND THE HERETIC: LEIBNITZ, SPINOZA, AND THE FATE OF GOD IN THE MODERN WORLD, by Matthew Stewart, Norton

7 BETRAYING SPINOZA, THE RENEGADE JEW WHO GAVE US MODERNITY, by Rebecca Goldstein, Nextbook/Shocken

8 THE BEST OF ALL POSSIBLE WORLDS, MATHEMATICS AND DESTINY, by Ivar Ekland, University of Chicago Press

9 PEACE BE UPON YOU, THE STORY OF MUSLIM, CHRISTIAN, AND JEWISH COEXISTENCE, by Zachary Karabell, Knopf

10 ENCYCLOPEDIA OF WOMEN AND RELIGION IN NORTH AMERICA, Rosemary Skinner Keller and Rosemary Radford Ruether, Editors, Indiana University Press

11 THE BATTLE OVER SCHOOL PRAYER HOW ENGEL v. VITALE CHANGED AMERICA, by Bruce J Dierenfield, University Press of Kansas

12 THE HUNT FOR THE DAWN MONKEY UNEARTHING THE ORIGINS OF MONKEYS, APES AND HUMANS, by Chris Beard, University of California Press

13 MORAL MINDS: HOW NATURE DESIGNED OUR UNIVERSAL SENSE OF RIGHT AND WRONG, by Marc D Hauser, Ecco Press

14 ADVERTISING SIN AND SICKNESS: THE POLITICS OF ALCOHOL AND TOBACCO MARKETING 1950-1990, by Pamela E Pennock, Northern Illinois Press

15 JOURNEY INTO ISLAM: THE CRISIS OF GLOBALIZATION, by Akbar Ahmed, Brookings Press

16 RELIGION AND FAMILY IN A CHANGING SOCIETY, by Penny Edgell, Princeton University Press

17 THE CLASH WITHIN: DEMOCRACY, RELIGIOUS VIOLENCE, AND INDIA'S FUTURE, by Martha C Nussbaum, Belknap Press/Harvard University Press

18 DEBATING MODERN ISLAM THE GEOPOLITICS OF ISLAM AND THE WEST, by M A Muqtedar Khan, Editor, Utah Press

19 THE VOICE, THE WORD, THE BOOKS: THE SACRED SCRIPTURE OF THE JEWS, CHRISTIANS AND MUSLIMS, by F E Peters, Princeton University Press

20 JESUS IN THE TALMUD, by Peter Schafer, Princeton University Press

21 THE ROMAN CATHOLIC CHURCH: AN ILLUSTRATED HISTORY, by Edward Norman, University of California Press

22 THE ATLAS OF RELIGION, by Joanne O'Brien and Martin Palmer, University of California Press
OPENING THE DOORS OF WONDER: REFLECTIONS ON RELIGIOUS RITES OF PASSAGE, by Arthur J Magida, University of California Press

23 MOTHER TERESA: COME BE MY LIGHT, by Mother Teresa (Agnes Bojaxhiu) edited by Brian Kolodiejchuk, Doubleday

24 THE NEW VOICES OF ISLAM: RETHINKING POLITICS AND MODERNITY, Mehran Kanirava, Editior, University of California Press

25 GODDESSES AND THE DIVINE FEMININE: A WESTERN RELIGIOUS HISTORY, by Rosemary Radford Ruether, University of California Press

26 THE LIFE OF HINDUISM, John Stratton Hawley and Vasudha Narayanan, Editors, the University of California Press

27 WHY INTELLIGENT DESIGN FAILS: A SCIENTIFIC CRITIQUE OF THE NEW CREATIONISM, by Matt Young and Taner Edis, Editors, Rutgers University Press

28 PREACHING FROM A PULPUT OF BONES: WE NEED MORALITY BUT NOT TRADITIONAL MORALITY, by Bob Avakian, Insight Press

29 GOD WITHOUT RELIGION: QUESTIONING CENTURIES OF ACCEPTED TRUTHS, by Sankara Saranam, The Pranayama Institute

30 CHRISTIANITY WITHOUT FAIRY TALES: WHEN SCIENCE AND RELIGION MERGE, by Jim Rigas, Pnevma Publications

31 NATURE VS. NURTURE, by Cynthia Chris, University of Minnesota Press

32 BARBARISM AND RELIGION: THE FIRST DECLINE AND FALL, by J G A Pocock, Cambridge University Press

33 WHY RELIGION MATTERS: THE FATE OF THE HUMAN SPIRIT IN AN AGE OF DISBELIEF, by Huston Smith

34 THE GLOBAL SPIRAL, Published by The Metanexus Institute, Bryn Mawr, Pennsylvania

35 THE EDGE OF EVOLUTION: THE SEARCH FOR THE LIMITS OF DARWINISM, by Michael J Behe, Free Press

36 EVOLUTION AND THE LEVELS OF SELECTION, by Samir Okasha, Clarendon Press Oxford

37 CENTERLINGS, A publication by The Center of Theological Inquiry, Princeton NJ

38 ALONE IN THE WORLD? HUMAN UNIQUENESS IN SCIENCE AND THEOLOGY, by J Wentzel van Huyssteen, W B Eerdmans Press

39 BLOOMFIELD AVENUE, A JEWISH-CATHOLIC JERSEY GIRL'S SPIRITUAL JOURNEY, by Linda Mercadante, Cowley Publications

40 GOD AND HUMAN DIGNITY, R Kendall Soulen and Linda Woodhead, Editors, W B Eerdmans Publishing Company

41 CYBER WORSHIP IN MULTIFAITH PERSPECTIVES, by Mohamic Taller, The Scarecrow Press/Rowman and Littlefield

42 HISTORICAL DICTIONARY OF SHAMANISM, by Graham Harvey and Robert Wallis, The Scarecrow Press/Rowman and Littlefield

43 THE A TO Z OF NEW RELIGIOUS MOVEMENTS, by George D Chryssides, The Scarecrow Press/Rowman and Littlefield

44 THE WESLEYAN HOLINESS MOVEMENT: A COMPREHENSIVE GUIDE, by Charles Edward Jones, The Scarecrow Press/Rowman and Littlefield

45 JOURNEY TO THE EAST: THE JESUIT MISSION TO CHINA 1579-1724, by Liam Matthew Brockey, Belnap Press/ Harvard University Press

46 DEMOCRACY AFFIRMED, Editorial, the New York Times, July 24, 2007

47 CROSSING THE DIVIDE, Science, February 22, 2008

48 THE PROVIDENT FORUM, Organizational letter, December 20, 2006

49 WHO'S BITTER NOW? Larry Bartels, New York Times, April 17, 2008

50 CONTRIBUTIONS TO METANEXUS, *INFQ@METANEXUS.NET*, December 15, 2006

51 AN EVEN POORER WORLD, New York Times Editorial, September 2, 2008

52 COUNCIL FOR SECULAR HUMANISM, January, 2008 Website, www. Secularhumanism.org, Buffalo. NY

53 CREATIVE TENSION, Michael Heller, 2008 Templeton Prize.

54 THE GLOBAL SPIRAL, A Publication of Metanexus Institute, July 2007

55 TEENAGE BIRTH RATE RISES FOR FIRST TIME SINCE '91, by Gardiner Harris, the New York Times, December 6, 2007

56 CHILDREN OF GOD, by Sara Corbett, the New York Times, July 27, 2008

57 AFFAIRS TO REMEMBER, Olivia Judson, the New York Times. May 28, 2006

58 SUICIDE BOMBER ON BIKE KILLS 29 IRAQI POLICEMEN, by Alissa J Rubin, the New York Times, October 30, 2007
59 THE BIGGEST ISSUE, by David Brooks, the New York Times, July 29, 2008
60 SOMETHING TO WORRY ABOUT, Editorial, the New York Times, June 29, 2008
61 THE POPE VS THE PILL, by John Allen, the New York Times, July 27, 2008
62 MILLION DOLLAR MEAT, Editorial, the New York Times, April 23, 2008
63 MADNESS AND SHAME, by Bob Herbert, the New York Times, July 22, 2008
64 THE MASK SLIPS, by William Kristol, the New York Times, April 24, 2008
65 THE COGNITIVE AGE, by David Brooks, the New York Times, May 2, 2008
66 NASA GOES DEEP, by Carolyn Porco, the New York Times, February 20, 2007
67 IRON FISTS: BRANDING THE 20TH CENTURY TOTALITARIAN STATE, by Steven Heller, Phaidron Press
68 FIGHTING AIDS BEHIND BARS, Editorial, the New York Times, July 18th 2007
69 LIVING FOR GOD: 18th CENTURY DUTCH PIETIST AUTOBIOGRAPHY, by Fred van Lieburg, The Scarecrow Press
70 HANDLE WITH CARE, by Cornelia Dean, the New York Times, August 12, 2008
71 THE CONS OF CREATIONISM, Editorial, the New York Times, June 7, 2008

72 THE EXECUTIONER'S HOOD, Editorial, the New York times, August 6, 2007

73 THE URGE TO END IT, by Scott Anderson, the New York Times, July 6, 2008

74 DAWKINS VERSUS THE GODS, by Eliot Marshall, Science December 1, 2007

75 THE OUTSOURCED BRAIN, David Brooks, the New York Times, October 26, 2007

76 COLLIDING WITH NATURE'S BEST-KEPT SECRETS, by Elizabeth Landau, 2008, CNN.COM

77 IN AT THE BIRTH OF DEATH, Editorial, the New York Times, June 2, 2008

78 PRIEST COSMOLOGIST WINS $1.6 MILLION TEMPLETON PRIZE, by Brenda Goodman, the New York Times, March 13, 2008

79 GOSPEL TRUTH, by April D Deconick, the New York Times, December 1, 2007

80 MARRIAGE PARTNERS, by Kenji Yoshino, the New York Times, June 1, 2008

81 DARWIN'S GOD, by Robin Marantz Henig, New York Times Magazine, March 4, 2007

82 A DIALOGUE OF CIVILIZATIONS: GULEN'S ISLAMIC IDEALS AND HUMANISTIC DISCOURSE, The Light Publishing

83 WHEN PEOPLE DRINK THEMSELVES SILLY, AND WHY, by Benedict Carey, the New York Times, March 4, 2008

84 THE HAPPINESS MYTH: WHY WHAT WE THINK IS RIGHT IS WRONG, by Jennifer Michael Hecht, Harper San Francisco

85 TEENAGE: THE CREATION OF YOUTH CULTURE, by Jon Savage, Viking

86 THE CANON: A WHIRLIGIG TOUR OF THE BEAUTIFUL BASICS OF SCIENCE, by Natalie Angier, Houghton Mifflin

87 THE JESUS MACHINE: HOW JAMES DOBSON, FOCUS ON THE FAMILY AND EVANGELICAL AMERICA ARE WINNING THE CULTURE WAR, by Dan Gilgoff, St Martin's Press

88 LOST CHRISTIANITES: CHRISTIAN SCRIPTURES AND THE BATTLE OVER AUTHENTICATION, Great Courses Lecture Series, The Teaching Company, Chantilly, VA

89 HOLY IMAGE, HALLOWED GROUND: ICONS FROM SINAI, Robert Nelson and Kristen Collins, Editors, Getty Publications

90 CARVED SPLENDOR: LATE GOTHIC ALTARPIECES IN SOUTH GERMANY, AUSTRIA AND SOUTH TYROL, by Ranier Kahsnitz, Getty Publications

91 WITH GOD ON THEIR SIDE, GEORGE W BUSH AND THE CHRISTIAN RIGHT, New Press

92 GOOD GOD! (AND OTHER FOLLIES) ESSAYS ON RELIGION, by Peter Heinegg, University Press of America

93 THE CHOICE PRINCIPLE: THE BIBLICAL CASE FOR LEGAL TOLERANCE, by Andy G Olree, University Press of America

94 SATAN'S CAULDRON: RELIGIOUS EXTREMISM AND THE PROSPECTS FOR TOLERANCE, by Charles Stewart Goodwin, University Press of America

95 AND GOD KNOWS THE SOLDIERS: THE AUTHORITATIVE AND AUTHORITARIAN IN ISLAMIC DISCOURSE, University Press of America

96 HOLY MURDER: ABRAHAM, ISAAC, AND THE RHETORIC OF SACRIFICE, by Larry Powell and William Self, University Press of America

97 BIBLICAL LAW AND ITS RELEVANCE: A CHRISTIAN UNDERSTANDING AND ETHICAL APPLICATION FOR TODAY OF THE MOSAIC REGULATIONS, by Joe M Sprinkle, University Press of America

98 TRUTH IN TRANSLATION: ACCURACY AND BIAS IN ENGLISH TRANSLATION OF THE NEW TESTAMENT, by Jason BeDuhn, University Press of America

99 THE MERGING OF THEOLOGY AND SPIRITUALITY: AN EXAMINATION OF THE LIFE AND WORK OF ALISTER McGRATH, University Press of America

100 ON THE BRINK OF ARTIFICIAL LIFE, by John Carey. Business Week, June 25, 2007

101 FAITH AND FORCE: A CHRISTIAN DEBATE ABOUT WAR, by David L Clough and Brian Stiltner, Georgetown University Press

102 THE DECLINE OF THE SECULAR UNIVERSITY, by C John Sommerville, Oxford University Press

103 THE LAW OF GOD, by Retni Brague, University of Chicago Press

104 BEARING WITNESS AGAINST SIN: THE EVANGELICAL BIRTH OF THE AMERICAN SOCIAL MOVEMENT, by Michael P Young, University of Chicago Press

105 PLAYING WITH GOD: RELIGION AND MODERN SPORT, by William J Baker, Harvard University Press

106 THE CURSE OF HAM: RACE AND SLAVERY IN EARLY CHRISTIANITY AND ISLAM, Princeton University Press

107 JOSEPH'S BONES: UNDERSTANDING THE STRUGGLE BETWEEN GOD AND MANKIND IN THE BIBLE, by Jerome M Segal, Riverhead/Penguin

108 QUANTUM PHYSICS AND THEOLOGY: AN UNEXPECTED KINSHIP, by John Polkinghorne, Yale University Press

109 KINGDOM COMING: THE RISE OF CHRISTIAN NATIONALISM, by Michelle Goldberg, W W Norton

110 THE ACCIDENTAL MIND: HOW THE BRAIN EVOLUTION HAS GIVEN US LOVE, MEMORY, DREAMS, AND GOD, by David J Linden, Harvard University Press

111 THE CASE AGAINST PERFECTION: ETHICS IN THE AGE OF GENETIC ENGINEERING, by Michael J Sandel, Harvard University Press

112 CHRISTIANITY AND THE TRANSFORMATION OF THE BOOK: ORIGEN, EUSEBIUS, AND THE LIBRARY OF CAESAREA, by Anthony Grafton and Megan Williams, Harvard University Press

113 THE MONK AND THE BOOK: JEROME AND THE MAKING OF CHRISTIAN SCHOLARSHIP, by Megan Hale Williams, University of Chicago Press

114 MOHAMMAD: PROPHET OF GOD, by Daniel C Peterson, W B Eerdmans

115 THE DIMINISHING DIVIDE: RELIGION'S CHANGING ROLE IN AMERICAN POLITICS, Brookings Institution Press

116 CQ: DEVELOPING CULTURAL INTELLIGENCE AT WORK, by P Christopher Earley, Soon Ang, and Jod-Seng Tan, Stanford University Press

117 A GODLY HERO: WILLIAM JENNINGS BRYAN, by Michael Kazin, Knopf

118 HIGHER GROUND: ETHICS AND LEADERSHIP IN THE MODERN UNIVERSITY, by Nannerl O Keohane, Duke University Press

119 SCANDALOUS KNOWLEDGE: SCIENCE, TRUTH, AND THE HUMAN, by Barbara H Smith, Duke University Press

120 LOST BODIES: INHIBITING THE BORDERS OF LIFE AND DEATH, by Laura E Tanner, Cornell University Press

121 SCIENCE AND THE TRINITY: THE CHRISTIAN ENCOUNTER WITH REALITY, by John Polkinghorne, Yale University Press

122 THE PSYCHOLOGY OF SCIENCE AND THE ORIGINS OF THE SCIENTIFIC MIND, by Gregory J Feist, Yale University Press

123 AFTER THE DEATH OF GOD, by John E Caputo and Gianni Vattimo, Columbia University Press

124 ZEN-BRAIN REFLECTIONS: REVIEWING RECENT DEVELOPMENTS IN MEDITATION AND STATES OF CONSCIOUSNESS, by James H Austin, MIT Press

125 AIMEE SEMPLE McPHERSON AND THE RESURRECTION OF CHRISTIAN AMERICA, by Matthew Avery Sutton, Harvard University Press

126 CORPORAL COMPASSION: ANIMAL ETHICS AND PHILOSOPHY OF BODY, by Ralph R Acampora, Pittsburgh University Press

127 THE ETHICS OF CREATIVITY: BEAUTY, MORALITY AND NATURE IN A PROCESSIVE COSMOS, by Brian G Henning, University of Pittsburgh Press

128 PERVERSION OF POWER: SEXUAL ABUSE IN THE CATHOLIC CHURCH, by Mary Gail Frawley-O'Dea, Vanderbilt University Press

129 THE QUEST FOR MEANING: A GUIDE TO SEMIJOTIC THEORY AND PRACTICE, by Marcel Danesi, University of Toronto Press

130 WHAT IS YOUR DANGEROUS IDEA? TODAY'S LEADING THINKERS ON THE UNTHINKABLE, by John Brockman, Harper Perennial

131 WHERE GOD WAS BORN: A DARING ADVENTURE THROUGH THE BIBLE'S GREAT STORIES, by Bruce Feiler. Harper Perennial

132 RECONCILIATION IN DIVIDED SOCIETIES: FINDING COMMON GROUND, by Erin Daly and Jeremy Sarkin, University of Pennsylvania Press

133 CHRISTIANITY AS A FAIRY TALE, by Ronald M Mazur, iUniverse Press

134 RETHINKING THE WORLD, by Peter Pogany, iUniverse Press

135 POPE PRAISES U.S. BUT WARNS OF SECULAR CHALLENGES, Laurie Goodstein and Sheryl Stolberg, the New York Times, April 17, 2008

136 ALTERNATIVE VALUES: THE PERENNIAL DEBATE ABOUT WEALTH, POWER, FAME, PRAISE, GLORY, AND PHYSICAL PLEASURE, Hunter Lewis, editor. Axios Press

137 WAYWARD CHRISTIAN SOLDIERS: FREEING THE GOSPEL FROM POLITICAL CAPTIVITY, by Charles Marsh, Oxford University Press

138 A COMPASS THAT CAN CLASH WITH MODERN LIFE, by Michael Slackman, the New York Times, June 12, 2007

139 SEEKING A SANCTUARY, SEVENTH-DAY ADVENTISM AND THE AMERICAN DREAM, by Malcom Bull and Keith Lockhart, Indiana University Press

140 RELIGIOUS EXPERIENCE AND THE NEW WOMAN: THE LIFE OF LILY DOUGALL, by Joanna Dean, Indiana University Press

141 DOGS OF GOD: COLUMBUS, THE INQUISITION, AND THE DEFEAT OF THE MOORS, by James Reston Jr, Anchor/Vintage

142 THE TRUTH ABOUT CONSERVATIVE CHRISTIANS: WHAT THEY THINK AND WHAT THEY BELIEVE, by Andrew Greeley and Michael Rout

143 AMERICAN THEOCRACY: THE PERILS AND POLITICS OF RADICAL RELIGION, OIL, AND BORROWED MONEY IN THE 21ST CENTURY, by Kevin Phillips, Viking Press

144 THE NEED FOR RELIGION: A UNIVERSAL CHRISTIANITY, by John Tomikel, Allegheny Press

145 JESUS CIRCLES: A WAY TO HEAL OUR WOUNDS, SUBVERT THE DOMINATION SYSTEM, AND BUILD AN ABUNDANT FUTURE, by Peter R Lawson, Xlibris Press

146 THE FAITHS OF THE FOUNDING FATHERS, by David L Holmes, Oxford University Press

147 ALL OUT OF FAITH: SOUTHERN WOMEN ON SPIRITUALITY, Wendy Reed and Jennifer Home, Editors, The University of Alabama Press

148 TOWARD A CIVIL DISCOURSE: RHETORIC AND FUNDAMENTALISM, by Sharon Crowley, University of Pittsburgh Press

149 SAINTS AND SINNERS: A HISTORY OF THE POPES, by Eamon Duffy, Yale University Press

150 CREDO: HISTORICAL AND THEOLOGICAL GUIDE TO CREEDS AND CONFESSIONS OF FAITH IN THE

CHRISTIAN TRADITION, Jaroslav Pelikan, Yale University Press

151 A VOICE FOR HUMAN RIGHTS, by Mary Robinson, University of Pennsylvania Press

152 TERRITORY, AUTHORITY, RIGHTS FROM MEDIEVAL TO GLOBAL ASSEMBLAGES, by Saskia Sassen, Princeton University Press

153 T D JAKES, AMERICA'S NEW PREACHER, by Shayne Lee, NYU Press

154 THE CENTRAL LIBERAL TRUTH, by Lawrence E Harrison, Oxford University Press

155 SECULARISM AND DEMOCRACY IN TURKEY, Editorial, the N Y Times, May 1, 2007

156 PLACEBO EFFECTS OF MARKETING ACTION: CONSUMERS MAY GET WHAT THEY PAY FOR, by Baba Shiv, Ziv Cannon and Dan Ariely, School of Social Science, The Princeton Institute for Advanced Study, October 2006

157 STILL AWAITING AN ACCOUNTING, Editorial, The New York Times, April 17, 2008

158 ELOHIM (GOD) CAME TO MY STREET WHERE I LIVE . . . . ! by Kenneth Lamar Williams, AuthorHouse

159 THE SHARIATIZATION OF PAKISTANI NATIONALISM, by Farzana Shaikh, Occasional Paper, School of Social Science, The Princeton Institute for Advanced Study, April 2007

160 GROWTH IN THIRD WORLD CHRISTIANITY TRANSFORMS MIDDLE-AMERICAN CHURCH, by Rachel Zoll, The (Trenton)Times, April 25, 2007

161 MASSIVE STELLAR EXPLOSION WOWS SCIENTISTS, by Ker Than, AOL News, May 8, 2007

162 GINGRICH WARNS AGAINST RADICAL SECULARISM: HE LAUDS FALWELL AT LIBERTY UNIVERSITY COMMENCEMENT, by Bob Lewis, AOL News, May 19, 2007

163 TERRORISM AND JUST WAR, Public Lecture by Michael Walzer, Princeton Institute for Advanced Study, May 4, 2007

164 A CHURCH IN SEARCH OF ITSELF: BENEDICT XVI AND THE BATTLE FOR THE FUTURE, by Robert Blair Kaiser, Anchor Vintage

165 OFFERING COMFORT TO THE SICK AND BLESSINGS TO THEIR HEALERS, by Jan Hoffman, the New York Times, July 17, 2007

166 ISLAMIC CREATIONIST AND A BOOK SENT AROUND THE WORLD, by Cornelia Dean, the New York Times July 17, 2007

167 DEBATE ON CHILD PORNOGRAPHY'S LINK TO MOLESTING, by Julian Sher and Benedict Carey, the New York Times, July 19, 2007

168 PORTRAIT OF A PRIESTESS: WOMEN AND RITUAL IN ANCIENT GREECE, by Joan Breton Connelly, Princeton University Press

169 IN A NEW SIGN OF THE TIMES, IRISH PARADES HELD PEACEFULLY, by Shawn Pogatchnik, The (Trenton) Times, July 13, 2007

170 IN STEPS BIG AND SMALL, SUPREME COURT MOVED RIGHT: THE FACE OF THE MAJORITY, by Linda Greenhouse, the New York Times, July 1, 2007

171 POPE APPEALS FOR RECONCILIATION AMONG CHINA'S CATHOLICS, by Elizabeth Rosenthal, the New York Times, July 1, 2007

172 SECULARISM CONFRONTS ISLAM, by Olivier Roy, Columbia University Press
173 THE FUTURE OF RELIGION, by Richard Rorty and Gianni Vattimo, Columbia University Press
174 ECOSPIRIT: RELIGIONS AND PHILOSOPHIES FOR THE EARTH, Edited by Laurel Kearns and Catherine Keller, Fordham University Press
175 BRAINPOWER MAY LIE IN COMPLEXITY OF SYNAPSES, Nicholas Wade, the New York Times, June 10, 2008
176 LIVING WITH DARWIN: EVOLUTION, DESIGN AND THE FUTURE OF FAITH, by Philip Kiwtcher, Oxford University Press
177 BIBLE-BASED COUNSELING, by Mark Chapman Jr, iUniverse
178 INTELLIGENT DESIGN IN SCIENCE, RELIGION, AND YOU, by Nickolas Bay, Xlibris
179 THE WHOLE PURPOSE OF GOD, by Dr Darren D Hulbert, Xlibris
180 TELL THEM: A MAN'S DISCOVERY OF GOD'S PLAN FOR HIM AND MANKIND, Xlibris
181 THE HEXATEUCH: AN AGNOSTIC REVIEW OF THE BOOKS OF GENESIS THROUGH JOSHUA, by F David Raymond Sr, iUniverse
182 THE LIFE OF MEANING: REFLECTIONS ON FAITH, DOUBT AND REPAIRING THE WORLD, by Bob Abernethy and William Bole, Seven Stories Press
183 WHY CHRISTIANITY MUST CHANGE OR DIE: A BISHOP SPEAKS TO BELIEVERS IN EXILE, by John Shelby Spong, Harper San Francisco

184 FLESH AND REASON: THE MODERN FOUNDATION OF BODY AND SOUL, by Roy Porter, W W Norton and Company
185 THE KORAN, Translated by N J Dawood, Penguin books
186 ETERNITY FOR ATHEISTS: IS GOD NECESSARY FOR IMMORTALITY? by Jim Holt, New York Times Magazine, July 29, 2007
187 IS RELIGION DANGEROUS? by Keith Ward, W B Eerdmans Publishing Company
188 THE END OF MEMORY: REMEMBERING RIGHTLY IN A VIOLENT WORLD, by Miroslav Volt W B Eerdmans Publishing Company
189 AT FERMILAB, THE RACE IS ON FOR THE 'GOD PARTICLE', by Dennis Overbye, the New York Times, July 24, 2007
190 TWO CHURCHES BECOME ONE IN MINISTRY, by Rose Y Colon, The (Trenton) Times, July 2, 2007
191 COLLECTION PLATE WAITS IN CYBERSPACE, by Ron Orozco, The (Trenton) Times, July 17, 2007
192 THE ASTONISHING HYPOTHESIS: THE SCIENTIFIC SEARCH FOR THE SOUL, by Francis Crick, Simon & Schuster
193 TOURING THE SPIRITUAL WORLD, by Ethan Todras-Whitehill, the New York Times, April 29, 2007
194 FIGMENTS OF REALITY: THE EVOLUTION OF THE CURIOUS MIND, by Ian Stewart and Jack Cohen, Cambridge University Press
195 WE NEVER REALLY TALK ANYMORE, Editorial, the New York Times, August 6, 2007
196 U S MARSHALS LET FUGITIVES COME TO THEM, IN CHURCH, by Theo Emery, the New York Times, August 6, 2007

197 SIX IMPOSSIBLE THINGS BEFORE BREAKFAST: THE EVOLUTIONARY ORIGINS OF BELIEF, by Lewis Wolpert, Norton

198 THE CLOSING OF THE AMERICAN MIND, by Allan Bloom, Simon and Schuster

199 ADVOCATES HAIL LUTHERAN ACT ON GAY MEMBERS. by Neela Banerjee, the New York Times, August 17, 2007

200 LEAVE RELIGION OUT OF IT, by Richard Cohen, the New York Times, August 15, 2007

201 THE EXPERIMENTAL INDUCTION OF OUT-OF-BODY EXPERIENCES, by H Hendrik Ehrsson, Science, August 24, 2007

202 IN PLACE OF SOLACE: FINDING FAITH AMONG THE SORROW, by Alexei Barrionuevo, the New York Times, August 20, 2007

203 A SURVEY REPORT ON TELEMETRY by John F Brinster. Prepared for The Applied Physics Laboratory, Johns Hopkins University, May 29, 1947

204 MONKS IN MYANMAR PROTEST FOR THIRD DAY, by Seth Mydans, the New York Times, September 21, 2007

205 VOICES RISE IN EGYPT TO SHIELD GIRLS FROM AN OLD TRADITION, by Michael Slackman, the New York Times, September 20, 2007

206 CALLS FOR A BREAKUP GROW EVER LOUDER IN BELGIUM, by Elaine Sciolino, the New York Times September 16, 2007

207 GOD IN SCIENCE, DOES SCIENCE REVEAL THE EXISTENCE OF GOD? Radio Station WGN News Talk, September 20, 2007

208 LOST IN A MILLION-YEAR GAP, SOLID CLUES TO HUMAN ORIGINS, by John Noble Wilford, the New York Times, September 18, 2007

209 CHRISTIAN DEBT TRUST, *www.christiandebttrust.com*

210 A SAINT'S DARK NIGHT, by James Martin, Op/Ed, the New York Times, August 20, 2007

211 CIRCLING MY MOTHER, by Mary Gordon, Pantheon Books

212 BIRTH WITHOUT BOTHER, by Nicholas D Kristof, the New York Times, July 23, 2007

213 POPE URGES EUROPEANS TO EMBRACE FAITH, The (Trenton) Times, September 9, 2007

214 WELCOME OR NOT, ORTHODOXY IS BACK IN RUSSIA'S PUBLIC SCHOOLS, by Clifford J Levy, the New York Times, September 23, 2007

215 THE WORLD COMES TO GEORGIA. AND AN OLD CHURCH ADAPTS, by Warren St. John, the New York Times, September 22, 2007

216 ARGENTINE CHURCH FACES 'DIRTY WAR' PAST. by Alexei Barrionuevo, the New York Times, September 17, 2007

217 MOTHER TERESA'S CRISIS OF FAITH, by David Van Biema, Time, August 23, 2007

218 DENTATE GYRUS NMDA RECEPTORS MEDIATE RAPID PATTERN SEPARATION IN THE HIPPOCAMPAL NETWORK, by Thomas J McHugh, et al, Science, July 6, 2007

219 GLOBAL PATTERN FORMATION AND ETHNIC/CULTURAL VIOLENCE, by May Lim, et al, Science, September 14, 2007

220 THOU SHALT NOT KILL, EXCEPT IN A POPULAR VIDEO GAME AT CHURCH, by Matt Richtel, the New York Times, October 10, 2007

221 FOR A TRUSTY VOTING BLOC, A FAITH SHAKEN, by Laurie Goodstein, the New York Times, October 7, 2007

222 HOW BABOONS THINK (YES, THINK) by Nicholas Wade, the New York. Times, October 9, 2007

223 HOUSE PANEL RAISES FUROR ON ARMENIAN GENOCIDE. by Steven Meyers and Carl Hulse, the New York Times, October 11, 2007

224 MUSICAL MYSTICISM IN SEARCH FOR GOD, by Allan Kozinn, the New York Times October 11, 2007

225 HOW CHINA GOT RELIGION, by Slavoj Zizek, the New York Times, October 11, 2007

226 RELATIONS SOUR BETWEEN SHIITES AND IRAQ MILITIA, by Sabrina Tavernise, the New York Times, October 12, 2007

227 LEGAL OR NOT, ABORTION RATES COMPARE, by Elisabeth Rosenthal, the New York Times, October 12, 2007

228 A CALL FOR HERESY: WHY DISSENT IS VITAL TO ISLAM AND AMERICA, by Anouar Majid, University of Minnesota Press

229 AN INTERNET JIHAD AIMS AT US VIEWERS, by Michael Moss and Souad Merhennet, the New York Times, October 18, 2007

230 RELIGIOUS RIGHT DIVIDES ITS VOTE AT SUMMIT, by Michael Luo, the New York Times, October 21, 2007

231 IRAQ'S PLAGUE: DEATH BY RELIGION, Associated Press, The (Trenton) Times, October 26,2007

232 COURT ALLOWS STEM-CELL QUESTION, by Kale Coscarelli, The (Trenton) Times, October 27, 2007

233 THE EVANGELICAL CRACKUP, by David D Kirkpatrick, the New York Times, October 28, 2007
234 SOCIAL DECISION-MAKING: INSIGHTS FROM GAME THEORY AND NEUROSCIENCE, by A G Sanfey, Science, October 26, 2007
235 DECISION THEORY: WHAT "SHOULD" THE NERVOUS SYSTEM DO? by K Kording, Science, October 26, 2007
236 DESCARTES' ERROR: EMOTION, REASON, AND THE HUMAN BRAIN, by Antonio R Damasio, G P Putnam's Sons
237 WEIGHING THE SOUL SCIENTIFIC DISCOVERY FROM THE BRILLIANT TO THE BIZARRE, by Len Fisher, Arcade Publishing
238 WHO WILL PRAISE THE LORD? by Harold Bloom, the New York Review of Books, November 22, 2007
239 PURIFY AND DESTROY: THE POLITICAL USES OF MASSACRE AND GENOCIDE, by Jacques Semelin, Columbia University Press
240 RELIGION: BEYOND A CONCEPT, Edited by Hent de Vries, Fordham University Press
241 FAITH IN THE HALLS OF POWER, by D Michael Lindsay, Oxford University Press
242 THE TURNING OF AN ATHEIST, by Mark Oppenheimer, the New York Times, November 4, 2007
243 NEW TERRORISM CASE CONFIRMS THAT DENMARK IS A TARGET, by Nicholas Kulish, the New York Times, September 17, 3007
244 IS THE COSMOS ALL THERE IS? By Owen Gingerich, Reflections, Spring 2002, a publication of the Center of Theological Inquiry, Princeton NJ

245 ESCHATOLOGY AND SCIENTIFIC COSMOLOGY: FROM CONFLICT TO INTERACTION, by Robert John Russell. Reflections, Spring 2006, a publication of the Center of Theological Inquiry, Princeton NJ

246 GAY MUSLIMS FIND FREEDOM OF A SORT IN THE U.S., by Neil MacFarquhar, the New York Times, November 7, 2007

247 PLUGGING IN TO MAKE A JOYFUL NOISE UNTO THE LORD, by Ben Ratcliff, the New York Times, November 7, 2007

248 IN DNA ERA, NEW WORRIES ABOUT PREJUDICE, by Amy Harmon, the New York Times, November 1, 2007

249 BISHOPS: SALVATION MAY REST AT POLLS, by MCT News Service. The (Trenton) Times, November 15, 2007

250 THROUGH GENETICS, TAPPING A TREE'S POTENTIAL AS A SOURCE OF ENERGY, by Andrew Pollack, the New York Times, November 20, 2007

251 SCIENTISTS BYPASS NEED FOR EMBRYO TO GET STEM CELLS, by Gina Kolata, the New York Times, November 21, 2007

252 A BATTLE RAGES IN LONDON OVER A MEGA-MOSQUE PLAN, by Jane Perlez, the New York Times, November 4, 2007

253 LAW AND RELIGION, Sixth Annual Continuing Legal Education Conference, Princeton University. May 29, 2008

254 LETTER TO A CHRISTIAN NATION, by Sam Harris

255 PRISONS TO RESTORE PURGED RELIGIOUS BOOKS, by Neela Banerjee, the New York Times. September 27, 2007

256 PRISONS PURGING BOOKS ON FAITH FROM LIBRARIES, by Laurie Goodstein, the New York Times, September 10, 2007

257 A TRUCE ON RELIGION, by Nicholas D Kristof, The (Trenton) Times December 6, 2006

258 THE GEOGRAPHY OF HATE, by Mark Potok, et al, the New York Times, November 25, 2007

259 EVERYTHING CONCEIVABLE: HOW ASSISTED REPRODUCTION IS CHANGING MEN, WOMEN AND THE WORLD, by Liza Munday, the New York Times

260 SCIENTIFIC PERSPECTIVES ON CHRISTIAN ANTHROPOLOGY, by Nancy Murphy, Reflections, Center of Theological Inquiry, Princeton New Jersey, Spring 2006

261 PRAYERS FOR A TROUBLED CITY, by Jeff Trently, The (Trenton) Times, September 4, 2007

262 OUT-OF-BODY EXPERIENCES ENTER THE LABORATORY, by Greg Miller, Science, August 24, 2007

263 GLOBAL PATTERN FORMATION AND ETHNIC/ CULTURAL VIOLENCE, by May Lim Richard Metzler, Yaneer Bar-Yam, Science, September 14, 2007

264 SAINTS THAT WEREN'T, by James Martin, the New York. Times, November 1, 2006

265 AS EXEMPTIONS GROW, RELIGION OUTWEIGHS REGULATION, by Diana Hendriques, the New York Times, October 8, 2006

266 GHOSTS IN THE MACHINE, by Deborah Blum. the New York Times, December 30, 2006

267 VEDIC GROUP HOPES TO TRANSCEND ISSUES, by Nyier Abdou, The (Trenton) Times. December 3, 2006

268 EVOLUTION AND CHRISTIAN FAITH: REFLECTIONS OF AN EVOLUTIONARY BIOLOGIST, by Joan Roughgarden, Island Press

269 NEARBY CLUSTER SHOWS EXTREMES OF STARDOM, Yudhijit Bhattacharjee, Science, August 18, 2006

270 ISLAM'S SILENT MODERATES, by Ayaan Hirsi Ali, the New York Times, December 7, 2007

280 BOY SCOUTS LOSE PHILADELPHIA LEASE IN GAY-RIGHTS FIGHT, by Ian Urbina, the New York Times, December 6, 2007

281 A CHURCH IS DIVIDED, AND HEADED FOR COURT, by Brenda Goodman, the New York Times, December 5, 2007

282 THE EVANGELICAL SURPRISE, by Frances FitzGerald, The New York Review of Books, April 26, 2007

283 GOD'S TROUBLE MAKERS: HOW WOMEN OF FAITH ARE CHANGING THE WORLD, by Katherine Henderson, Continuum Books

284 SUICIDE BOMBERS IN IRAQ: THE STRATEGY AND IDEOLOGY OF MARTYRDOM, by Mohammed Hafez, USIP Press

285 THE HISTORICAL MUHAMMAD, by Irving Zeitlin, Polity

286 FOR THE LOVE OF GOD: THE BIBLE AS AND OPEN BOOK, by Alicia Ostriker, Rutgers Press

287 LATTER-DAY REPUBLICANS vs, THE CHURCH OF OPRAH, by Frank Rich, the New York Times, December 16, 2007

288 THE HUCKABEE FACTOR, by Zev Chat, the New York Times, December 12, 2007

289 WHAT'S THE BIG IDEA? By Merrell Noden, The Princeton Alumni Weekly, June 6, 2007

290 PEOPLE ARE EVOLVING FASTER, STUDY SAYS, by Randolph E Schmid, Science News, Associated Press, December 11, 2007

291 A MIDNIGHT SERVICE HELPS AFRICAN IMMIGRANTS COMBAT DEMONS, by Neela Banerjee, The New York. Times, December 18, 2007

292 LAWS OF NATURE, SOURCE UNKNOWN, by Dennis Overbye, the New York Times, December 18, 2007

293 RELIGION, CULTURE, CONFLICT . . . . AND COMEDY, A Presentation featuring Bob Alper, Asher Usman and Nazareth, by Drew University Center on Religion, Culture & Conflict, February 7, 2008

294 WHERE BOYS WERE KINGS, A SHIFT TOWARD BABY GIRLS, by Choe Sang-hun, the New York Times, December 23, 2007

295 MOB SETS KENYA CHURCH ON FIRE BEFORE KILLING DOZENS, by Jeffrey Gettleman, the New York Times, January 2, 2008

296 POLL DELVES INTO HOW WE VIEW GOD, by Jeff Diamant, The (Trenton) Times, September 12, 2007

297 WHAT'S SO GREAT ABOUT CHRISTIANITY? by Dinesh D'Souza, Regnery

298 COURT BARS STATE EFFORT USING FAITH IN PRISONS, by Neela Banerjee, the New York Times, December 4, 2007

299 SEX, SCIENCE AND SAVINGS, Editorial, The New York Times, December 2, 2007

300 NEURO-IMMUNOLOGY AND MENTAL HEALTH, by Ljubisa Vitkovik and Stephen Koslow, National Institute of Mental Health

301 CLOTHES MAKE THE WOMAN, by Vicki Hyman, The (Trenton) Times, January 8, 2008

302 INSTITUTE FOR HUMANIST STUDIES, The Humanist Center, Albany New York

303 A CUTTING TRADITION, by Sara Corbett, the New York Times, January 20, 2008

304 THE BLIGHT THAT IS STILL WITH US, by Bob Herbert, the New York Times, January 22, 2008

305 POLITICAL ANIMALS (YES, ANIMALS), by Natalie Angier, the New York Times, January 22, 2008

306 THE AGE OF AMBITION, by Nicholas D Kristof, the New York. Times, January 27, 2008

307 A BAPTIST COALITION AIMS FOR MODERATE IMAGE, by Neela Banerjee. the New York Times, January 27, 2008

308 GENDER AND THE BRAIN, by Ronald Kotulak, Chicago Tribune, On-line Edition, April 30, 2006

309 RED AND GOOD, by Hagit Benbaji, An Occasional Paper of the Princeton Institute for Advanced Study, October, 2007

310 EVANGELICALS A LIBERAL CAN LOVE, by Nicholas KristoL the New York Times February 5, 2008

311 THE COLD WAR AS ANCIENT HISTORY, by Roger Cohen, the New York Times, February 4, 2008

312 ENDOWMENTS WIDEN A HIGHER EDUCATION GAP, by Karen Arenson. the New York Times, February 4, 2008

313 AN ALTAR BEYOND OLYMPUS FOR A DEITY PREDATING ZEUS, by John Wilford, the New York Times, February 5, 2008

314 FOOD POLITICS, HALF-BAKED, by James McWilliams, the New York Times, February 5, 2008

315 A VICTORY FOR SAME-SEX MARRIAGE, Editorial, the New York Times, February 5, 2008

316 SPEAKING OF FAITH, WHY RELIGION MATTERS AND HOW TO TALK ABOUT IT, by Krista Tippett, Kindle Books

317 CONSERVATIVE RABBIS TO VOTE ON RESOLUTION CRITICIZING POPE'S REVISION OF PRAYER, by Neela Banerjee, the New York Times, February 9, 2008

318 THE BIBLE AS GRAPHIC NOVEL, WITH A SAMURAI STRANGER CALLED CHRIST, by Neela Banerjee, the New York Times, February 20, 2008

319 THE FIRST ACHE, by Annie Murphy Paul, the New York Times, February 10, 2008

320 NEW WEIGHT IN ARMY MANUAL ON STABILIZATION, by Michael Gorman, the New York Times, February 8, 2008

321 AMID STORM DEBRIS, A NEED FOR MIRACLES SMALL OR LARGE, by J Michael Kennedy, the New York Times, February 8, 2008

322 SEX AND THE TEENAGE GIRL, by Caitlin Flanagan, the New York Times, January 13, 2008

323 THE MORAL INSTINCT, by Steven Pinker, the New York Times, January 13, 2008

324 TEAM CREATES RAT HEART USING CELLS OF BABY RATS, by Lawrence Altman, ie New York Times, January 14, 2008

325 A NATION OF CHRISTIANS IS NOT A CHRISTIAN NATION, by Jon Meacham, the New York Times, October 7, 2007

326 OPTIMISM IN EVOLUTION, by Olivia Judson, the New York Times, August 13, 2008

327 DIOCESE TO SELL HEADQUARTERS TO HELP SETTLE ABUSE CLAIMS, by Neela Banerjee, the New York Times, May 26, 2006

328 GOD AND MAN ON SCREEN: BIG QUESTIONS AS ENTERTAINMENT, by Caryn James, the New York Times May 27, 2006

329 WHY AMERICAN COLLEGE STUDENTS HATE SCIENCE, by Brent Staples, the New York Times, May 25, 2006

330 COURT REJECTS EVANGELICAL PRISON PLAN OVER STATE AID, by Neela Banerjee, the New York Times June 3, 2006

331 IN POLAND, POPE SEEKS TO STRENGTHEN BOND WITH FAITHFUL AND HONOR HIS PREDECESSOR, by Ian Fisher, the New York Times, May 26, 2006

332 OPUS DEI'S BOX OFFICE TRIUMPH, Paul Fortunato, the New York Times, June 2, 2006

333 JUDGING WHETHER A KILLER IS SANE ENOUGH TO DIE, by Ralph Blumenhal and Adam Liptak, the New York Times, June 2, 2006

334 ON PUBLIC LAND, SUNDAY IN THE PARK WITH PRAYER, by Neela Banerjee, the New York Times, July 24, 2006

335 THE HISTORICAL JESUS, Course by Bart D Ehrman, The Teaching Company

336 FREE SPEECH CASE DIVIDES BUSH AND RELIGIOUS RIGHT, by Linda Greenhouse, the New York Times March 18, 2007

337 A VEIL CLOSES FRANCE'S DOOR TO CITIZENSHIP, by Katrin Bennhold, the New York Times, July 19, 2008

338 PLAYING GOD, THE HOME GAME, by Seth Schiesel, the New York Times, September 5, 2008

339 TAKING MARRIAGE PRIVATE, by Stephanie Coonz, the New York Times, November 26, 2007

340 PASTOR DEFENDS INVITATION TO OBAMA, by Nedra Pickier, A P (AOL News) November 30, 2006

341 MIND OVER MANUAL, by Sally Saatel, the New York Times, September 13, 2007

342 SHIITE REFUGEES FEEL FORSAKEN IN THEIR HOLY CITY, by Alissa Rubin, the New York times, October 19, 2007

343 PARENTING INHIBITS NEUROGENESIS IN CALIFORNIA MICE, by Ashley Pavlic, Senior thesis 2007, Princeton University

344 COLLABORATORS IN A QUEST FOR HUMAN PERFECTION, comments by Abigal Zuger M D on David Friedman book, the New York Times, August 28, 2007

345 THE UNITED STATES IN A GLOBALIZED WORLD-HOW WILL WE EARN OUR LIVING? A talk by Peter Wodtke, The Old Guard of Princeton., January 9, 2008

346 DEMOCRACY'S IDENTITY PROBLEM: IS CONSTITUTIONAL PATRIOTISM THE ANSWER? By Clarissa Hayward, School of Social Science, Princeton Institute for Advanced Study

347 THE OLD TESTAMENT, Course by Amy-Jill Levine, The Teaching Company

348 REINVENTING THE SACRED, A NEW VIEW OF SCIENCE, REASON, AND RELIGION, by Stuart A Kaufman, Basic Books

349 LIFE AFTER DEATH: NUCLEAR POWER IS CLEAN, BUT CAN IT OVERCOME ITS IMAGE PROBLEM? The Economist, June 22, 2008

350 WHERE SCIENCE MEETS RELIGION, Letter to the Editor, The Princeton Alumni Weekly by John F Brinster

351 FEDERAL GRANT FOR A MEDICAL MISSION GOES AWRY, by Diana Henriques and Andrew Lehren, the New York Times, June 13, 2008

352 LOVE'S RULES VEX AND ENTRANCE YOUNG SAUDIS, by Michael Slaclunan, the New York Times, May 12, 2008
353 DNA IS TAKEN FROM SECT'S CHILDREN, by Dirk Johnson, the New York Times, April 22, 2008
354 CYBER WORSHIP IN MULTIFAITH PERSPECTIVES, by Mohamed Taher, Scarecrow Press
355 THE MERGING OF THEOLOGY AND SPIRITUALITY: AN EXAMINATION OF THE LIFE AND WORK OF ALISTER E McGRATH, by Larry McDonald, University Press of America
356 A SUMMER OF MADNESS, by Oliver Sacks, the New York Review of Books, September 25, 2008
357 THE DECLINING SIGNIFICANCE OF GENDER, by Francine Blau, Mary Brinton and David Gruslcy, Editors, Russell Sage Foundation
358 OF WINE, HASTE AND RELIGION, by Roger Cohen, the New York Times, April 21, 2008
359 EXPLORATIONS IN NEUROSCIENCE, PSYCHOLOGY, AND RELIGION, by Kevin Seybold, Ashgate Publishing Company
360 OFFICIAL LEAVES POST AS TEXAS PREPARES TO DEBATE SCIENCE EDUCATION STANDARDS, by Ralph Blumenthal, the New York Times, December 3, 2007
361 THE WORLD OF UNDERGRADUATE EDUCATION: COUNTRIES IN PROFILE, Science, July 6, 2007
362 THIS IS YOUR (FATHER'S) BRAIN ON DRUGS, by Mike Males, the New York Times, September 17, 2007
363 AT ST PATRICK'S CATHEDRAL, POPE CALLS FOR UNITY, by Ian Fisher and Sewell Chan, the New York Times, April 28, 2008
364 TRUTH IN TRANSLATION: ACCURACY AND BIAS IN ENGLISH TRANSLATIONS OF THE NEW TESTAMENT, by Jason BeDuhn, University Press of America

365 INTERVIEW OF ALISON SODEN, DEAN OF RELIGIOUS LIFE, PRINCETON UNIVERSITY, by Brett Tomilson. Princeton Alumni Weekly, September 26, 2007

366 BIBLICAL LAW, A CHRISTIAN UNDERSTANDING AND ETHICAL APPLICATION FOR TODAY OF THE MOSAIC REGULATIONS, by Joe Sprinkle, University Press of America

367 A PERSON COULD DEVELOP OCCULT, by Alessandra Stanley, the New York Times, October 14, 2007

368 EPISCOPAL PARISHES IN VIRGINIA VOTE TO SECEDE, by Laurie Goodsein, the New York times, December 18, 2006

369 CALLS FOR A BREAKUP GROW EVER LOUDER IN BELGIUM, by Elaine Sciolino, the New York Times September 21, 2007

370 VATICAN HINTS AT CHANGES IN CHURCH LAWS ON ABUSE, by Laurie Goodstein, the New York Times, April 19, 2008

371 RELIGION OF FEAR: THE POLITICS OF HORROR IN CONSERVATIVE EVANGELICALISM, by Jason C Bivins, Oxford University Press

372 RELIGIONS: MYTH OR REALITY, by Ibrahim Abdel-Motaleb, iUniverse.

373 THE SHAPE OF THE WORLD TO COME: CHARTING THE GEOPOLITICS OF A NEW CENTURY, by Laurent Cohen-Tanugi, Columbia University Press

374 DECK THE SCHOOL WITH SIGNS OF WINTER, by Jeff Treutly, The (Trenton) Times, December 15, 2006.

375 WHAT IS INTERNATIONAL HUMAN RIGHTS LAW'? THREE APPLICATIONS OF A DISTRIBUTIVE ACCOUNT,

by Patrick Macklem, Occasional Paper, Princeton Institute for Advanced Study, February, 2008

376 SEXUALITY AND THE TRINITY, Mark Jordan ed., Princeton University Press

377 GOD AND THE GAVEL: RELIGION AND THE RULE OF LAW, by Marci Hamilton, Cambridge University Press

378 MISSION AND MONEY: UNDERSTANDING THE UNIVERSITY, by Burton Weisbrod, Jeffrey Balton and Evelyn Asch, Cambridge Press

379 BIBLICAL INTERPRETATION IN AFRICAN PERSPECTIVE, by David Adamo, editor, The University Press of America

380 DALAI LAMA ARRIVES FOR A FIVE DAY CONFERENCE IN SEATTLE, VERY MUCH HIS KIND OF TOWN, by William Yardley, the New York Times, April 14, 2008

381 THE COST OF SMARTS, by Verlyn Klinkenborg, the New York Times, May 7, 2008

382 PASSIONATE UPRISINGS: IRAN'S SEXUAL REVOLUTION, Pardis Mandavi, Stanford University Press

383 AN INITIATIVE ON READING IS RATED INEFFECTIVE, by Sam Dillon, the New York Times, May 2, 2008

384 RELIGION AND THE POST-METAPHYSICAL SELF-UNDERSTANDING OF MODERNITY, Juergen Habermas, 2008 Castle Lectures, Yale University

385 THE STRANGE CASE OF JUDGE ALITO, by Ronald Dwarkin, The New York Review of Books, February 23, 2006

386 THE OTHER HALF OF THE BRAIN, by R Douglas Fields, Scientific American, April 2004

387 ECUMENSUS-THE NEXT VISION, by Clifford Mark, i Universe

388 A DISEASE THAT ALLOWED TORRENTS OF CREATIVITY, by Sandra Blakeslee, the New York Times, April 8, 2008

389 GLOBAL CATASTROPHES AND TRENDS-THE NEXT FIFTY YEARS, by Vaclav Smil, MIT Press

390 AIMEE SEMPLE McPHERSON AND THE RESURRECTION OF CHRISTIAN AMERICA, by Matthew Avery Sutton, Harvard University Press

391 VARIATION ON A THEME, by Charles Krauthammer, the New York Times, 2008

392 LET'S TALK ABOUT SEX, by Charles Blow, the New York Times, September 6, 2008

393 A GUIDE TO ORTHODOX PSYCHOTHERAPY, by Archbishop Chrysostomos, University Press of America

394 RELIGION JOINS CUSTODY CASES TO JUDGES' UNEASE. by Neela Banerjee, the New York Times, February 13, 2008

395 THE FAITHFUL GET THE CALI, by Linda Seein, The (Trenton) Times, February 13, 2008

396 SMALLER VERSION OF THE SOLAR. SYSTEM IS DISCOVERED, by Dennis Overbye, the New York Times. February 15, 2008

397 WHAT PEOPLE OWE FISH: A LOT, by Natalie Angier, the New York Times, February 19, 2008

398 THE GLOBAL SPIRAL, A Publication of the Metanexus Instsitute, February, 2008

399 FREEDOM & NEUROBIOLGY: REFLECTIONS ON FREE WILL LANGUAGE AND POLITICAL POWER, by John Searle, Columbia University Press

400 WHY RADICAL ISLAM JUST WON'T DIE, by Paul Berman, the New York Times, March 23, 2008

401 BLIND TO CHANGE, EVEN AS IT STARES US IN THE FACE, by Natalie Angier, the New York Times, April 1, 2008

402 GROWING GULF DIVIDES CHINA AND OLD FOE, by Howard W French, the New York Times, March 29, 2008

403 REPORT SKETCHES CRIME COSTING BILLIONS: THEFT FROM CHARITIES, by Stephanie Strom, the New York Times, March 29, 2008

404 SEX INFECTIONS FOUND IN QUARTER OF TEENAGE GIRLS, by Lawrence K Altman, the New York Times, March 12, 2008

405 EPISCOPAL CHURCH VOTES TO OUST BISHOP WHO SECEDED, by Neela Banerjee, the New York Times, March 13, 2008

406 MANY MUSLIMS TURN TO HOME SCHOOLING, by Neil MacEarquhar, the New York Times, March 26, 2008

407 TWO VIEWS OF LIFE, ENDURING, UNYIELDING, by William Grimes, the New York Times, March 26, 2008

408 SPAIN'S MANY MUSLIMS FACE DEARTH OF MOSQUES, by Victoria Burnet, the New York Times, March 16, 2008

409 BLACK RABBI REACHES OUT TO MAINSTREAM OF HIS FAITH, by Niko Koppel, the New York Times March 16, 2008

410 GOD AND THE NEW ATHEISM, by John Haught, Westminster John Knox Press

411 OBAMA, NOW ON THE DEFENSIVE, CALLS BITTER WORDS ILL-CHOSEN, by Katherine Q Seelye and Jeff Zeleny, the New York Times, April 13, 2008

412 INMATE COUNT IN U.S. DWARFS OTHER NATIONS, by Adam Liptak, the New York Times, April 23, 2008

413 NO SHOES, NO SHIRT, NO WORRIES, by Michelle Higgins, the New York Times, April 27, 2008

414 THE QUEST OF MICHEL DE CERTEAU, by Natalie Davis. New York Review of Books, May 15, 2008

415 MISTAKES WERE MADE (BUT NOT BY ME) WHY WE JUSTIFY FOOLISH BELIEFS, BAD DECISIONS, AND HURTFUL ACTS, by Carol Tavris and Elliot Aronson, Powells Books

416 BEAUTIFUL MINDS: THE PARALLEL LIVES OF GREAT APES AND DOLPHINS, by Maddalina Bearzi and Craig Stanford, Harvard University Press

417 BETWEEN NATURALISM AND RELIGION, by Jurgen Habermas, Polity Press

418 SURVEY SHOWS U S RELIGIOUS TOLERANCE, by Neela Banerjee, the New York Times, June 24, 2008

419 FOR ALIEN LIFE-SEEKERS, NEW REASON TO HOPE, Natalie Angier, the New York Times, June 24, 2008

420 TAKING THEIR FAITH, BUT NOT THEIR POLITICS, TO THE PEOPLE, by Neela Banerjee, the New York Times, June 1, 2008

421 OPPONENTS OF EVOLUTION ADOPTING A NEW STGRATEGY, by Laura Beil, the New York Times, June 4, 2008

422 THE CHOICE THEY MADE, by William Kn'stol, the New York Times, June 30, 2008

423 YOUR BRAIN LIES TO YOU, by Sam Wang and Sandra Aamodt, the New York Times, June 27, 2008

424 CHURCH OF ENGLAND ENDORSES WOMEN AS BISHOPS, by John F Bums, the New York Times, July 8, 2008

425 WHAT'S NEXT IN THE LAW? THE UNALIENABLE RIGHTS OF CHIMPS, by Adam Cohen, the New York Times, July 14, 2008

426 THE UNITY OF SCIENCE AND THE HUMANITIES IN THE EDUCATIONAL SYSTEM, by John F Brinster, A private communication, Princeton University Class of 1943 as member of the SCAM Committee, January, 1994

427 THE STRUGGLE TO MEASURE COSMIC EXPANSION, by Dennis Overbye, the New York Times, August 19, 2008

428 CRUSADER SEES WEALTH AS CURE FOR INDIA CASTE BIAS, by Somini Sengupta, the New York Times, August 30, 2008

429 GO AHEAD RATIONALIZE, MONKEYS DO IT, TOO, by John Tierney, the New York Times, November 6, 2007

430 THE ORIGINS OF THE UNIVERSE: A CRASH COURSE, by Brian Greene, the New York Times, September 12, 2008

431 PUBLISHER WHO FOUGHT PURITANISM, AND WON, by Charles McGrath, the New York Times, September 24, 2008

432 A HEROINE FROM THE BROTHELS, by Nicholas Kristof the New York Times, September 25, 2008

433 CHRISTIAN THOUGHT AND PRACTICE, COGNITION AND RELIGION, Public Theology, Faculty directed conferences and lecture series, Princeton Center for the Study of Religion, www.Princeton.edu/csr.

434 A CURE FOR GREED, by Eduardo Porter, the New York Times, September 29, 2008

435 SECULAR DEFEATS ULTRA-ORTHODOX IN JERUSALEM, by Isabel Kershner, the New York Times, November 12, 2008

436 RAISING THE WORLD'S I.Q., by Nicholas Kristof, New York Times, December 4, 2008

437 PRESERVING THE CITY: CHURCH AND STATE, by Robin Pogrebin, the New York Times, December 1, 2008

438 THE WORLD WE WANT: NEW DIMENSIONS IN PHILANTHROPY AND SOCIAL CHANGE, by H Peter Karoff with Jane Maddox, AhaMira Press

439 LAWMAKER IN KENTUCKY MIXES PIETY AND POLITICS, by Ian Urbina, the New York Times, January 4, 2009

440 YOUR SPIRITS WALK BESIDE US: THE POLITICS OF BLACK RELIGION, by Barbara Dianne Savage, Harvard University Press

441 HOPES AND HABITS PERSEVERE AT CHURCHES GONE, BUT NOT DESTROYED, by Paul Vitello and Christine Haughney, the New York Times, May 17, 2009

# PRINCIPAL REFERENCES USED FOR THIS WORK BY AUTHOR

(All first authors, editors and other prime sources are listed alphabetically with numbers in parentheses corresponding to those shown listed in the text.)


| | |
|---|---|
| Abdel-Motaleb, Ibrahim | 372 |
| Abernathy, Bob | 182 |
| Abdou, Nyier | 267 |
| Acampora, Ralph | 126 |
| Adamo, David | 379 |
| Ahmad, Munir | (125) |
| Ahmed, Akbar | 15 |
| Ali, Ayaan | 270 |
| Allen, John | (236), 61 |
| Alper, Bob | (135), 293 |
| Alsanea, Kajaa | (135) |
| Altman, Lawrence | 324, 404 |
| Anderson, Lars | (108) |
| Anderson, Per | (70) |
| Anderson, Scott | 73 |
| Angier, Natalie | (80), 86, 305, 397, 401, 419 |
| Antony, Louise | (89) |
| AOL News | (48) |
| Arenson, Karen | 312 |
| Armstrong, Karen | (63) |
| Atran, Scott | (187) |
| Austin, James | 124 |
| Avakian, Bob | 28 |

# Index

## B

## C

**H**

### L



### M

**X**



**Y**



**Z**

# About the Author

Born a Roman Catholic, John F. Brinster spent his early years in support of a traditional neighborhood church associated with a Catholic teaching monastery. His experience included the study of Christian doctrine for which he received several awards. He wore its robes, carried its banners and crosses, sang its praises, and administered to officiating clergy. He manufactured wafers, poured the wine and produced the entrancing mist of incense. He bowed and knelt and responded in Latin.

His religious experience was followed by awakening to the reality of the broader world as he left the influence of the Church and advanced to fields of factual science and technology. The imaginative character of the church doctrine became clear and the basis of realism became fixed in his mind with new understanding and conviction. His interest expanded to analysis of other belief systems, first attending school with a Methodist seminary and later Princeton University, situated in the shadow of a Presbyterian seminary. His Princeton education in physics was supplemented by a special interest in psychology and neuroscience in the subfield of learning and memory. His education included diverse disciplines in the humanities with a wide range of literature, philosophy and religion. He graduated with magna cum laude honors, was elected to Phi Bet Kappa, and awarded membership in the Sigma Xi scientific honor society. In curtailed wartime graduate study he benefited by the presence of gifted colleagues such as Albert Einstein, Wolfgang Pauli, Richard Feynman, John Wheeler, Eugene Wigner, Robert Oppenheimer and he worked with Wehrner Von Braun in space vehicle development and testing for the United States. He subsequently founded and managed several hi-tech companies following which he contributed to the expansion of mind study at Princeton University and is now retired. He is a member of the Author's League, the New York Academy of Science, the American Association for the Advancement of Science, and an active supporter of the Princeton Institute for Advanced Study.

Among his technical contributions was the development of systems for radio telemetry, instrumentation systems with which a multiplicity of data could be acquired in real time from missiles, spacecraft and other moving vehicles for recording and analysis at fixed stations. It marked the beginning of international telemetry conferences for comparing and exchanging related ideas. In 1946 he was assigned by the Navy Bureau of Ordnance to make a theoretical study of data transmission systems by radio which concluded that the most efficient system would be based on a new concept of digital coding of data. It also marked the birth of digital technology as suitable solid state components and chip circuitry became generally available. Up until that time only vacuum tubes were possible using relatively high power and producing an excessive amount of heat. His report titled *A Survey Report on Telemetry*, 203, not only became the basis for radio data transmission but, because of its reproduction fidelity, digital technology quickly expanded to many other electronic applications.

This book is his fifth in a series that reflects a similar theme including a 2005 satire titled *The Man Who Created God* (15). His most recent fictional work, *The Abduction*, is currently under Hollywood consideration. Other books are listed in the reference section. He has frequently written about the trend to eventual secularism in a natural mechanistic world and has encouraged factual enlightenment and responsibility with respect to fellow man during his short stay on planet Earth.